An Introduction to the
Physics
Particle *of* Accelerators
Second Edition

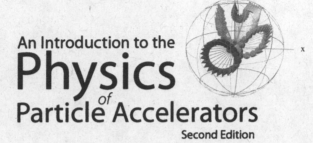

Second Edition

An Introduction to the
Physics
Particle *of* Accelerators

Mario Conte
INFN, Italy

William W MacKay
Brookhaven National Laboratory, USA

 **World Scientific**

NEW JERSEY · LONDON · SINGAPORE · BEIJING · SHANGHAI · HONG KONG · TAIPEI · CHENNAI

Published by

World Scientific Publishing Co. Pte. Ltd.

5 Toh Tuck Link, Singapore 596224

USA office: 27 Warren Street, Suite 401-402, Hackensack, NJ 07601

UK office: 57 Shelton Street, Covent Garden, London WC2H 9HE

British Library Cataloguing-in-Publication Data
A catalogue record for this book is available from the British Library.

ISBN-13 978-981-277-960-1
ISBN-10 981-277-960-4
ISBN-13 978-981-277-961-8 (pbk)
ISBN-10 981-277-961-2 (pbk)

Printed in Singapore by World Scientific Printers

We would like to dedicate
this book to the memory of
Benjamin Franklin
whose arbitrary selection of polarity
has confounded myriads of physics students.

Contents

Preface to the 2nd Edition

In the second edition, we have made a number of corrections of minor typographical errors in the first edition. More problems, and several new chapters and sections have been added, in particular:

1 Several sections on Lie groups and algebras with symplectification and integration algorithms to Chapter 3.

2 Expanded discussions of linear coupling, eigenvalues, and stability in chapters 3, 6, and 10.

3 A new chapter on spin dynamics with a detailed discussion of spin, spin precession, spinors, depolarizing resonances, effect of crossing a spin resonance, Siberian snakes, and the invariant spin field.

4 A new chapter which discusses how beam position measurements are made, and the power spectra of bunched beams.

5 An expanded appendix calculating luminosity of colliding beams.

6 A new appendix on the leap-frog integration technique.

We would like to thank our many colleagues over the years. In particular, we would like to thank D. Barber, M. Blaskiewicz, P. Cameron, E. D. Courant, W. Fischer, E. Forest, L. Hand, H. Huang, R. Huson, V. Litvinenko, A. U. Luccio, S. Mane, P. McIntyre, C. Montag, S. Ohnuma, S. Peggs, S. Peck, M. Pusterla, D. Rice, D. Rubin, T. Roser, T. Satogata, R. Talman, L. Teng, and D. Trbojevic for very fruitful discussions.

Preface to the 1st Edition

Our purpose in writing this book has been to provide a concise and coherent introduction to the physics of particle accelerators, with attention being paid to the design of an accelerator for use as an experimental tool. The book has grown from the lecture notes for courses given at the Texas Accelerator Center, Texas A&M University, and the Università di Genova. The book is intended for study at the graduate level in physics, and may be used as a text for either a one semester course or self-study. It is hoped that the reader will also find this work to be a useful reference. We have assumed that the student has an undergraduate knowledge of electrodynamics, special relativity, and a knowledge of classical mechanics including basic Hamiltonian formalism.

This subject brings together several areas of classical physics (particularly classical mechanics, special relativity, electrodynamics, and statistical mechanics) by applying them to a concrete example which is nearly linear. As such it provides an excellent pedagogical framework to gain greater insight into these basic fields.

The book is organized into twelve chapters, the first six of which deal mainly with transverse motion of single particles:

1. An introduction describing various types of accelerators, a discussion of the end parameters needed to make the accelerator a useful tool, and a short review of relativity.

2. Weak focusing with a simple treatment of particle motion.

3. A more general study of particle trajectories using Hamiltonian formalism, symplectic transformations, and Liouville's theorem.

4. An introduction to field expansions of various beam transport magnets.

5. Strong focusing with the concepts of emittance, dispersion, and the Twiss parameters.

6. Applications demonstrating calculation techniques, optimization of compound optical elements, coupled motion, and chromatic effects.

The next three chapters introduce longitudinal motion of single particles:

7. Synchrotron oscillations.

8. Effects of synchrotron radiation on beam trajectories.

9. Linear accelerators.

Chapter 10 is an introduction to resonances:

10. A study of resonance behavior including coupled motion and nonlinear effects.

The last two chapters examine some collective effects of beams:

11 Space charge effects with self-beam and beam-beam interactions.

12 Two methods for circumventing Liouville's theorem: electron and stochastic cooling.

Problems of varying difficulty are included at the ends of each chapter to give the reader an opportunity to obtain a deeper understanding of the topics. Several appendices are included either to give background material or to elaborate a specific point. The first appendix is included to minimize confusion by reconciling incompatible definitions and symbols of different authors (note especially emittance.)

The authors are very grateful to the students who have given feedback and enabled us to improve the text and problems. We would especially like to thank Professor Sho Ohnuma for his comments.

1

Introduction

1.1 Motivation and overview

The reasons for studying accelerator physics fall into two basic categories: first, to design, build, and improve real accelerators for use as tools, and second, to study the dynamics of particle motion in a non-linear system.

Some of the tools provide energy to process matter such as electron beam welding, x-ray lithography, cancer therapy, preparation of radioisotopes for medicine, food sterilization and Star Wars. Other accelerators are used as microscopes to study matter and fundamental interactions of nature. Examples of these are cathode ray tubes, electron microscopes, accelerators for studying nuclear and high energy physics.

A particle accelerator can also be used as an analog computer to test models of non-linear dynamics. The two mechanical systems in the physical universe, that are closest to being linear systems, are accelerators, and the solar system. Of the two, accelerators provide the more linear case. They can be observed over many more cycles, since the particles in an accelerator usually travel much faster than planets, and their orbits are much smaller than planets.

In this book we only consider accelerators that accelerate microscopic particles such as electrons, protons, and ions under the influence of electromagnetic fields. This influence can be divided into two effects: longitudinal acceleration due to electric fields along the direction of motion of the particle, and transverse bending of the trajectory due to transverse electric and magnetic fields. The motion of charged particles is determined by the relativistic extensions to Newton's laws and the Lorentz force,

$$\vec{F} = q(\vec{E} + \vec{v} \times \vec{B}).$$

(1.1)

For particles with spin, there is also the the Stern-Gerlach force which in the particle's rest frame is

$$\vec{F} = \nabla(\vec{\mu} \cdot \vec{B}),$$

(1.2)

where $\vec{\mu}$ is the magnetic moment of the particle. This spin force is much smaller (of order $\hbar$) than the Lorentz force for a charged particle, but for a neutral particle of

1

low momentum, it can be a useful effect.

Particle accelerators for the most part are divided into linear accelerators that accelerate the beam in a single pass, and circular accelerators that recirculate the beam through the accelerating voltage many times. Some examples of the linear type are the Van de Graaff generator, the Cockcroft-Walton cascade generator, the radio frequency quadrupole (RFQ), and the drift tube linac (DTL). The circular accelerators include cyclotrons, synchrocyclotrons, betatrons, microtrons, and synchrotrons.

Some simple parameters of importance to the user are: input power, the type of particle accelerated, the average beam energy, the distribution of energy of the particles in the beam, the angular divergence, intensity, duty factor, repetition rate, background rates, the polarization of the beam and target, and the cost. These are fairly obvious concepts; however, some of the definitions used for intensity deserve a little clarification.

For example let us consider the requirements for a high energy physics experiment. The experiment is trying to measure some esoteric process like[1]

$$e^+ + e^- \rightarrow \Upsilon(4s) \rightarrow B^+ + B^-. \qquad (1.3)$$

This type of experiment would probably be done with colliding electron and positron beams in a storage ring. This type of reaction has a certain production cross section, which when multiplied by something called the luminosity gives the rate of production of the B mesons. The usual backhanded definition of luminosity is: the number, that when multiplied by the cross section yields the interaction rate. The units for cross section are [cm^2] or sometimes the [barn $= 10^{-24}$ cm^2]. The instantaneous luminosity (see Appendix B) has units of [cm$^{-2} \cdot$ s^{-1}], and is proportional to the overlap integral of the densities of the two beams,

$$\mathcal{L} = |\vec{v}_+ - \vec{v}_-| f_0 \iiiint \rho_+(\vec{x} - \vec{v}_+ t, t) \, \rho_-(\vec{x} - \vec{v}_- t, t) \, dx \, dy \, dz \, dt, \qquad (1.4)$$

where v_+ and v_- are the velocities of the beams in the lab, f_0 is the frequency of beam crossings, and the integrations are carried out for a single crossing of bunches. (Note that for extremely relativistic head-on collisions in the center of mass, the factor $|\vec{v}_+ - \vec{v}_-| = 2c$. Simply put, it takes half as long for two beams to pass each other if they are moving towards each other than if one is stationary.) The densities, ρ_- and ρ_+ are the densities of the respective electron and positron bunches, and will vary in z with the shape of the beam envelope. The total number of such interactions in an experiment is given by the cross section times the integrated luminosity, which is defined as the instantaneous luminosity integrated over the total time of the experiment. If there is more than one bunch per beam, the number of bunch crossings at a given interaction point (i. e., for a single experiment) is increased by the number of bunches, N_b.

Of course for a useful number, we must account for any dead time of the experimental apparatus. In order to identify the B mesons in this experiment, the detector must identify the tracks of the particles from the decays of the B mesons. The detector takes a certain amount of time to accumulate and log the data, producing a period of time in which the detector is unable to identify a new event. This is called the dead time of the detector.

This type of event produces many daughter particles, and if there are two simultaneous events, the data is usually too confused to be useful. Because of this confusion, it is useless to have the average number of interactions per beam crossing greater than some value (usually one or less for most experiments.)

For an experiment with a single beam incident on a fixed target, the intensity is traditionally quoted as the number of beam particles hitting the target per second, rather than as a luminosity.

Two other terms used for defining intensities are used with synchrotron light sources: brightness and brilliance. Both are proportional to the number of photons hitting the target per second, but they have the added feature of being inversely proportional to the bandwidth, or energy spread of the photon beam. Brightness is defined as the number dn, of photons per time interval dt, passing through a solid angle $d\Omega$, and divided by 0.1% of the bandwidth $d\lambda/\lambda$,

$$d\Phi_\Omega = 1000 \frac{d^4 n}{dt \, d\Omega \, (d\lambda/\lambda)}.$$
(1.5)

Brilliance is defined as the brightness per area, s, of the source,

$$B = \frac{d\Phi_\Omega}{ds}.$$
(1.6)

In the rest of this chapter we review some basic concepts and briefly discuss a few of the early types of particle accelerators, which illustrate the different techniques used to accelerate charged particles.

1.2 Direct–voltage accelerators

The simplest type of elementary particle accelerator is a source of electrons or ions, and a pair of electrodes, activated by a potential drop ΔV, as shown in Fig. 1.1.

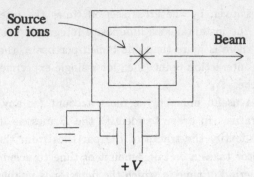

Figure. 1.1 A simple accelerator for charged ions of charge q. The kinetic energy of the beam is approximately qV.

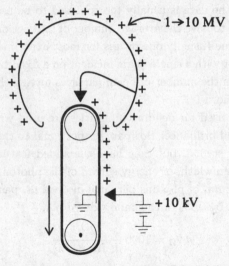

Figure. 1.2 A Van de Graaff generator. Electrons are pulled off the belt by a corona discharge at the bottom. The net positive charge moves up with the belt inside the dome, where electrons from the dome are pulled onto the belt through another corona discharge. As a result, the dome can reach a potential of several million volts relative to the lower corona points which are at ground.

Indeed such an apparatus is the prototype of a class of devices (cathode ray tubes, electron microscopes, etc.) which come under the branch of "Electron Optics." All direct current accelerators are variations on this theme, e. g., the electrostatic generator constructed by Van de Graaff[2] and the cascade generator developed by Cockcroft and Walton,[3] who first succeeded in disintegrating nuclei with accelerated particles.

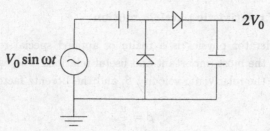

Figure. 1.3 A simple cascade circuit for doubling the voltage of an input generator.

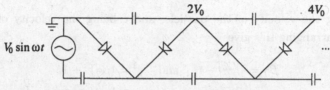

Figure. 1.4 A multistage cascade Cockcroft–Walton circuit, which rectifies and multiplies the input voltage.

Fig. 1.2 shows the sketch of the Van de Graaff electrostatic generator: a belt of insulating material runs between ground and a high-voltage generator ($\simeq 10$ kV); corona discharge provides charge to the belt which, in its turn, induces electrostatic charges in the "hot" terminal (1—10 MV); another corona discharge neutralizes the belt. Notice how the drive of the engine is one of the best examples of electromotive force!

Mixtures of high pressure gases (N_2 and CO_2, for example) provide insulation and material for corona discharges. This machine can accelerate charged particles of either polarity.

A further improvement is the tandem generator, where negative ions are accelerated from the ground to the terminal, then are stripped of most of their electrons by a thin foil, hence the resulting positive ions are accelerated back to ground potential. In principle the energy gain can be increased (twice for protons) with respect to a simple acceleration.

The cascade generator is an extension of the doubling circuit, shown in Fig. 1.3, where two rectifying diodes and two capacitors, applied to an ac generator $V(t) = V_0 \sin \omega t$, give an almost dc voltage, $2V_0$.

Fig. 1.4 shows a sketch of a multistage cascade generator, capable of reaching a few million volts of potential. In both these high voltage generators the voltage must be distributed along the accelerating tube via either capacitive or resistive partitions, in order to avoid electric disruptions.

1.3 A review of relativistic particle motion

Particle accelerator physics is a realm of applied special relativity. In this section we review the most important and useful relations. Following the standard method we define the relativistic velocity β, and the Lorentz factor γ as

$$\beta = \frac{v}{c} \tag{1.7}$$

$$\gamma = (1 - \beta^2)^{-\frac{1}{2}} \tag{1.8}$$

with v being the velocity of the particle, and c being the velocity of light in a vacuum. Rearranging this gives

$$\beta\gamma = \sqrt{\gamma^2 - 1}, \quad \text{and} \quad \gamma^2 = (\beta\gamma)^2 + 1. \tag{1.9}$$

The total energy, momentum, and kinetic energy for a particle of rest mass, m, are, respectively:

$$U = \gamma mc^2, \tag{1.10}$$

$$p = \beta\gamma mc = \beta\frac{U}{c}, \quad \text{and} \tag{1.11}$$

$$W = (\gamma - 1)mc^2. \tag{1.12}$$

The relation between energy and momentum is

$$U = \sqrt{(pc)^2 + (mc^2)^2} = \sqrt{p^2 + m^2}, \tag{1.13}$$

where we have used the ever popular set of units, with $c = 1$, in the last expression. The most frequently accelerated particles are the electron with mass $m_e = 0.510999$ MeV, and the proton with mass, $m_p = 938.272$ MeV.

The following divisions are frequently used

$$\gamma \simeq 1 \quad \text{Non-relativistic} \quad \text{N. R.}$$
$$\gamma > 1 \quad \text{Relativistic} \quad -$$
$$\gamma \gg 1 \quad \text{Ultra-Relativistic} \quad \text{U. R.}$$

The non-relativistic case can be checked by expanding Eq. (1.8) for small β and obtaining $\gamma \simeq 1 + \frac{1}{2}\beta^2$, which when inserted into Eq. (1.12) produces

$$W \simeq \frac{1}{2}mc^2\beta^2 = \frac{1}{2}mv^2. \tag{1.14}$$

In the ultra-relativistic case the mass becomes negligible and Eqs. (1.10, 1.11, and 1.12) collapse into the simpler relation $U \simeq W \simeq pc$.

Now a set of relations, particularly useful to accelerator physics, will be deduced. Differentiating respectively (1.8) and (1.9), we obtain:

$$d\gamma = \beta(1 - \beta^2)^{-\frac{3}{2}} d\beta = \beta\gamma^3 d\beta \qquad (1.15)$$

$$d(\beta\gamma) = \gamma d\beta + \beta d\gamma = \gamma(1 + \beta^2\gamma^2) d\beta = \gamma^3 d\beta = \frac{d\gamma}{\beta}. \qquad (1.16)$$

Squaring and differentiating Eq. (1.13) gives $2U dU = 2p dp$ or

$$\frac{dU}{U} = \frac{p^2}{U^2}\frac{dp}{p} = \beta^2\frac{dp}{p}. \qquad (1.17)$$

By dividing Eq. (1.15) by $\beta^2\gamma^3$, the fractional change in velocity can be found as

$$\frac{d\beta}{\beta} = \frac{1}{(\beta\gamma)^2}\frac{d\gamma}{\gamma} = \frac{1}{(\beta\gamma)^2}\frac{dU}{U} = \frac{1}{\gamma^2}\frac{dp}{p}. \qquad (1.18)$$

For acceleration in one dimension, Newton's second law becomes

$$F = \frac{dp}{dt} = mc\frac{d}{dt}(\beta\gamma) = \gamma^3 m\frac{dv}{dt} = m^*\frac{dv}{dt}, \qquad (1.19)$$

having considered Eq. (1.16), and defining as effective mass,

$$m^* = \frac{dp}{dv} = \frac{d(\gamma mv)}{dv} = m\gamma^3. \qquad (1.20)$$

The electromagnetic force is what accelerates charged particles, and is described mathematically by the Lorentz equation,

$$\vec{F} = q(\vec{E} + \vec{v} \times \vec{B}). \qquad (1.21)$$

If there is no electric field and only a uniform magnetic field, then the force equation may be written as

$$\vec{F} = q\vec{v} \times \vec{B} = \frac{d}{dt}(\gamma m\vec{v}) = m(\gamma\frac{d\vec{v}}{dt} + \frac{d\gamma}{dt}\vec{v}) = \gamma m\frac{d\vec{v}}{dt}, \qquad (1.22)$$

since $\beta = |\vec{\beta}|$ is a constant, which implies that $(d\gamma/dt) = 0$. The velocity $\vec{v} = \vec{\omega} \times \vec{\rho}$, with the angular velocity, $\vec{\omega}$ being constant for a central force of constant magnitude.

The *cyclotron radius* ρ is just the radius of the particle's orbit. Eq. (1.22) now becomes

$$q\vec{v} \times \vec{B} = \gamma m \vec{\omega} \times \frac{d\vec{\rho}}{dt} = \gamma m \vec{\omega} \times \vec{v}, \qquad (1.23)$$

or for a particle moving in a plane perpendicular to $\vec{B}$,

$$qvB = \gamma m \omega v = \gamma m \frac{v^2}{\rho}, \qquad (1.24)$$

i. e., the Lorentz force is the centripetal force which keeps the particle of charge q and mass m on a circular orbit. Dividing Eq. (1.24) by v/ρ, we get a relation for the momentum in terms of the orbit radius, magnetic field and charge of the particle:

$$p = \beta \gamma m c = qB\rho. \qquad (1.25)$$

For a particle with same charge as the electron, it is useful to remember

$$p[\text{GeV/c}] \simeq 0.3 B[\text{T}] \, \rho[\text{m}]. \qquad (1.26)$$

Another popular formula is the one for the angular velocity or cyclotron frequency:

$$\omega = \frac{qB}{\gamma m}. \qquad (1.27)$$

It may also be useful to remember the Lorentz transformations of the electromagnetic field from the lab system to the rest system:

$$\vec{E}_{\perp}^{*} = \gamma(\vec{E}_{\perp} + \vec{v} \times \vec{B}_{\perp}), \qquad (1.28)$$
$$\vec{E}_{\parallel}^{*} = \vec{E}_{\parallel}, \qquad (1.29)$$
$$\vec{B}_{\perp}^{*} = \gamma(\vec{B}_{\perp} - \vec{v} \times \vec{E}_{\perp}), \quad \text{and} \qquad (1.30)$$
$$\vec{B}_{\parallel}^{*} = \vec{B}_{\parallel}, \qquad (1.31)$$

where the $\parallel$ designates the component of the field parallel to the boost velocity $\vec{v}$, $\perp$ indicates the component perpendicular to the boost, and the asterisks indicate quantities in the rest system.

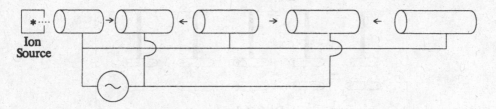

Figure. 1.5 A schematic of the Wideröe linac structure. The arrows indicate the directions of the accelerating electric field at a given instant of time.

1.4 Linear accelerators with oscillating electric fields

Since it is very difficult to produce dc voltages more than a few million volts, it was necessary to find a new method for acceleration to energies beyond a few MeV. In 1928 Wideröe proposed an accelerating structure using a series of cylindrical tubes, called drift tubes, which were alternately connected to a high frequency oscillator, as shown in Fig. 1.5. Charged particles from the source are accelerated in the gaps between tubes. They then drift in the field free region inside the tube. While the particles are inside the tube the direction of the field is reversed so that when the particles reach the next gap, they again see an accelerating electric field. If a constant frequency generator is used, the tubes must increase in length as the particle velocity increases, so that the particles will always arrive at the next gap with the correct phase of the accelerating voltage in the gap. Particles will only leave the source when the voltage in the first gap has the correct sign to accelerate them, thus the accelerated beam will have a pulsed structure to it.

A cell is usually defined as the region from the midplane of one drift tube to the next. (Sometimes it is convenient to offset this slightly, but the length of the cell remains the same.) The Wideröe structure is called a $\beta\lambda/2$ or π-mode structure, since the electric field configuration repeats every two cells. The product $\beta\lambda$ is the distance that the particle travels during one rf cycle. As the particles' velocities increase, the cell lengths must increase. Bunches of particles cannot be accelerated in every gap during a half cycle, but must be spaced with a free gap between every pair of bunches.

The Alvarez structure, shown in Fig. 1.6, is a $\beta\lambda$ or 2π-mode structure and can accelerate particles simultaneously in each gap. Since this is a 2π-mode structure, the rods connecting the tank and drift tubes are only necessary for support. In this structure the charges oscillate between the ends of the drift tubes. Unlike the

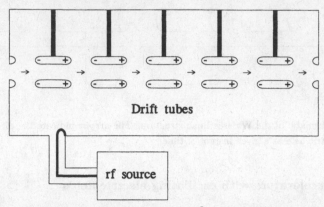

Figure. 1.6 The Alvarez drift tube structure. This is a 2π-mode structure with the field pattern repeating in every cell. The arrows indicate the direction of the electric field at one instant.

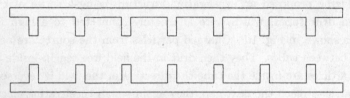

Figure. 1.7 A traveling wave disk and washer structure.

Wideröe structure with the charges actually traveling from one tube to the next by passing through the rf generator, the Alvarez structure is a resonant structure which is inductively coupled through a transformer consisting of a one turn primary inserted through the wall of the resonant tank containing the drift tubes.

Clearly, as the particles become more relativistic, the length of a cell increases. To counteract this requires either a much longer cell, or a source of higher frequency. Conventional triode and tetrode tubes were unable to operate at high frequencies in the microwave regime. Another type of tube called a klystron[5] can produce very high power at frequencies from a few hundred megahertz to several tens of gigahertz. The klystron is really more like a small linear accelerator than an electron vacuum tube. It uses a driven rf cavity to modulate a dc beam by varying the velocity of the particles with respect to the time of passage through the cavity. The particles drift for some distance and accumulate into bunches which appear as a pulsed current at a second resonant output cavity. The output power can be coupled by an inductive loop to a waveguide that pipes the power to an accelerating structure.

At about the same time as the invention of the klystron, it was realized that

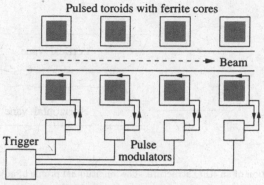

Figure. 1.8 An induction linac. This operates in the same manner as a transformer. The primary turns are toroidal pulsed electromagnets, and the secondary is the beam.

a traveling wave could be used to accelerate relativistic particles. A cylindrical waveguide propagates waves with phase velocities greater than the speed of light. Since the charged particles must be traveling at less than the speed of light, they will not obtain any net acceleration, because they cannot keep in phase with the wave. If the waveguide is loaded by corrugating its walls, so that the induced charges have a longer path length (as in Fig. 1.7), the phase velocity of the wave can be slowed down to a usable value (or even slower.) The particles may then "surf" along the wave with a phase yielding an accelerating force. This type of structure is called a *traveling wave structure*.

A *standing wave* structure is a structure which has two traveling waves moving in opposite directions. This type of structure is necessary for accelerating oppositely charged particles (e^+e^- or $p\bar{p}$) in opposite directions in the same accelerator.

Another type of linac is the induction linac (see Fig. 1.8.) This linac uses a series of toroidal electromagnets coaxially placed along the beam axis. By successively pulsing each magnet, large peak values of emf can be produced which will accelerate the beam. Typical currents of several kiloamperes can be achieved.

More recently, the radio frequency quadrupole (RFQ) has been developed for preliminary acceleration of protons and heavier ions. This uses four parallel electrodes around the beam axis as shown in Fig. 1.9. It is a resonant structure with adjacent electrodes having opposite charges. From the end, they look like an electric quadrupole. This arrangement of electric fields focuses the beam in one plane and defocuses the beam in the other plane. Since the electric field oscillates, a net focusing effect can be obtained. If the electrodes are scalloped with a curve somewhat like a sinusoid, and with the curves of the adjacent electrodes differing

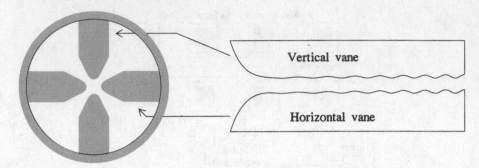

Figure. 1.9 A cross section of an RFQ structure. The longitudinal profile of adjacent vanes is shown a the right.

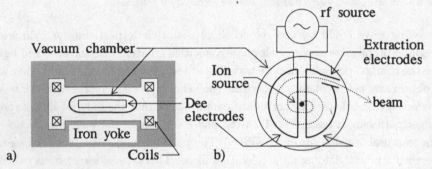

Figure. 1.10 a) Side view of a cyclotron. b) Top view of a cyclotron, with "D" electrodes shown inside the vacuum chamber.

by 180° in phase, then a component of longitudinal field will be produced, which may be used to accelerate the particles.

1.5 Circular machines

The cyclotron is the first example of a circular machine.[6,7] A homogeneous magnetic field, supplied by an H-shaped magnet, as in Fig. 1.10a, bends back the particles to the same rf gap between the two D-shaped electrodes shown in Fig. 1.10b, twice each period of the radio frequency oscillation. If the rf is set equal to the cyclotron frequency (a resonance condition) given by Eq. (1.27) with $\gamma = 1$ (N.R. ions or protons), the particles will continue to pass near the peak of the rf voltage twice per turn, gaining kinetic energy, and then increasing the radius of their orbits by Eq. (1.25), till reaching some extraction device.

Usually $\Delta E \leq 200$ keV/turn, then for $W_{\max} \simeq 20$—25 MeV, one can infer

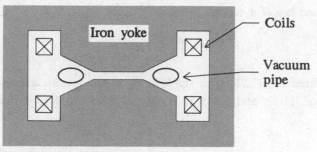

Figure. 1.11 Side cross section of a betatron.

that some 100 to 125 turns suffice to achieve the wanted acceleration. The required frequency can be calculated from

$$\frac{\nu_{\mathrm{rf}}}{B} = \frac{e}{2\pi m} \simeq 15\,\mathrm{MHz} \cdot \mathrm{T}^{-1}, \text{ for protons.} \tag{1.32}$$

Typical beam currents would be about 1mA.

However, Eq. (1.27) contains the Lorentz factor γ, which, as soon as the kinetic energy is significantly increased, begins to deviate from one, causing a decrease of the radio-frequency. A remedy to this consists of just varying the rf according to a $1/\gamma$ law; this means that only particles in phase with this varying rf can be accelerated. That is, a synchronous acceleration principle[8,9] is now required, and only one bunch of particles will reach the final kinetic energy (600—700 MeV for protons), due to technical limitations on magnet dimensions and rf modulation. The current is of the order of a few microamperes, but still a rather small number (a few thousand) of revolutions are required to accomplish the full cycle of acceleration.

This attention paid to the number of revolutions, run by particles from rest to their goal-energy, is dictated by the need of having orbits with limited dimensions. In fact, since particles leave the ion source with nonzero angles and energy spread to form a beam of nonzero size, a very high number of turns would imply a broadening of the beam, unless some focusing mechanism plays a role.

This problem first arose for the betatron,[10] a rather different machine based on acceleration of electrons by induction. The working principle of this accelerator will be illustrated briefly: the magnetic flux, crossing a circle of radius R, is $\phi = \pi R^2 \bar{B}$, where $\bar{B}$ is some average magnetic induction. If ϕ varies with time, an electron orbiting at this radius will experience a force,

$$F = -eE \simeq (-e)\frac{V}{2\pi R} = \frac{(-e)}{2\pi R}(-\frac{d\phi}{dt}) = \frac{1}{2}eR\frac{d\bar{B}}{dt}. \tag{1.33}$$

Newton's second law of dynamics gives

$$F = \frac{dp}{dt} = eR\frac{dB_g}{dt}, \qquad (1.34)$$

having considered Eq. (1.25) with the field B_g in the magnet gap (see Fig. 1.11). From both Eqs. (1.33, and 1.34), it is easy to infer the 2-to-1 rule or Wideröe condition,[11]

$$B_g = \frac{1}{2}\bar{B}, \qquad (1.35)$$

typical of betatrons, which gives rise to the rather cumbersome structure of these machines.

Working out a few realistic numbers, one obtains $W_{max} \simeq 50$ MeV, which for $B_g = 0.5$ T, gives $R = 33$ cm and $\bar{\tau}_{rev} = (2\pi R)/\bar{v} \simeq 14$ ns, since $\bar{v} \simeq c/2$. For $\Delta t_{cycle} = (1/4f_{main}) = 5$ ms, we get

$$N_{rev} = \frac{\Delta t_{cycle}}{\bar{\tau}} \simeq 357,000 \text{ turns.} \qquad (1.36)$$

Now the problem of focusing the circulating beam, mildly tackled in cyclotrons, becomes of primary importance. In the next chapter, we will derive the trajectory equations for a charged particle traveling through various shaped magnetic fields. Also, we will see that several effects will cause focusing in the dimensions transverse to the beam.

1.6 Momentum compaction and the synchronous particle

If a particle of fixed energy moves along a closed trajectory, the integral of the curvature, $1/\rho$, around the closed path must be

$$\oint \frac{ds}{\rho} = 2\pi. \qquad (1.37)$$

Since $1/\rho = qB_\perp/p$, this means that the momentum

$$p = \frac{q}{2\pi} \oint B_\perp \, ds. \qquad (1.38)$$

Leaving the magnetic guide field unchanged, consider another particle also with a closed trajectory but a slightly different momentum. This second particle's closed orbit must differ from the path of the first particle. By varying the momentum, the path length, $L = \oint ds$, of the closed trajectory can be varied. The fractional deviation of this path length divided by the fractional deviation of the momentum is frequently called *momentum compaction*

$$\alpha_p = \frac{dL}{L} \bigg/ \frac{dp}{p} = \frac{p}{L}\frac{dL}{dp}. \tag{1.39}$$

The name momentum compaction (see Appendix A) is reviled by some authors since an increase in momentum implies a lengthening of the orbit if $\alpha_p > 0$, i. e., a dilation rather than a compaction.

For example, it is easy to show that the momentum compaction for a satellite of mass m in a circular orbit* around the earth at a radius r has a momentum compaction of -2, ignoring elliptical orbits:

$$F = \frac{GMm}{r^2} = \frac{p^2}{mr}, \tag{1.40}$$

$$\frac{dr}{dp} = -\frac{2pr^2}{GMm^2}, \tag{1.41}$$

and thus

$$\alpha_p = \frac{p}{r}\frac{dr}{dp} = -\frac{2p^2r}{GMm^2} = -2. \tag{1.42}$$

In order to understand the acceleration of a particle due to an rf field, let us first consider a proton synchrotron with a fixed magnetic guide field and a single rf cavity driven by an oscillating electric field of constant amplitude and frequency ω_{rf}. The electric field on the axis of the cavity can be described by

$$E_s(x = 0, y = 0, s, t) = E_0(s)\cos(\omega_{rf}t), \tag{1.43}$$

where $E_0(s)$ is the s-component of the amplitude of the electric field along the axis. For this simple example, let us ignore Maxwell and approximate $E_0(s)$ by a constant inside the cavity and zero everywhere else. This constant would then be the maximum voltage across the gap divided by the length of the gap, V_0/g.

As a particle of velocity v crosses the gap it will experience a varying acceleration. The change in energy can be calculated by

$$\Delta U = \int_{t_0}^{t_0+T} \frac{eV_0}{g}\cos(\omega_{rf}t)\,v\,dt, \tag{1.44}$$

---

* Here, it is assumed the increment dp of momentum is applied in such a way as to keep the satellite in a circular orbit. If this were not done, then the momentum would not be constant, and our definition of momentum compaction would not be applicable.

where the particle enters the gap at time t_0 and position s_0. The transit time T through the cavity must satisfy the condition

$$g = \int_{t_0}^{t_0+T} v \, dt. \qquad (1.45)$$

Particles arriving at different times will accrue different energy shifts on passing through the cavity, some positive, some negative. There will be a specific phase relative to the rf oscillations, for which $\Delta U = 0$. In this simple case, a *synchronous particle* is defined as a hypothetical particle that moves along the design trajectory and passes through the rf cavity with a phase, such that there is no change in energy. The synchronous particle must have a path length which is an integral number of $\beta\lambda$, where λ is the wavelength of the rf field, and $\beta = v/c$ for the particle.

When the accelerator is ramped up in energy (by ramping up the fields of the guide magnets), the meaning of the synchronous particle changes slightly. The ramping is assumed to be slow, so that the particles gain only a very small increase in energy per revolution. It is then possible to consider the ramping to increase as a step function, so that on each revolution the momentum corresponding to the closed orbit increases in steps. The synchronous particle is now defined as the hypothetical particle whose momentum increases exactly by this amount. Clearly the phase of this synchronous particle with respect to the rf must be shifted slightly from the case of no acceleration, i. e., the particle must see a net electric potential averaged over the time that it spends in the gap of the cavity.

This concept of synchronous particle may also be extended to a linac. A linac is designed to give a specific increase in energy for each cell. These increments typically vary from cell to cell. For the linac, the synchronous particle is defined as the hypothetical particle that obtains exactly the increment of energy of the design at each cell.

The next few chapters will study the transverse component of motion in beam lines and accelerators. The concept of phase stability and acceleration of particles will be discussed starting in Chapter 7.

Problems

1–1   a) If the bunches can be described by Gaussian ellipsoids with

$$\rho \propto \exp\left(-\left(\frac{x^2}{2\sigma_x^2} + \frac{y^2}{2\sigma_y^2} + \frac{z^2}{2\sigma_z^2}\right)\right),$$

show that the luminosity reduces to

$$\mathcal{L} = f_0 N_b \frac{N_+ N_-}{\pi(4\sigma_x \sigma_y)},$$

where it is assumed that the beams move along the z-axis. The number of particles per bunch in the electron and positron beams are N_- and N_+, respectively. The number of bunches in each beam is given by N_b. (Assume the bunches do not change shape either due to the accelerator optics or the interaction of the beams passing through each other.) b) An $e^+ e^-$ storage ring (CESR) operates at 5.3 GeV with 7 bunches of e^+ and 7 bunches of e^- orbiting in opposite directions. Assume the current per bunch is initially 8 mA, and the ring circumference is 768 m. For $\sigma_x = 8.4 \times 10^{-4}$ m, $\sigma_y = 3.5 \times 10^{-5}$ m, and $\sigma_z = 2.2$ cm, what is the initial luminosity in one of the experiments? What is the integrated luminosity of this experiment for a 3 hr run if the beam lifetimes are both 2 hr? (Assume that the beam currents decay exponentially.)

1–2   Calculate the brightness of a NdYAG laser with the following parameters:

$$\lambda = 1.064\mu \text{ m}$$

$$\text{Power} = 20 \text{ W}$$

$$\text{Bandwidth } \frac{\Delta\omega}{2\pi} = 120 \text{ GHz}$$

$$\text{Beam divergence} = 10 \text{ mrad.}$$

1–3   In a fixed Cartesian coordinates system, show that the equations of motion for a charged particle moving in a magnetic field may be written as

$$x'' = \frac{q}{p}(1 + x'^2 + y'^2)^{\frac{1}{2}}[y'B_z - (1 + x'^2)B_y + x'y'B_x], \quad \text{and}$$

$$y'' = -\frac{q}{p}(1 + x'^2 + y'^2)^{\frac{1}{2}}[x'B_z - (1 + y'^2)B_x + x'y'B_y],$$

where the primes denote derivatives with respect to z (i. e., $x' = dx/dz$). Here it has been assumed that the electric field is zero and that $dz/dt \neq 0$.

1–4   Consider a charged pion decaying into a muon plus an antineutrino:

$$\pi^- \to \mu^- + \bar{\nu}_\mu.$$

Use $M_{\pi\pm} = 140 \text{ eV}/c^2$, $m_\mu = 106 \text{ MeV}/c^2$, and $m_{\bar{\nu}} = 0$.

    a) In the rest system of the pion, what are the energies and momenta of the muon and antineutrino?

    b) For a moving pion with total energy $U_\pi = \gamma M_\pi c^2$ find an expression for the direction, θ_μ of the muon relative to the pion in the lab in terms of the angle θ_μ^* in the in the pion's rest system.

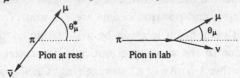

1–5   The Tevatron collides protons ($m_p = 0.938$ GeV) at 1 TeV per beam. What is the equivalent proton beam energy required to produce the same center-of-mass energy with a stationary hydrogen target? How fast would you have to drive your new 1.3 ton VW Beetle to have the same kinetic energy as a bunch of 10^{13} protons with this energy? (The speed of sound in air is 330 m/s.)

1–6   HERA collides 920 GeV protons with 27.5 GeV electrons with zero crossing angle.

    a) What is the center-of-mass energy?

    b) What is the velocity of the center of mass in the lab system?

1–7   The Stanford Linear Accelerator is 3.05 km long and can accelerate electrons up to 50 GeV.

    a) What is the average accelerating gradient of the rf cavities?

    b) For bunches of 4×10^{10} electrons per bunch and a duty cycle of 100 Hz, what is the power transferred to the beam?

1–8   An experiment has a 10 cm long liquid hydrogen target with a density of

$$\rho = .063 \text{ g/cm}^3.$$

Estimate the interaction rate for p+p collisions for a beam of 10^{13} protons every two minutes. Assume the total cross section is 40 mb. Note: 1 barn $= 10^{-24}$ cm^2 and Avogadro's number is 6×10^{23}.

References for Chapter 1

[1] S. Behrends et al., Phys. Rev. Lett., 50, 881 (1983).

[2] R. J. Van de Graaff, Phys. Rev., 38, 1919 (1931).

[3] J. D. Cockcroft and E. T. S. Walton, Proc. Royal Soc., A136, 619 (1932).

[4] J. Le Duff, "Dynamics and Acceleration in Linear Structures", *Proc. of the CERN Accelerator School, General Accelerator Physics, Gif-sur-Yvette*, CERN 85-19, Geneva (1985).

[5] S. Y. Liao, *Microwave Electron-Tube Devices*, Prentice-Hall Inc., Englewood Cliffs, N. J. (1988).

[6] E. O. Lawrence and N. E. Edlefsen, Science, 72, 376 (1930).

[7] E. O. Lawrence and M. S. Livingston, Phys. Rev. 40, 19 (1932).

[8] E. M. McMillan, Phys. Rev. 68, 143 (1945).

[9] V. Veksler, Journal of Physics USSR, 9, 153 (1945).

[10] D. W. Kerst, Phys. Rev., 60, 47 (1941).

[11] R. Wideröe, Arch. Electrotech., 21, 400 (1928).

[12] R. Kose, "Status of the HERA-Project", *1987 IEEE Particle Accelerator Conference*, March 1987, Washington, D. C., IEEE, Piscataway, NJ.

[13] K. G. Steffen, *High Energy Beam Optics*, John Wiley and Sons, New York, (1965).

2

Equations of Motion for Weak Focusing

2.1 Reference system for a circular machine

Consider an accelerator which is designed so that a specially prepared particle of charge q will orbit in a circle of radius ρ. Such an accelerator could be a betatron with the field being held constant in time. The special preparation requires that the particle have a momentum of $qB_0\rho$, where B_0 is the field at ρ pointing perpendicular to the plane of the circle. It also requires that the particle initially starts somewhere on the circle with a velocity vector tangent to the circle. This orbit is called the *design orbit*. The position of this special particle may be parametrized in terms of an angle θ relative to a fixed point on the circle. The arc length to a different point would then be $s = \rho\theta$.

Define the local Cartesian coordinate system as shown in Fig. 2.1 moving along with this special particle, such that the z–axis points in the direction of motion, the y–axis points up, and the x–axis points radially outward from the center of the circle so that $x = R - \rho$ for some point a distance R from the central axis of the circle. The trajectory of a generic particle traveling on some trajectory slightly different from the design particle may be parametrized in the traveling coordinate system as a function of the special design particle's position θ.

Notice that all these considerations are valid for a positive elementary particle. Little change occurs for a negative one. (Frequently we must define a left-handed coordinate system for one of the beams when we have counter-rotating beams in the same machine.)

2.2 Equations of motion

Let Eq. (1.22) be projected over the x and y axes to obtain

$$F_x = \frac{d}{dt}(\gamma m\dot{x}) = -q\beta c B_y, \quad \text{and} \tag{2.1}$$

$$F_y = \frac{d}{dt}(\gamma m\dot{y}) = q\beta c B_x, \tag{2.2}$$

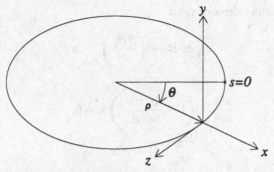

Figure. 2.1 Reference orbit and coordinate system.

since $\vec{B} \equiv (B_x, B_y, 0)$. The longitudinal motion is governed by other aspects than the ones considered in this chapter.

Eqs. (2.1, and 2.2) are respectively the horizontal (radial) motion equation, and the vertical (axial) motion equation. Since the energy variation due to the acceleration process can be considered very small with respect to $\ddot{x}$ and $\ddot{y}$ (adiabatic approximation), we may write

$$\frac{d}{dt}(\gamma m \dot{x}) \simeq \gamma m \left(\ddot{x} - \frac{v^2}{R} \right) = \gamma m \left(\frac{\beta^2 c^2}{R^2} \frac{d^2 x}{d\theta^2} - \frac{v^2}{R} \right) \qquad (2.3)$$

with

$$\theta = \frac{s}{R} = \frac{\beta c t}{R}. \qquad (2.4)$$

Then Eqs. (2.1, and 2.2) can be simplified to

$$\frac{d^2 x}{d\theta^2} + \left(\frac{qB_y}{\beta \gamma m c} R - 1 \right) R = 0, \quad \text{and} \qquad (2.5)$$

$$\frac{d^2 y}{d\theta^2} - \frac{qB_x}{\beta \gamma m c} R^2 = 0. \qquad (2.6)$$

By applying the *paraxial approximation*, i. e.,

$$x, y \ll \rho, R, \quad \text{and} \qquad (2.7)$$

$$\beta \simeq \beta_0 \quad \text{(relativistic speed of the reference particle),} \qquad (2.8)$$

and Eq. (1.25), we get

$$\frac{d^2 x}{d\theta^2} + \left[\left(1 + \frac{1}{B_0} \frac{\partial B_y}{\partial x} x \right) \left(1 + \frac{x}{\rho} \right) - 1 \right] \rho \left(1 + \frac{x}{\rho} \right) = 0. \qquad (2.9)$$

Expanding about the reference orbit

$$B_y \simeq B_0 + \left(\frac{\partial B_y}{\partial x}\right)_{x=0} x, \tag{2.10}$$

gives

$$\frac{d^2x}{d\theta^2} + \left(1 + \frac{\rho}{B_0}\frac{\partial B_y}{\partial x}\right) x = 0, \quad \text{or} \tag{2.11}$$

$$\frac{d^2x}{d\theta^2} + (1-n)x = 0, \tag{2.12}$$

where we have neglected terms of order $(x/\rho)^2$ or higher, and have defined the *field index*

$$n = -\frac{\rho}{B_0}\left(\frac{\partial B_y}{\partial x}\right)_{x=0}. \tag{2.13}$$

Similarly in the vertical plane, Eq. (2.6) becomes

$$\frac{d^2y}{d\theta^2} + \left(-\frac{\rho}{B_0}\frac{\partial B_y}{\partial x}\right) y \left(1 + 2\frac{x}{\rho}\right) = 0, \quad \text{or} \tag{2.14}$$

$$\frac{d^2y}{d\theta^2} + ny = 0. \tag{2.15}$$

This follows from Ampere's law $\nabla \times B = 0$ (in the vacuum chamber and in the quasi-stationary regime):

$$\frac{\partial B_y}{\partial x} = \frac{\partial B_x}{\partial y} \simeq \frac{B_x}{y}. \tag{2.16}$$

Since the field index is proportional to the opposite of the gradient, i. e., $n \propto -\nabla B_y$, one can devise the simple qualitative explanation of the role of the field gradient, shown in Fig. 2.2. Therefore a simultaneous focusing of both motions cannot occur unless

$$0 < n < 1. \tag{2.17}$$

This condition, called weak focusing, allows stable solutions for motion in both transverse dimensions (Eqs. (2.12 and 2.15)) and requires a further semiquantitative interpretation.

Bearing in mind that $n > 0$ means vertical stability, a few words are to be spent about horizontal motion[1], where the centrifugal and the Lorentz forces balance each other,

$$F_{res} = \frac{mv^2}{R} - qvB_y \simeq \frac{mv^2}{\rho}\left(\frac{\rho}{R}\right) - qvB_0\left(\frac{\rho}{R}\right)^n, \tag{2.18}$$

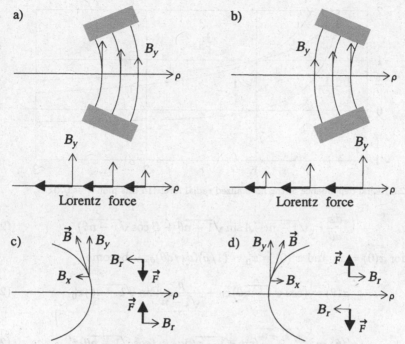

Figure. 2.2 Forces for a positive particle coming out of the page: a) Radial defocusing force from radially opening poles. b) Radial focusing gradient from radially closing poles. c) Vertical focusing forces from radially opening poles. d) Vertical defocusing forces from radially closing poles.

having integrated Eq. (2.13); Eq. (2.18) can be transformed into

$$\phi = \frac{1}{\zeta} - \frac{1}{\zeta^n}, \tag{2.19}$$

where

$$\phi = \frac{F_{res}}{qvB_0}, \quad \text{and} \tag{2.20}$$

$$\zeta = \frac{R}{\rho}. \tag{2.21}$$

Fig. 2.3 illustrates how F_{res} varies with R, showing that $F_{res} < 0$ for $R > \rho$, and $F_{res} > 0$ for $R < \rho$. This means the resulting force acts to push the particles towards the reference or equilibrium orbit.

2.3 Solutions of the motion equations and transfer matrices

Equation (2.12) with the condition $0 < n < 1$ yields

$$x(\theta) = A \cos \sqrt{1 - n}\theta + B \sin \sqrt{1 - n}\theta \tag{2.22}$$

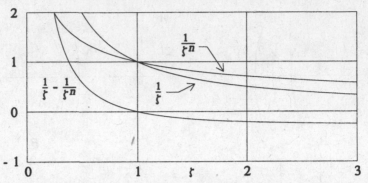

Figure. 2.3 Radial dependence of the normalized radial force. In this plot $n = 0.5$.

$$\frac{dx}{d\theta} = \sqrt{1-n}(-A\sin\sqrt{1-n}\theta + B\cos\sqrt{1-n}\theta) \qquad (2.23)$$

which for $x(0) = x_0$ and $x'(0) = x_0' = (1/\rho)(dx/d\theta)_{\theta=0}$, become

$$x(\theta) = (\cos\sqrt{1-n}\theta)x_0 + \frac{\rho}{\sqrt{1-n}}(\sin\sqrt{1-n}\theta)x_0' \qquad (2.24)$$

$$x'(\theta) = -\frac{\sqrt{1-n}}{\rho}(\sin\sqrt{1-n}\theta)x_0 + (\cos\sqrt{1-n}\theta)x_0', \qquad (2.25)$$

or

$$\begin{pmatrix} x \\ x' \end{pmatrix} = \begin{pmatrix} \cos\sqrt{1-n}\theta & \frac{\rho}{\sqrt{1-n}}\sin\sqrt{1-n}\theta \\ -\frac{\sqrt{1-n}}{\rho}\sin\sqrt{1-n}\theta & \cos\sqrt{1-n}\theta \end{pmatrix} \begin{pmatrix} x_0 \\ x_0' \end{pmatrix}$$

$$= \mathbf{M}_H(\theta)\begin{pmatrix} x_0 \\ x_0' \end{pmatrix}, \qquad (2.26)$$

and similarly in the vertical plane

$$\begin{pmatrix} y \\ y' \end{pmatrix} = \begin{pmatrix} \cos\sqrt{n}\theta & \frac{\rho}{\sqrt{n}}\sin\sqrt{n}\theta \\ -\frac{\sqrt{n}}{\rho}\sin\sqrt{n}\theta & \cos\sqrt{n}\theta \end{pmatrix} \begin{pmatrix} y_0 \\ y_0' \end{pmatrix} = \mathbf{M}_V(\theta)\begin{pmatrix} y_0 \\ y_0' \end{pmatrix}. \qquad (2.27)$$

If we look at the motion in the x–direction, we see a sinusoidal behavior. This type of transverse oscillation is called the *betatron oscillation*. By defining a phase $\phi(s) = \sqrt{1-n}\theta = \rho^{-1}\sqrt{1-n}\,s$, we can write

$$x(s) = x_0\cos\phi(s) + \frac{\rho}{\sqrt{1-n}}\sin\phi(s)x_0'. \qquad (2.28)$$

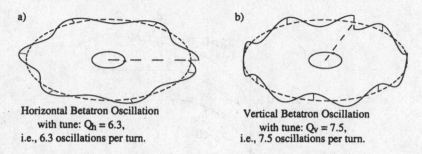

Horizontal Betatron Oscillation
with tune: $Q_h = 6.3$,
i.e., 6.3 oscillations per turn.

Vertical Betatron Oscillation
with tune: $Q_v = 7.5$,
i.e., 7.5 oscillations per turn.

Figure 2.4 a) Horizontal betatron oscillation. b) Vertical betatron oscillation. The tune is the number of wavelengths of the oscillation in one complete orbit around the ring. Note that for these figures the tunes are much higher than would be found in a weak focusing ring.

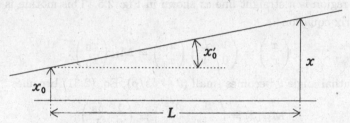

Figure. 2.5 Trajectory of a particle in a field free region.

The phase $\phi(s)$ is called the betatron phase, and the *betatron tune* is defined as the number of cycles of the betatron oscillation made by a particle during one full orbit (see Fig. 4),

$$Q_H = \frac{1}{2\pi}\sqrt{1-n}\,2\pi = \sqrt{1-n}. \qquad (2.29)$$

Similarly in the vertical direction, the betatron phase may be defined as $\phi(s) = \sqrt{n}\,\theta = \rho^{-1}\sqrt{n}\,s$, with a betatron tune of

$$Q_V = \sqrt{n}. \qquad (2.30)$$

It is interesting to note that $Q_H{}^2 + Q_V{}^2 = 1$, for the special case of weak focusing with a θ-independent guide field.

For $n = 0$, the matrix in Eq. (2.26) becomes simply

$$\mathbf{M_H}(n=0) = \begin{pmatrix} \cos\theta & \rho\sin\theta \\ -\frac{1}{\rho}\sin\theta & \cos\theta \end{pmatrix}. \qquad (2.31)$$

The vertical matrix becomes

$$\mathbf{M_V}(n=0) = \begin{pmatrix} 1 & \rho\,\underset{\sqrt{n}\to 0}{\lim}\frac{\sin\sqrt{n}\theta}{\sqrt{n}} \\ 0 & 1 \end{pmatrix} = \begin{pmatrix} 1 & \rho\theta \\ 0 & 1 \end{pmatrix}. \qquad (2.32)$$

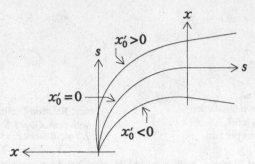

Figure. 2.6 Effect of divergence on a beam from a point source.

This is just the matrix of a free flight of length $\rho\theta$. The trajectory of a particle in a field free region is a straight line as shown in Fig. 2.5. This motion is described by the matrix equation

$$\begin{pmatrix} x \\ x' \end{pmatrix} = \begin{pmatrix} 1 & L \\ 0 & 1 \end{pmatrix} \begin{pmatrix} x_0 \\ x_0' \end{pmatrix} = \mathbf{M_L} \begin{pmatrix} x_0 \\ x_0' \end{pmatrix}. \tag{2.33}$$

If the azimuthal angle θ becomes small $(\theta \to \delta s/\rho)$, Eq. (2.31) becomes

$$\mathbf{M_H}(\theta) = \begin{pmatrix} 1 & \rho\frac{\delta s}{\rho} \\ \frac{-\delta s}{\rho^2} & 1 \end{pmatrix} \simeq \begin{pmatrix} 1 & \delta s \\ 0 & 1 \end{pmatrix}, \tag{2.34}$$

which coincides with Eq. (2.33) when $L = \delta s$. As a check, let the matrix in Eq. (2.31) be considered for $\theta = \pi/2$,

$$\begin{pmatrix} x \\ x' \end{pmatrix} = \begin{pmatrix} 0 & \rho \\ -\frac{1}{\rho} & 0 \end{pmatrix} \begin{pmatrix} 0 \\ x_0' \end{pmatrix} = \begin{pmatrix} \rho x_0' \\ 0 \end{pmatrix}, \tag{2.35}$$

which is shown graphically in Fig. 2.6.

If the sign of the field index is changed, the matrix in Eq. (2.27) changes to

$$\mathbf{M_V} = \begin{pmatrix} \cos i\sqrt{|n|}\,\theta & \frac{\rho}{i\sqrt{|n|}}\sin i\sqrt{|n|}\,\theta \\ -i\frac{\sqrt{|n|}}{\rho}\sin i\sqrt{|n|}\,\theta & \cos i\sqrt{|n|}\,\theta \end{pmatrix}$$

$$= \begin{pmatrix} \cosh\sqrt{|n|}\,\theta & \frac{\rho}{\sqrt{|n|}}\sinh\sqrt{|n|}\,\theta \\ \frac{\sqrt{|n|}}{\rho}\sinh\sqrt{|n|}\,\theta & \cosh\sqrt{|n|}\,\theta \end{pmatrix}, \tag{2.36}$$

where $i = \sqrt{-1}$.

A similar result can be obtained for $\mathbf{M_H}$ of Eq. (2.26) when $n > 1$,

$$\mathbf{M_H} = \begin{pmatrix} \cosh\sqrt{n-1}\,\theta & \frac{\rho}{\sqrt{n-1}}\sinh\sqrt{n-1}\,\theta \\ \frac{\sqrt{n-1}}{\rho}\sinh\sqrt{n-1}\,\theta & \cosh\sqrt{n-1}\,\theta \end{pmatrix}. \tag{2.37}$$

2.4 Momentum dispersion

So far all the particles have been considered as monoenergetic. Indeed a momentum spread does exist, and this implies a further radial broadening of the beam, while the vertical oscillations remain unaffected. In fact, if $\beta\gamma mc = p$ in Eq. (2.5) is replaced by $p + \delta p$, the horizontal motion is described by

$$\frac{d^2x}{d\theta^2} + \left[\left(1 - \frac{\delta p}{p}\right)\left(1 + \frac{1}{B_0}\frac{\partial B_y}{\partial x}x\right)\left(1 + \frac{x}{\rho}\right) - 1\right]\rho\left(1 + \frac{x}{\rho}\right) = 0, \qquad (2.38)$$

which simplifies to

$$\frac{d^2x}{d\theta^2} + (1 - n)x = \rho\frac{\delta p}{p}. \qquad (2.39)$$

The vertical motion equation with dispersion becomes

$$\frac{d^2y}{d\theta^2} + \left(1 - \frac{\delta p}{p}\right)ny\left(1 + \frac{2x}{\rho}\right) \simeq \frac{d^2y}{d\theta^2} + ny = 0, \qquad (2.40)$$

as before.

The solution of Eq. (2.39) is

$$x = A\cos\sqrt{1-n}\,\theta + B\sin\sqrt{1-n}\,\theta + \frac{\rho}{1-n}\frac{\delta p}{p}, \qquad (2.41)$$

$$x' = \frac{dx}{ds} = \frac{\sqrt{1-n}}{\rho}(-A\sin\sqrt{1-n}\,\theta + B\cos\sqrt{1-n}\,\theta). \qquad (2.42)$$

For $x(0) = x_0$, and $x'(0) = x_0'$ these can be written

$$x = (\cos\sqrt{1-n}\,\theta)x_0 + \left(\frac{\rho}{\sqrt{1-n}}\sin\sqrt{1-n}\,\theta\right)x_0'$$
$$+ \frac{\rho}{1-n}(1 - \cos\sqrt{1-n}\,\theta)\left(\frac{\delta p}{p}\right)_0 \qquad (2.43)$$

$$x' = -\left(\frac{\sqrt{1-n}}{\rho}\sin\sqrt{1-n}\,\theta\right)x_0 + (\cos\sqrt{1-n}\,\theta)x_0'$$
$$+ \left(\frac{1}{\sqrt{1-n}}\sin\sqrt{1-n}\,\theta\right)\left(\frac{\delta p}{p}\right)_0 \quad \text{with} \qquad (2.44)$$

$$\frac{\delta p}{p} = \left(\frac{\delta p}{p}\right)_0. \qquad (2.45)$$

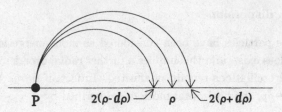

Figure. 2.7 Effects of momentum spread in a beam from a point source at P.

It is convenient to write this in a 3×3 matrix form

$$
\begin{pmatrix} x \\ x' \\ \frac{\delta p}{p} \end{pmatrix} = \begin{pmatrix} \cos\sqrt{1-n}\theta & \frac{\rho}{\sqrt{1-n}}\sin\sqrt{1-n}\theta & \frac{\rho}{1-n}(1-\cos\sqrt{1-n}\theta) \\ -\frac{\sqrt{1-n}}{\rho}\sin\sqrt{1-n}\theta & \cos\sqrt{1-n}\theta & \frac{1}{\sqrt{1-n}}\sin\sqrt{1-n}\theta \\ 0 & 0 & 1 \end{pmatrix}
$$
$$
\begin{pmatrix} x_0 \\ x'_0 \\ \left(\frac{\delta p}{p}\right)_0 \end{pmatrix}.
$$
(2.46)

This matrix is frequently written with, $n = 0$, as

$$
\mathbf{M_H}(\theta) = \begin{pmatrix} \cos\theta & \rho\sin\theta & \rho(1-\cos\theta) \\ -\frac{1}{\rho}\sin\theta & \cos\theta & \sin\theta \\ 0 & 0 & 1 \end{pmatrix}.
$$
(2.47)

For $\theta = \pi$, we have

$$
\mathbf{M_H}(\pi) = \begin{pmatrix} -1 & 0 & 2\rho \\ 0 & -1 & 0 \\ 0 & 0 & 1 \end{pmatrix},
$$
(2.48)

which when applied to a vector of initial conditions

$$
\begin{pmatrix} 0 \\ 0 \\ \pm\frac{\delta p}{p} \end{pmatrix}
$$
(2.49)

yields

$$
x_+ = 2\rho\frac{\delta p}{p},
$$
(2.50)

$$
x_- = -2\rho\frac{\delta p}{p},
$$
(2.51)

$$
x'_\pm = 0.
$$
(2.52)

The orbit deviation after a bend of 180° is $\delta x = x_+ - x_- = 4\rho\,\delta p/p$, which may be checked by simple geometry as shown in Fig. 2.6

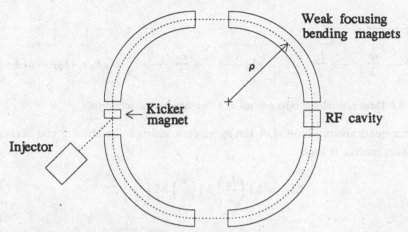

Figure. 2.8 Lay-out of an "old-fashioned" weak focusing synchrotron.

2.5 Weak focusing synchrotron

The next step after (synchro)cyclotrons was the implementation of the weak focusing synchrotron: a machine at constant radius, with its magnetic field varying in time together with the increasing momentum of particles (usually either electrons or protons).

Every turn, particles receive an energy-kick by an rf cavity, and, if the synchronism is correct, the goal-energy is reached after several revolutions. A straight-section is needed for locating this rf equipment.

Besides, the initial momentum must not be too small, because it is hard to realize a low magnetic field with a very good uniformity. Therefore an injector, of the direct-voltage type described in § 1.2, can provide a beam with enough kinetic energy to be captured in the synchrotron. This requires a second straight section free of the normal bending magnets in order to match the injected beam with the optics of the cyclic ring.

Since some other equipment (e. g., control systems, internal targets, etc.) need more room, another two straight sections are frequently introduced, producing the common four-fold symmetric structure (see Fig. 2.8) with four bends and four straight sections.

By definition, a *periodic-element* or *cell* is any block of items (bending magnets and straight sections for the present chapter) which can span the whole machine by repeating itself. Fig. 2.9 shows three possible ways of choosing a periodic cell

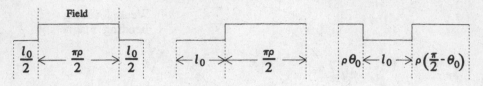

Figure. 2.9 Three examples of cells referred to a four-quadrant synchrotron.

in a four-quadrant machine. Let the symmetric example be chosen; the horizontal oscillation matrix is then

$$\mathbf{M_H} = \mathbf{M}\left(\frac{l_0}{2}\right)\mathbf{M}\left(\frac{\pi}{2}\right)\mathbf{M}\left(\frac{l_0}{2}\right), \tag{2.53}$$

or

$$\mathbf{M_H} = \begin{pmatrix} 1 & \frac{l_0}{2} \\ 0 & 1 \end{pmatrix}\begin{pmatrix} \cos\frac{\sqrt{1-n}\pi}{2} & \frac{\rho}{\sqrt{1-n}}\sin\frac{\sqrt{1-n}\pi}{2} \\ -\frac{\sqrt{1-n}}{\rho}\sin\frac{\sqrt{1-n}\pi}{2} & \cos\frac{\sqrt{1-n}\pi}{2} \end{pmatrix}\begin{pmatrix} 1 & \frac{l_0}{2} \\ 0 & 1 \end{pmatrix}. \tag{2.54}$$

Carrying out the multiplication gives

$$\mathbf{M_H} = \left(\begin{matrix} \cos\frac{\sqrt{1-n}\pi}{2} - \frac{l_0}{2\rho}\sqrt{1-n}\sin\frac{\sqrt{1-n}\pi}{2} \\ -\frac{\sqrt{1-n}}{\rho}\sin\frac{\sqrt{1-n}\pi}{2} \\[1em] l_0\cos\frac{\sqrt{1-n}\pi}{2} + \frac{\rho}{\sqrt{1-n}}\left[1 - \frac{l_0^2(1-n)}{4\rho^2}\right]\sin\frac{\sqrt{1-n}\pi}{2} \\ \cos\frac{\sqrt{1-n}\pi}{2} - \frac{l_0}{2\rho}\sqrt{1-n}\sin\frac{\sqrt{1-n}\pi}{2} \end{matrix} \right). \tag{2.55}$$

For $l_0/2\rho \ll \pi/2$ one can write:

$$\cos\frac{\sqrt{1-n}\pi}{2} - \frac{l_0}{2\rho}\sqrt{1-n}\sin\frac{\sqrt{1-n}\pi}{2} \simeq \cos\left[\left(1+\frac{l_0}{\rho\pi}\right)\frac{\sqrt{1-n}\pi}{2}\right]$$

$$= \cos\mu_H = \frac{1}{2}\mathrm{tr}\mathbf{M} \tag{2.56}$$

with

$$\mu_H = \left(1 + \frac{l_0}{\rho\pi}\right)\frac{\sqrt{1-n}\pi}{2} = \text{phase} - \text{advance per cell}, \tag{2.57}$$

and

$$Q_H = \frac{\mu}{\pi/2} = \left(1 + \frac{l_0}{\rho\pi}\right)\sqrt{1-n} = \text{horizontal betatron tune.} \tag{2.58}$$

As an example let us show that the matrix Eq. (2.55) can be written in the form:

$$\mathbf{M_H} = \begin{pmatrix} \cos\mu_H & \beta_H\sin\mu_H \\ -\frac{1}{\beta_H}\sin\mu_H & \cos\mu_H \end{pmatrix}. \tag{2.59}$$

Solution: Let the following ancillary quantities be defined:

$$\psi = \frac{\sqrt{1-n}\pi}{2} \tag{2.60}$$

$$\lambda = \frac{\rho}{\sqrt{1-n}}. \tag{2.61}$$

Then taking into account Eq. (2.57), we have:

$$\cos\mu_H \simeq \cos\psi - \frac{l_0}{2\lambda}\sin\psi, \tag{2.62}$$

$$\sin\mu_H \simeq \sin\psi + \frac{l_0}{2\lambda}\cos\psi, \tag{2.63}$$

or

$$\cos\psi \simeq \cos\mu_H + \frac{l_0}{2\lambda}\sin\psi, \tag{2.64}$$

$$\sin\psi \simeq \sin\mu_H - \frac{l_0}{2\lambda}\cos\psi. \tag{2.65}$$

We may now write Eq. (2.55) as

$$\mathbf{M_H} = \begin{pmatrix} \cos\psi - \frac{l_0}{2\lambda}\sin\psi & \lambda\left(\sin\psi + \frac{l_0}{\lambda}\cos\psi - \frac{l_0^2}{4\lambda^2}\sin\psi\right) \\ -\frac{\sin\psi}{\lambda} & \cos\psi - \frac{l_0}{2\lambda}\sin\psi \end{pmatrix}. \tag{2.66}$$

It is easy to see that the diagonal elements are just $\cos\mu_H$.

$$\mathbf{M_{12}} = \lambda\left[\sin\psi + \frac{l_0}{2\lambda}\cos\psi + \frac{l_0}{2\lambda}\left(\cos\psi - \frac{l_0}{2\lambda}\sin\psi\right)\right]$$

$$\simeq \lambda\left(\sin\mu_H + \frac{l_0}{2\lambda}\cos\mu_H\right). \tag{2.67}$$

In order to have $\mathbf{M_{12}} = \beta_H\sin\mu_H$, it suffices to set

$$\beta_H \simeq \lambda\left(1 + \frac{l_0}{2\lambda}\cot\mu_H\right). \tag{2.68}$$

On the other hand, $\mathbf{M_{21}} = -\sin\psi/\lambda$ should be equal to $-\sin\mu_H/\beta_H$,

$$\beta_H = \lambda\frac{\sin\mu_H}{\sin\psi} \simeq \lambda\frac{\sin\mu_H}{\sin\mu_H - \frac{l_0}{2\lambda}\cos\mu_H} = \lambda\left(1 - \frac{l_0}{2\lambda}\cot\mu_H\right)^{-1} \tag{2.69}$$

or

$$\beta_{\mathrm{H}} \simeq \lambda \left(1 + \frac{l_0}{2\lambda} \cot \mu_{\mathrm{H}}\right) \quad \text{again.} \tag{2.70}$$

Comparing Eq. (2.59) to Eq. (2.26), we may state that the real quasi-circular synchrotron can be replaced by a dummy circular machine, whose circumference and radius are respectively

$$L = 2\pi\rho + 4l_0 \tag{2.71}$$

$$R = \rho\left(1 + \frac{2l_0}{\pi\rho}\right). \tag{2.72}$$

Notice that, if $\cot \mu_{\mathrm{H}} \simeq 1/\mu_{\mathrm{H}}$, Eq. (2.70) becomes

$$\beta_{\mathrm{H}} \simeq \frac{\rho}{\sqrt{1-n}} \left[1 + \frac{l_0}{\pi\rho}\left(1 - \frac{l_0}{\pi\rho}\right)\right] \tag{2.73}$$

or

$$\beta_{\mathrm{H}} \simeq \frac{\rho}{\sqrt{1-n}} \left(1 + \frac{l_0}{\pi\rho}\right) \tag{2.74}$$

having neglected terms of order $(l_0/(\pi\rho))^2$ or higher. We could also start by defining $\beta_{\mathrm{H}} = R/Q_{\mathrm{H}}$, with Q_{H} and R given by Eqs. (2.58 and 2.72), respectively.

Similar results can be obtained for the vertical oscillations:

$$\mu_{\mathrm{V}} = \left(1 + \frac{l_0}{\rho\pi}\right) \sqrt{n} \frac{\pi}{2} \tag{2.75}$$

$$Q_{\mathrm{V}} = \left(1 + \frac{l_0}{\rho\pi}\right) \sqrt{n} \tag{2.76}$$

$$\beta_{\mathrm{V}} \simeq \frac{\rho}{\sqrt{n}} \left(1 + \frac{l_0}{\rho\pi}\right). \tag{2.77}$$

2.6 Momentum compaction factor

Particles with slightly different momenta travel along reference orbits, that vary according to the "golden relation" Eq. (1.25) between momentum and bending radius. Then Eq. (2.71) gives $dL = 2\pi\, d\rho$ or $dL/L = (1 + 2l_0/(\pi\rho))^{-1}\, d\rho/\rho$, since paths in the straight sections are the same for particles with different momenta. In addition Eqs. (1.25 and 2.13) yield

$$\frac{\delta p}{p} = \frac{d\rho}{\rho} + \frac{dB}{B} = \left(1 + \frac{\rho}{B}\frac{\partial B}{\partial\rho}\right) \frac{d\rho}{\rho} = (1-n)\frac{d\rho}{\rho}. \tag{2.78}$$

We may now calculate the momentum compaction, defined in Eq. (1.38), as the fractional difference in circumference of the reference orbit divided by the fractional difference in the particle momenta,

$$\alpha_p = \frac{dL/L}{dp/p} = \left(1 + \frac{2l_0}{\pi\rho}\right)^{-1} (1-n)^{-1} \simeq \frac{1}{Q_H^2} \qquad (2.79)$$

as one can prove by squaring the inverse of Eq. (2.58).

Reintroducing time as the independent variable into Eq. (2.12) produces

$$\frac{d^2x}{dt^2} + \omega_H^2 x = 0, \qquad (2.80)$$

with $\omega_H = Q_H\omega_s = Q_H v/R \simeq v/\beta_H$, where ω_s is the angular frequency of revolution for the particle. But ω_H is just $2\pi/\tau_H = 2\pi v/\lambda_H$, therefore $\lambda_H = 2\pi/\beta_H = L/Q_H$. Similarly, $\lambda_V = 2\pi\beta_V = L/Q_V$.

This means that particles undergo betatron oscillations with wavelengths λ_H, and λ_V slightly longer than the machine circumference. This implies that the horizontal and vertical dimensions of the vacuum chamber must be proportional to the machine radius, i. e., to the maximum momentum of particles. Then the volume of the synchrotron is proportional to the *third power* of the goal–energy, making unpractical high–energy weak focusing synchrotrons.

In Chapter 5, the principle of *strong focusing* will be discussed thoroughly. In simple terms, this method amounts to making Q_H and Q_V arbitrarily large. In doing this, the vacuum chamber cross–section is practically independent of particle momenta; the volume, weight and, last but not least, the cost are roughly proportional to the energy. It is worthwhile to mention that Eq. (2.79) is still a good approximation for the strong focusing case.

Before proceeding to strong focusing, we must first examine separated function magnets: particularly, the dipole bending magnet, and the quadrupole focusing magnet.

Problems

2–1   Derive Eqs. (2.39 and 2.40), but keep terms to second order in x, y, and δ.

2–2   Work out the equivalent for Eq. (2.55) with dispersion.

2–3   Calculate the momentum compaction factor, α_p, for a betatron with a field index, n.

2–4   Show by explicit multiplication of matrices that the transfer matrices (both vertical and horizontal) for a combination of two sector magnets with identical bending radii, but with different bend angles θ_1 and θ_2, is equivalent to a sector magnet with a bend of $\theta_1 + \theta_2$, if there is no drift between the two magnets.

2–5   The RHIC collider collides fully stripped gold ions ($A = 197$, $Z = 79$) at a total energy of 100 GeV/nucl. per beam. The circumference of each ring is 3834 m. (Assume the mass of a gold ion is 197×0.93113 GeV/c^2.)

a) If the injection energy is 10.5 GeV/nucleon, what is the required swing in revolution frequency during acceleration?

b) If we assume that there are 192 identical dipoles per ring, what is the field at top field? Assume each dipole is 10 m long.

References for Chapter 2

[1] J. J. Livingood, **Cyclic Particle Accelerators**, Van Nostrand, New York (1961).

3

Mechanics of Trajectories

In this chapter, the general theory of particle trajectories moving in a six dimensional phase space is examined in terms of generalized transformations and Hamiltonian dynamics. For small deviations about a design trajectory, the linearized transformation yields a matrix which is the Jacobian matrix for the transformation. Liouville's theorem[1,2], which requires the conservation of particle density in phase space for nondissipative systems, is shown to be equivalent to the requirement that the Jacobian matrix have a unit determinant. The trajectories are studied in terms of Hamiltonian dynamics, and the transformations are shown to be symplectic* for a canonical choice of spatial and momentum coordinates. Finally, the standard coordinates of Chapter 2 are obtained using the paraxial approximation, and their limitations are studied.

3.1 Liouville's theorem

For a large group of particles, a density function $f(x, y, z, p_x, p_y, p_z, t)$ may be defined for the number of particles per volume of six dimensional phase space with coordinates given by the three spatial coordinates x, y, and z, and their corresponding momentum coordinates p_x, p_y, and p_z. We are assuming that the number of particles is large enough so we may treat f as a continuous function of the phase space coordinates and ignore the fact that the particles are really point objects. In relation to accelerator physics, Liouville's theorem may be stated as follows:

> In the local region of a particle, the particle density in phase space is constant, provided that the particles move in a general field consisting of magnetic fields and of fields whose forces are independent of velocity.

Here we are ignoring the effect of radiation due to the acceleration of the charges. Liouville's theorem does not hold for systems with dissipation of energy, which is generally driven by velocity dependent forces.

---

* The term *symplectic* was coined by H. Weyl[3] to indicate the symplectic group. It is derived from the Greek word $\sigma\upsilon\mu\pi\lambda\epsilon\gamma\mu\alpha$ meaning "complex" or "braided together".

We can define a six dimensional current of the particles moving in phase space by

$$\vec{J}_6 = (f\dot{x}, f\dot{y}, f\dot{z}, f\dot{p}_x, f\dot{p}_y, f\dot{p}_z) = (f\vec{v}, f\vec{F}), \qquad (3.1)$$

where the subscript six implies a six dimensional vector. The velocity $\vec{v}$ is the time derivative of the spatial coordinates, and the force $\vec{F}$ is the time derivative of the momentum coordinates. The continuity equation for this current is then

$$\frac{\partial f}{\partial t} + \nabla_6 \cdot \vec{J}_6 = 0, \qquad (3.2)$$

which just reflects the fact that the total number of particles in the beam is constant.* In order to prove Liouville's theorem, we will show that the left hand side of this equation is just the total derivative of f with respect to time; hence

$$\frac{df}{dt} = 0. \qquad (3.3)$$

The gradient

$$\nabla_6 \cdot \vec{J}_6 = \nabla \cdot (f\vec{v}) + \nabla_p \cdot (f\vec{F})$$
$$= (\nabla f) \cdot \vec{v} + f(\nabla \cdot \vec{v}) + (\nabla_p f) \cdot \vec{F} + f(\nabla_p \cdot \vec{F}), \qquad (3.4)$$

where ∇_p is the gradient with respect to the three momentum coordinates. Relativistically,

$$\vec{v} = (p^2 c^2 + m^2 c^4)^{-\frac{1}{2}} \vec{p} c^2, \qquad (3.5)$$

which is independent of the spatial coordinates, and thus $\nabla \cdot \vec{v} = 0$.

The force may be written as

$$\vec{F} = \frac{d\vec{p}}{dt} = \vec{g}(\vec{r}) + q\vec{v} \times \vec{B}(\vec{r}), \qquad (3.6)$$

---

* The rate of change of particles inside an arbitrary volume, V, is equal to the negative of the total flux leaving the volume:

$$\frac{\partial}{\partial t} \int_V f \, d^6 X = \int_V \frac{\partial f}{\partial t} d^6 X = -\int_{\partial V} \vec{J}_6 \cdot d\vec{S} = -\int_V (\nabla_6 \cdot \vec{J}_6) \, d^6 X,$$

where the last equality is the six-dimensional version of the divergence theorem which is a special case of Stokes' theorem in n-dimensions.[4] Since this is true for any arbitrary volume, the integrands of the second and fourth integrals must be identically equal, thus proving the continuity equation.

with $\vec{r} = (x, y, z)$. The function $\vec{g}$ is some general force which is independent of $\vec{v}$, and $\vec{B}$ is some external magnetic field through which the beam particles of charge q are passing. Since $\vec{g}$ is independent of $\vec{p}$, we have $\nabla_p \cdot \vec{g} = 0$, and we may write

$$\nabla_p \cdot \vec{F} = q\nabla_p \cdot (\vec{v} \times \vec{B}) = q\vec{B} \cdot (\nabla_p \times \vec{v}) - q\vec{v} \cdot (\nabla_p \times \vec{B}). \tag{3.7}$$

The curl in momentum space of $\vec{B}$ is zero since the magnetic field only depends on spatial coordinates, hence

$$\nabla_p \cdot \vec{F} = q\vec{B} \cdot (\nabla_p \times \vec{v}). \tag{3.8}$$

Looking at the p_z-component of $\nabla_p \times \vec{v}$, we find

$$\left[\nabla_p \times \left(\frac{\vec{p}}{\sqrt{p^2 c^2 + m^2 c^4}} \right) \right]_z$$
$$= \frac{\partial}{\partial p_x} \left(\frac{p_y}{\sqrt{p^2 c^2 + m^2 c^4}} \right) - \frac{\partial}{\partial p_y} \left(\frac{p_x}{\sqrt{p^2 c^2 + m^2 c^4}} \right)$$
$$= 0. \tag{3.9}$$

By cyclic permutation of the coordinates, it is obvious that

$$\nabla_p \times \vec{v} = 0, \tag{3.10}$$

and

$$\nabla_6 \cdot \vec{J}_6 = (\nabla f) \cdot \vec{v} + (\nabla_p f) \cdot \vec{F}. \tag{3.11}$$

Substituting this back in the continuity equation gives

$$\frac{\partial f}{\partial t} + (\nabla f) \cdot \frac{d\vec{r}}{dt} + (\nabla_p f) \cdot \frac{d\vec{p}}{dt} = \frac{df}{dt} = 0, \tag{3.12}$$

thus proving the theorem.

This theorem seems to imply that once you have a beam of particles, there is no way you can increase the density of particles in the beam. This is true if the initial assumptions are rigidly observed; however, in reality they are only approximately true. By assuming a continuous distribution for the beam, we are eliminating any possible interactions between individual particles. In a real beam, single particles may come very close together, resulting in a repulsive Coulomb force much larger than the force calculated from the continuous distribution. This two body force must depend on the velocities of the individual particles, since particles of different

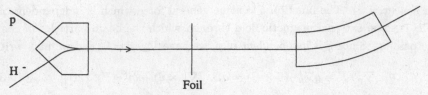

Figure. 3.1 Two oppositely charged beams of the same momentum can be made to travel along the same trajectory through a straight section since they can be bent in opposite directions.

momenta will result in different distances of closest approach. These collisions can provide large momentum transfers which decrease the number of particles remaining in the beam. When these single particle collisions are ignored, we call this model of the beam a *collisionless* beam. The *space-charge force*, as calculated from the continuous distribution of charge, may be obtained from a velocity independent potential,

$$V(\vec{x}) = \frac{q}{4\pi\epsilon_0} \int \frac{f(\vec{x}', \vec{p})}{|\vec{x} - \vec{x}'|} d^3x' \, d^3p, \qquad (3.13)$$

and thus will satisfy the conditions for Liouville's theorem. More will be said about this space-charge force in Chapter 11.

We will mention a few more ways to circumvent Liouville's theorem by the use of dissipative forces. In circular electron accelerators, synchrotron radiation is emitted in the forward direction with an opening angle of about $1/\gamma = mc^2/E$. The momentum is thus decreased in both the longitudinal and transverse directions. When the beam passes through the rf structures, it is accelerated in only the longitudinal direction, causing the transverse beam size to shrink. The energy loss due to synchrotron radiation is caused by a dissipative force, which implies that Liouville's theorem does not hold. This is called *radiation damping* and will be covered in more detail in Chapter 8.

In high energy proton accelerators, H^- ions are initially accelerated in a linac and can then be injected into a circular accelerator, so that they are placed in the same volume with an existing bunch of protons that are circulating in the ring (see Figure 3.1). It is possible to have the two oppositely charged bunches traveling together in a straight section, since they can be bent in opposite directions by the same magnet. Two bunches of protons of the same energy would not be able to follow the same trajectory. In this straight section, the beams are passed through a thin foil which strips the electrons from the H^- ions, leaving a single proton beam of higher density in phase space. The interaction of the H^- beam with the foil is a dissipative force; therefore, Liouville's theorem can not be invoked.

Another trick is *electron cooling*, which uses a "cold" beam of electrons, i. e., a beam with very little spread in phase space. This cold beam can be made to travel in a straight section along with a hot beam of antiprotons whose average velocity is the same as the electron beam. The hot antiproton beam will interact with the cold electron beam via the Coulomb force and lose some of its volume in phase space to the electron beam. The total phase space volume of the two beams is conserved, but the volume of the antiprotons is reduced. At the end of the straight section the two beams are separated by a bending magnet since they have different momenta.

Stochastic cooling uses the fact that the density of the beam is not a continuous function, but is really a bunch of spikes due to the point-like nature of the particles. In the space between the particles, the density is zero. This method basically removes some of the holes between the particles by statistically averaging out density fluctuations, thus increasing the particle density. This clearly works better for smaller numbers of particles. Electron cooling and stochastic cooling are covered in Chapter 12.

3.2 General transformations

In a section of an accelerator or a beam line, particle trajectories may be studied as deviations from the trajectory of the design particle. At the input end, $s = s_0$, of the section the trajectory of a particular particle can be defined relative to the design particle by using the difference in the six dimensional phase space coordinates of the two particles,

$$\vec{X} = (\Delta x_0, \Delta p_{x0}, \Delta y_0, \Delta p_{y0}, \Delta z_0, \Delta p_{z0}). \tag{3.14}$$

Similarly at the output end $s = s_1$ of the section, we may write

$$\vec{Y} = (\Delta x_1, \Delta p_{x1}, \Delta y_1, \Delta p_{y1}, \Delta z_1, \Delta p_{z1}). \tag{3.15}$$

When the design particle has moved to s_1, we may parametrize the location of the output particle by a vector function of the input coordinates,

$$\vec{Y} = \vec{T}(\vec{X}). \tag{3.16}$$

This map can be a non–linear function of the input coordinates, and only requires that the path length, s, of the design particle is a monotonically increasing function of time, i. e., the design particle does not backtrack.

For small deviations of the input particle from the design trajectory, we may expand $\vec{T}(\vec{X})$ about the design trajectory,

$$\Delta \vec{Y} = \vec{T}(\vec{0}) + \sum_{j=1}^{6} \frac{\partial \vec{T}}{\partial \Delta X_j}(\vec{0}) \, \Delta X_j + \frac{1}{2} \sum_{j,k=1}^{6} \frac{\partial^2 \vec{T}}{\partial \Delta X_j \, \partial \Delta X_k}(\vec{0}) \Delta X_j \, \Delta X_k + \cdots. \quad (3.17)$$

Clearly $\vec{T}(\vec{0}) = 0$, since this is just the path of the design particle. Keeping only linear terms yields

$$\Delta Y_i = \sum_{j=1}^{6} M_{ij} \, \Delta X_j, \quad (3.18)$$

where we defined the matrix $\mathbf{M}$ to be the Jacobian matrix of the transformation from initial to final coordinates with

$$M_{ij} = \frac{\partial T_i}{\partial X_j}(\vec{0}). \quad (3.19)$$

The determinant of this matrix gives the ratio of the volume element of the new coordinates, $\vec{Y}$, to the volume element of the old coordinates, $\vec{X}$. Since the coordinates chosen were the coordinates of six dimensional phase space, Liouville's theorem requires that

$$\det \mathbf{M} = 1. \quad (3.20)$$

If we parametrize the transformation $\vec{T}(\vec{X})$ in terms of canonical coordinates and momenta we shall find that an even stronger requirement than Liouville's theorem is obtained.

3.3 Canonical momentum and vector potential

For a conservative force, the work done by the force in moving from point P_1 to point P_2 is independent of the path taken:

$$\int_{P_1}^{P_2} \vec{F} \cdot d\vec{r} \quad \text{is invariant,} \quad (3.21)$$

or more succinctly

$$\oint \vec{F} \cdot d\vec{r} = 0 \quad (3.22)$$

from which by use of Stoke's theorem we get in differential form

$$\nabla \times \vec{F} = 0. \quad (3.23)$$

From basic mechanics we learned that forces which only depend on position and not the velocity of the particle being worked upon are conservative. Examples of such forces are those from gravitational and static electric fields. When there are magnetic fields present the Lorentz force can depend on velocity:

$$\frac{d\vec{p}}{dt} = \vec{F} = q(\vec{E} + \vec{v} \times \vec{B}), \tag{3.24}$$

and is not always conservative. Taking the curl of the Lorentz force yields:

$$\begin{aligned}
\nabla \times \vec{F} = \nabla \times \frac{d\vec{p}}{dt} &= q(\nabla \times \vec{E} + \nabla \times (\vec{v} \times \vec{B})) \\
&= -q\frac{\partial \vec{B}}{\partial t} + q\left[(\vec{B}\cdot\nabla)\vec{v} - (\vec{v}\cdot\nabla)\vec{B} + (\nabla\cdot\vec{B})\vec{v} - (\nabla\cdot\vec{v})\vec{B}\right] \\
&= -q\frac{\partial \vec{B}}{\partial t} - q\left[(\vec{v}\cdot\nabla)\vec{B}\right] \\
&= -q\left[\frac{\partial \vec{B}}{\partial t} + \frac{\partial \vec{B}}{\partial x}\frac{dx}{dt} + \frac{\partial \vec{B}}{\partial y}\frac{dy}{dt} + \frac{\partial \vec{B}}{\partial z}\frac{dz}{dt}\right] \\
&= -q\frac{d\vec{B}}{dt} \\
&= -\frac{d}{dt}(\nabla \times q\vec{A}), \tag{3.25}
\end{aligned}$$

where $\vec{A}$ is vector potential for the magnetic field $\vec{B}$. Moving terms to the left side produces

$$\nabla \times \frac{d\vec{p}}{dt} + \frac{d}{dt}(\nabla \times q\vec{A}) = 0, \tag{3.26}$$

which after reordering the differentiation becomes

$$\nabla \times \left[\frac{d}{dt}\left(\vec{p} + q\vec{A}\right)\right] = 0. \tag{3.27}$$

If we define a new *canonical momentum* by

$$\vec{P} = \vec{p} + q\vec{A}, \tag{3.28}$$

then the corresponding *canonical force*

$$\vec{F}_{\text{can}} = \frac{d\vec{P}}{dt} \tag{3.29}$$

is conservative. We should note that the choice of vector potential is not unique, since the gradient of any scalar function $\phi(x, y, z)$ of spatial coordinates may be added to $\vec{A}$ without changing $\vec{B}$, i. e.,

$$\nabla \times (\nabla\phi) = 0. \tag{3.30}$$

3.4 Hamiltonian formalism and canonical coordinates

In this section, we use several results from Hamiltonian dynamics without proof. For a more detailed discussion of these we recommend a graduate level classical mechanics book such as that of Goldstein.[6]

For a free particle with rest mass, m, and charge, q, and kinetic momentum $\vec{p} = \gamma m \vec{v}$, we may write the relativistic Hamiltonian as

$$H = \sqrt{p^2 c^2 + m^2 c^4}. \tag{3.31}$$

In an electromagnetic field with potentials $\vec{A}(\vec{x}, t)$ and $\phi(\vec{x}, t)$ such that

$$\vec{B} = \nabla \times \vec{A}, \tag{3.32}$$

and

$$\vec{E} = -\nabla \phi - \frac{\partial \vec{A}}{\partial t}, \tag{3.33}$$

the Hamiltonian can be given by

$$H = \sqrt{(\vec{P} - q\vec{A})^2 c^2 + m^2 c^4} + q\phi, \tag{3.34}$$

where the canonical momentum

$$\vec{P} = \vec{p} + q\vec{A}(\vec{x}, t). \tag{3.35}$$

This canonical momentum shows the coupling of the kinetic momentum of the particle to the momentum associated with the magnetic field. The result of the coupling is seen in the spiraling of charged particles in magnetic fields, i. e., the Lorentz force.

Using Hamilton's equations, the equations of motion can be written as

$$\frac{d\vec{P}}{dt} = -\nabla H, \quad \text{and} \quad \frac{d\vec{x}}{dt} = \nabla_P H. \tag{3.36}$$

We would like to write the Hamiltonian in terms of the local coordinate system, defined in Chapter 2, which was really a cylindrical system of coordinates with the origin at the center of curvature of the design trajectory. The position vector, $\vec{x} = (r, \theta, y)$, is the position of the particle relative to a fixed Cartesian coordinate system, (ξ, η, y), whose origin coincides with the center of curvature of the design

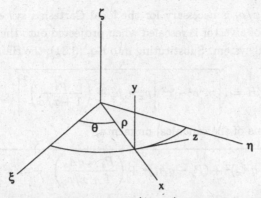

Figure. 3.2 Relative to a fixed coordinate system, (ξ, η, y), centered on the center of curvature of the design trajectory, the particle has cylindrical coordinates (r, θ, y). Note that the local system is right-handed if the coordinates are ordered as (x, z, y). The origin of the local system moves with the design particle.

trajectory as shown in Figure 3.2. Care must be taken when transforming from the cylindrical system to the local system. The velocity of the particle is

$$\vec{v} = \frac{d\vec{x}}{dt} = \frac{d}{dt}(r\hat{r} + y\hat{y})$$
$$= \dot{r}\hat{r} + r\dot{\theta}\hat{\theta} + \dot{y}\hat{y}. \tag{3.37}$$

The kinetic momentum is then

$$\vec{p} = \gamma m \vec{v} = \gamma m(\dot{r}\hat{r} + r\dot{\theta}\hat{\theta} + \dot{y}\hat{y})$$
$$= p_r\hat{r} + p_\theta\hat{\theta} + p_y\hat{y}. \tag{3.38}$$

In terms of the design trajectory radius, ρ, we may write

$$r = \rho + x, \tag{3.39}$$

and

$$s = \rho\theta, \tag{3.40}$$

with x being the excess distance of our particle from the origin, as in Chapter 2. The momentum conjugate to x is just $p_x = p_r$, since ρ may be taken to be constant in piecewise sections of the accelerator. It can be shown (see Problem 3-1) that a momentum coordinate which is conjugate to the coordinate s is given by

$$p_s = \left(1 + \frac{x}{\rho}\right)(\vec{p} \cdot \hat{s}). \tag{3.41}$$

The factor of $(1 + x/\rho)$ is necessary for the local Cartesian system, since the tangential component of a vector is rescaled when projected onto the design trajectory using the cylindrical system. Substituting into Eq. (3.31), the Hamiltonian becomes

$$H = \sqrt{m^2c^4 + c^2 \left[p_x^2 + p_y^2 + \left(\frac{p_s}{1 + x/\rho} \right)^2 \right]}. \tag{3.42}$$

Writing this in terms of the canonical momenta,

$$H = c\sqrt{(P_x - qA_x)^2 + (P_y - qA_y)^2 + \left(\frac{P_s - qA_s}{1 + x/\rho} \right)^2 + m^2c^2} + q\phi, \tag{3.43}$$

where we define

$$A_s = \left(1 + \frac{x}{\rho} \right) (\vec{A} \cdot \hat{\theta}), \tag{3.44}$$

as in Eq. (3.41).

Provided that the design particle moves without backtracking, we may change the role of the s and t coordinates, i. e., s now becomes* our independent parameter and t becomes one of the canonical coordinates. The conjugate momentum corresponding to t is now $-H$, which is just the negative of the total energy, and the new Hamiltonian is

$$\mathcal{H} = -P_s(x, P_x, y, P_y, t, -U; s)$$

$$= -qA_s - \left(1 + \frac{x}{\rho} \right) \sqrt{\left(\frac{U - q\phi}{c} \right)^2 - m^2c^2 - (P_x - qA_x)^2 - (P_y - qA_y)^2}, \tag{3.45}$$

---

* A canonical transformation from the variables $(\vec{q}, \vec{p})$ to the variables $(\vec{Q}, \vec{P})$ preserves the *integral invariant of Poincaré-Cartan*[4]: $\vec{p} \cdot d\vec{q} - H dt = \vec{P} \cdot d\vec{Q} - K dt$, where $K(\vec{Q}, \vec{P}; t)$ is the Hamiltonian expressed in terms of the new coordinates. If some coordinate q_j increases in time, then we may rewrite this invariant as

$$\left(\sum_{i \neq j} p_i dq_i + (-H) dt \right) - (-p_j) dq_j.$$

Note that this method of defining canonical transformations does not allow a simple rescaling of coordinates such as $q \to \bar{q}$ and $\vec{p} \to a\vec{p}$ for some constant a. This transformation amounts to a change of dimensions and will result in a new Hamiltonian in which the new momenta are still conjugate to the new coordinates; however, the volume element in phase space will have to be rescaled.

where we have replaced H by the total energy U.

In the cylindrical coordinates (r, θ, y), the curl of $\vec{A}$ is

$$\nabla \times \vec{A} = \left(\frac{1}{r} \frac{\partial A_y}{\partial \theta} - \frac{\partial A_\theta}{\partial y} \right) \hat{r} + \left(\frac{\partial A_r}{\partial y} - \frac{\partial A_y}{\partial r} \right) \hat{\theta}$$
$$+ \frac{1}{r} \left(\frac{\partial}{\partial r}(r A_\theta) - \frac{\partial A_r}{\partial \theta} \right) \hat{y}. \tag{3.46}$$

Transforming to the local coordinate system, (x, y, s), gives

$$\nabla \times \vec{A} = \frac{1}{1 + x/\rho} \left(\frac{\partial A_s}{\partial y} - \frac{\partial A_y}{\partial s} \right) \hat{x} + \frac{1}{1 + x/\rho} \left(\frac{\partial A_x}{\partial s} - \frac{\partial A_s}{\partial x} \right) \hat{y}$$
$$+ \left(\frac{\partial A_y}{\partial x} - \frac{\partial A_x}{\partial y} \right) \hat{s}. \tag{3.47}$$

Note that an overall minus sign is added because we have interchanged the position of the θ-like coordinate with the vertical coordinate. This gives the usual formula for a curl in Cartesian coordinates when the bending radius goes to infinity. For transverse magnetic fields which are static, we may write $\phi = 0$, $A_x = 0$, and $A_y = 0$. This lets the canonical momenta in the transverse dimensions revert to the normal kinetic momenta, and the energy just reverts to $\sqrt{(pc)^2 + (mc^2)^2}$. The longitudinal component of the vector potential must satisfy the requirement that

$$B_x = \frac{1}{1 + x/\rho} \frac{\partial A_s}{\partial y}, \tag{3.48}$$

$$B_y = -\frac{1}{1 + x/\rho} \frac{\partial A_s}{\partial x}, \tag{3.49}$$

where we have modified the curl operator in cylindrical coordinates to account for the definition of A_s in Eq. (3.44).

For longitudinal fields such as in a solenoid, the transverse canonical momenta must contain a coupling to the momentum of the solenoidal field. This effect forces the usual paraxial approximation to break down and become non-symplectic, since the transverse components of the vector potential are not negligible compared to the small transverse components of the kinetic momentum. In the next section we will examine what this symplectic condition means and how it arises.

3.5 Symplectic transformations and matrices

In this section we demonstrate that a nondissipative Hamiltonian system leads to a symplectic behavior of the canonical variables. Look at a transformation $\vec{T}$ of trajectories through a fixed section of beam line. A particle's coordinate in phase space will be a six dimensional vector whose components are made up of the three canonical coordinates and their conjugate momenta, $\vec{X} = (\vec{x}, \vec{P})$. Although the results of this section will be true for any Hamiltonian function and its canonical coordinates, we will use the coordinates arrived at for the Hamiltonian function in Eq. (3.45) as a concrete example:

$$H = H(x, P_x, y, P_y, t, -U; s). \tag{3.50}$$

If the six dimensional phase space vector is written with the ordering

$$\vec{X} = (x, P_x, y, P_y, t, -U), \tag{3.51}$$

then Hamilton's equations,

$$\frac{dx_i}{ds} = \frac{\partial H}{\partial P_i}, \quad \text{and} \quad \frac{dP_i}{ds} = -\frac{\partial H}{\partial x_i}, \tag{3.52}$$

become in matrix form

$$\frac{dX_i}{ds} = \sum_{j=1}^{6} S_{ij} \frac{\partial H}{\partial X_j}, \tag{3.53}$$

where the matrix

$$\mathbf{S} = \begin{pmatrix} 0 & 1 & 0 & 0 & 0 & 0 \\ -1 & 0 & 0 & 0 & 0 & 0 \\ 0 & 0 & 0 & 1 & 0 & 0 \\ 0 & 0 & -1 & 0 & 0 & 0 \\ 0 & 0 & 0 & 0 & 0 & 1 \\ 0 & 0 & 0 & 0 & -1 & 0 \end{pmatrix}. \tag{3.54}$$

A particle entering the beginning of the beam line at $s = s_0$ with coordinates $\vec{x}_0 = (x_0, y_0, t_0)$, and momenta $\vec{P}_0 = (P_{0x}, P_{0y}, -U_0)$, would exit the end of the beam line at $s = s_1$ with new coordinates $\vec{x}_1 = (x_1, y_1, t_1)$, and momenta $\vec{P}_1 = (P_{1x}, P_{1y}, -U_1)$. Keeping the ends of the section of beam line fixed, the final coordinates may be written as functions of the initial coordinates:

$$\vec{x}_1 = \vec{x}_1(\vec{x}_0, \vec{P}_0), \tag{3.55}$$

$$\vec{P}_1 = \vec{P}_1(\vec{x}_0, \vec{P}_0), \tag{3.56}$$

so that

$$\vec{X}_1 = \vec{X}_1(\vec{X}_0). \tag{3.57}$$

These functions form a canonical transformation from the beginning of the beam line to the end of the beam line, since the particle trajectories must obey Hamilton's equations. This is actually a *restricted canonical transformation*, since we have restricted the functions to be independent of s by fixing the ends of the beam line.

From Eq. (3.18) the slopes of the components of $\vec{X}_1$ become

$$\frac{dX_{1i}}{ds} = \sum_j \mathbf{M}_{ij} \frac{dX_{0j}}{ds}, \tag{3.58}$$

where $\mathbf{M}_{ij} = \partial X_{1i}/\partial X_{0j}$ is the Jacobian matrix for the transformation $\vec{T}$ from s_0 to s_1. Since the motion of these particles is reversible, the functions in Eqs. (3.55, and 3.56) may be inverted to give $\vec{x}_0 = \vec{x}_0(\vec{x}_1, \vec{P}_1)$, and $\vec{P}_0 = \vec{P}_0(\vec{x}_1, \vec{P}_1)$.

Using Eq. (3.53) we can write

$$\frac{dX_{1i}}{ds} = \sum_j \mathbf{S}_{ij} \frac{\partial \mathrm{H}}{\partial X_{1j}} = \sum_{j,k} \mathbf{S}_{ij} \frac{\partial \mathrm{H}}{\partial X_{0k}} \frac{\partial X_{0k}}{\partial X_{1j}}. \tag{3.59}$$

The last derivative in Eq. (3.59) is the Jacobian matrix of the inverse of the transformation $\vec{T}$ from s_0 to s_1,

$$\left(\mathbf{M}^{-1}\right)_{kj} = \frac{\partial X_{0k}}{\partial X_{1j}}, \tag{3.60}$$

where

$$\mathbf{M}_{jk} = \frac{\partial X_{1j}}{\partial X_{0k}}. \tag{3.61}$$

Substituting Eq. (3.60) into Eq (3.59) gives

$$\frac{dX_{1i}}{ds} = \sum_{jk} \mathbf{S}_{ij} \left(\left(\mathbf{M}^{-1}\right)^{\mathrm{T}}\right)_{jk} \frac{\partial \mathrm{H}}{\partial X_{0k}}, \tag{3.62}$$

where the superscript T indicates the transpose of the matrix. But since $\mathbf{S}^2 = -\mathbf{I}$, the derivative of H with respect to X_{0k} may be written as

$$\frac{\partial \mathrm{H}}{\partial X_{0k}} = -\sum_l \mathbf{S}_{kl} \frac{dX_{0l}}{ds} \tag{3.63}$$

which yields

$$\frac{dX_{1i}}{ds} = -\sum_{jkl} \mathbf{S}_{ij} \left(\left(\mathbf{M}^{-1}\right)^{\mathrm{T}}\right)_{jk} \mathbf{S}_{kl} \frac{dX_{0l}}{ds}. \tag{3.64}$$

Comparing this with Eq. (3.58), requires that

$$\mathbf{M} = -\mathbf{S}\left(\mathbf{M}^T\right)^{-1}\mathbf{S}, \tag{3.65}$$

which may be rearranged to give

$$\mathbf{S} = \mathbf{M}^T\mathbf{S}\mathbf{M}. \tag{3.66}$$

It will be convenient to define the symplectic conjugate of a general $2n \times 2n$-matrix $\mathbf{N}$ as

$$\tilde{\mathbf{N}} = \mathbf{S}^T\mathbf{N}^T\mathbf{S} = \mathbf{S}\mathbf{N}^T\mathbf{S}^T,$$

where $\mathbf{S}$ is the symplectic matrix with 2×2-blocks of

$$\begin{pmatrix} 0 & 1 \\ -1 & 0 \end{pmatrix} \tag{3.67}$$

along the diagonal and zeros everywhere else. If $\mathbf{N}$ is symplectic, then $\tilde{\mathbf{N}} = \mathbf{N}^{-1}$.

A symplectic matrix, $\mathbf{M}$, is defined as any matrix that satisfies Eq. (3.66), with $\mathbf{S}$ being the matrix defined by Eq. (3.54). All the $2n$-dimensional matrices which are symplectic form an algebraic group.[7,8]* Several of the interesting properties of these matrices will be examined in the problems at the end of the chapter. The most important of these is

$$\det(\mathbf{M}) = 1, \tag{3.68}$$

which would be expected from Liouville's theorem.

In general, a transformation $\vec{X}_1 = \vec{T}(\vec{X}_0)$ is called symplectic if its Jacobian matrix is a symplectic matrix. More specifically, the transformation of a particle through a section of beam line is symplectic if it is described in terms of a restricted canonical transformation of canonical coordinates and their conjugate momentum coordinates. It is possible to show that this symplectic condition applies to canonical transformations which are not restricted, by breaking the transformation into an infinite series of infinitesimal transformations. An example of this will be shown in the next section .

---

* We have demonstrated this for a system with three coordinates and three conjugate momenta. This can be generalized to any number n of coordinates and the same number of conjugate momenta. The corresponding $\mathbf{S}$ would be an $2n \times 2n$-matrix with the same structure as in the 6×6 case, but with the repetition given n times rather than three times.

3.6 The standard canonical coordinates

The usual coordinates for the linear approximation of most modern day accelerators are not those found in Eq. (3.50); however, a few simple transformations and the paraxial approximation will yield the familiar transverse coordinates previously obtained in Chapter 2. If we divide Eq. (3.50) by the design momentum, p_0, we get

$$\frac{H}{p_0} = -\frac{qA_s}{p_0} - \left(1 + \frac{x}{\rho}\right) \sqrt{\left(\frac{U - q\phi}{p_0 c}\right)^2 - \left(\frac{mc}{p_0}\right)^2 - \left(\frac{P_x - qA_x}{p_0}\right)^2 - \left(\frac{P_y - qA_y}{p_0}\right)^2}. \quad (3.69)$$

For static magnetic fields with no electric field, $\phi = 0$, and

$$\frac{H}{p_0} = -\frac{qA_s}{p_0} - \left(1 + \frac{x}{\rho}\right) \sqrt{\left(\frac{U}{p_0 c}\right)^2 - \left(\frac{mc}{p_0}\right)^2 - \left(\frac{P_x - qA_x}{p_0}\right)^2 - \left(\frac{P_y - qA_y}{p_0}\right)^2}. \quad (3.70)$$

If the magnetic field is perpendicular to the design orbit, we may require that $A_x = A_y = 0$ without any loss of generality. This means that $P_x = p_x$, and $P_y = p_y$, implying

$$\frac{H}{p_0} = -\frac{q}{p_0} A_s - \left(1 + \frac{x}{\rho}\right) \sqrt{\left(\frac{p}{p_0}\right)^2 - \left(\frac{p_x}{p_0}\right)^2 - \left(\frac{p_y}{p_0}\right)^2}. \quad (3.71)$$

For small deviations from the conditions of the design trajectory, the transverse slopes are

$$x' = \frac{dx}{ds} \simeq \frac{p_x}{p_0}, \quad \text{and} \quad y' = \frac{dy}{ds} \simeq \frac{p_y}{p_0}, \quad (3.72)$$

and the fractional momentum deviation is

$$\delta = \frac{\Delta p}{p_0} = \frac{p - p_0}{p_0} = -\frac{1}{v} \frac{U_0 - U}{p_0}. \quad (3.73)$$

We may use δ as a canonical momentum variable provided that we change the corresponding coordinate to something like

$$z = s - vt, \quad (3.74)$$

where $t_0 = s/v$ and t is the time at which the observed particle passes the position s. Eq. (3.71) may be rewritten using the paraxial approximation as

$$H_1(x, x', y, y', t, -U/p_0; s) = -\frac{q}{p_0} A_s$$
$$- \left(1 + \frac{x}{\rho}\right) \sqrt{\left(\frac{U}{p_0 c}\right)^2 - \left(\frac{mc}{p_0}\right)^2 - x'^2 - y'^2}. \quad (3.75)$$

(As previously mentioned, this establishes a new set of conjugate momenta, even though this is not really a canonical transformation.)

Next we canonically transform from the old longitudinal canonical coordinates $(t, -U/p_0)$ to a Hamiltonian in terms of new canonical coordinates (z, δ):

$$\mathcal{H}(x, x', y, y', z, \delta; s) = H_1(x, x', y, y', t, -U/p_0; s) + \frac{\partial F_2(t, \delta; s)}{\partial s}, \quad (3.76)$$

where F_2 is a generating function for the canonical transformation (See Appendix C.) with

$$z = \frac{\partial F_2}{\partial \delta}, \quad \text{and} \quad -\frac{U}{p_0} = \frac{\partial F_2}{\partial t}. \quad (3.77)$$

Since

$$\delta = \frac{\Delta p}{p_0} = \frac{U_0^2}{p_0^2 c^2} \frac{\Delta U}{U_0} = \frac{1}{\beta_0^2} \frac{\Delta U}{U_0}, \quad (3.78)$$

the relation between $\frac{U}{p_0}$ and δ is

$$\frac{U}{p_0} = \frac{c}{\beta}(1 + \beta^2 \delta). \quad (3.79)$$

To select an acceptable generating function for the canonical transformation, we must have

$$-\frac{U}{p_0} = \frac{\partial F_2}{\partial \Delta t}, \quad (3.80)$$

so F_2 should be the product of the old momentum and old coordinate t plus some function of s. Here Δt rather than t is used to shift the time coordinate origin to that of the synchronous particle t_0. In other words we are expanding all coordinates (transverse and longitudinal) about those of the design trajectory. The new canonical coordinate will then be

$$z = \frac{\partial F_2}{\partial \delta}. \quad (3.81)$$

A good candidate for the generating function is

$$F_2(t, \delta; s) = \frac{U}{p_0}(t_0 - t) - \frac{s}{\beta_0^2} + s$$

$$= \frac{c}{\beta_0}(1 + \beta_0^2 \delta)(t_0 - t) - \frac{s}{\beta_0^2} + s$$

$$= \frac{c}{\beta_0}(1 + \beta_0^2 \delta)\left(\frac{s}{v_0} - t\right) - \frac{s}{\beta_0^2} + s, \qquad (3.82)$$

since $s = v_0 t_0$. Evaluating for z then gives

$$z = s - v_0 t,$$

which is like Eq. (3.74).

The fractional momentum deviation is only truly canonical to z, if v is constant. This is true in the ultra-relativistic regime, or when the only external fields are static magnetic fields. If the fields are static, the variable z is not terribly interesting; although, it may be necessary to keep track of both z and δ if the particle then goes through an accelerating element with $\phi = \phi(t, s)$. For example, a high energy synchrotron with an rf accelerating cavity can be modeled with a simple longitudinal momentum kick which is proportional to the difference of the time at which the tracked particle passes the cavity minus the time that the hypothetical synchronous particle passes the cavity.

Expanding the square root and keeping the low order terms gives

$$\mathcal{H} = -\frac{q}{p_0}A_s - \left(1 + \frac{x}{\rho}\right)\left(1 + \frac{1}{2}[2\delta + \delta^2 - (x'^2 + y'^2) + \ldots]\right.$$

$$\left. - \frac{1}{8}(4\delta^2 + \ldots) + \ldots\right) + \frac{\partial F_2}{\partial s}$$

$$= -\frac{q}{p_0}A_s - \left(1 + \frac{x}{\rho}\right)\left(1 + \delta - \frac{1}{2}(x'^2 + y'^2) + \ldots\right) + 1 + \delta. \qquad (3.83)$$

We will use this in the next section to obtain the 6×6 transfer matrix for a sector magnet, i. e., a bending magnet with a uniform field that is transverse to the trajectory of the design particle.

For the energy oscillations associated with the longitudinal coordinate, the usual canonical momentum that is used to describe a particle is the deviation of the particle's momentum from the momentum of the synchronous particle. The corresponding conjugate coordinate is the difference of the times that the particle in question and the synchronous particle pass the longitudinal position s. This time deviation is frequently scaled to a length, by multiplying by the velocity of the synchronous particle, to obtain the longitudinal distance between the particle of interest and the synchronous particle. Another version of the longitudinal coordinate is derived by dividing the longitudinal path difference by the wavelength of the accelerating rf field. (This is the same as multiplying the time difference by the frequency of the accelerating field.) Energy oscillations will be covered in detail in Chapter 7.

3.7 Symplectic generators

In this section, we will show how it is possible to construct a transfer matrix along some path of finite length, by using an infinite sequence of infinitesimal transformations. This is really just a simple integration of the equations of motion and is identical to the Lie algebraic formalism used in quantum mechanics to construct the spin rotation matrices of SU(2) or more properly SU(2,$\mathbb{C}$) . (The "$\mathbb{C}$" indicates that the matrices elements are complex numbers.)

Simply put, for a field independent of s, this can be demonstrated by finding the group generator $\mathbf{G}$ of the transformation matrix $\mathbf{M}(s)$, which maps a particle from a starting point at $s = 0$ to a new point at s;

$$\begin{pmatrix} x \\ x' \\ y \\ y' \\ z \\ \delta \end{pmatrix} = \mathbf{M}(s) \begin{pmatrix} x_0 \\ x'_0 \\ y_0 \\ y'_0 \\ z_0 \\ \delta_0 \end{pmatrix}. \tag{3.84}$$

The transformation along an infinitesimal ds expands to

$$\mathbf{M}(ds) = 1 + \mathbf{G}\, ds. \tag{3.85}$$

The total transformation may be written as

$$\mathbf{M}(s) = \lim_{n \to \infty} \left(\mathbf{I} + \mathbf{G}\frac{s}{n} \right)^n = e^{\mathbf{G}s}, \tag{3.86}$$

where we have approximated ds by s/n. (It should be noted that this is effectively just an integral, even though we have not explicitly written an integral sign.)

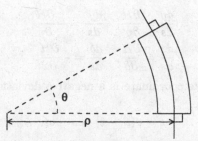

Figure. 3.3 In a sector magnet, the design trajectory enters and exits perpendicularly to the field, because the ends of the magnet lie along two radii of a circle whose center is the radius of curvature of the design trajectory.

As an example, let us find the transfer matrix for a sector dipole magnet in the ultrarelativistic limit, while ignoring the fringe fields of the ends of the magnet. This amounts to a field given by

$$\vec{B} = B_0\hat{y}, \quad \text{for } 0 < s < L, \quad \text{and} \tag{3.87}$$

$$\vec{B} = 0 \quad \text{elsewhere.} \tag{3.88}$$

The use of the word sector implies a wedge-shaped section whose edges intersect at the center of curvature, thus the design trajectory enters and exits perpendicularly to the ends of the magnet (see Fig. 3.3.) The path length of the design trajectory will be $L = \rho\theta$ if the radius of curvature is ρ, and the bending angle is θ. We will also assume that the particles are ultra-relativistic, so that δ may be considered to be a good canonical variable.

The method of attack to find **G** will be to find A_s for this magnetic field, use Eq. (3.83) as the Hamiltonian, and then linearize the equations of motion with respect to s. From Eq. (3.48), we find that

$$A_s = -B_0 \left(x + \frac{x^2}{2\rho} \right) \tag{3.89}$$

gives the correct value for the field inside the magnet. Our Hamiltonian is now

$$\mathcal{H} = \left(\frac{x}{\rho} + \frac{x^2}{2\rho^2} \right) - \left(1 + \frac{x}{\rho} \right) \left(1 + \delta - \frac{1}{2}(x'^2 + y'^2) + \dots \right) + \delta + 1, \tag{3.90}$$

remembering that $\rho = p_0/qB_0$. Hamilton's equations for our canonical variables are now

$$\frac{dx}{ds} = \frac{\partial \mathcal{H}}{\partial x'}, \quad \frac{dx'}{ds} = -\frac{\partial \mathcal{H}}{\partial x},$$

$$\frac{dy}{ds} = \frac{\partial \mathcal{H}}{\partial y'}, \quad \frac{dy'}{ds} = -\frac{\partial \mathcal{H}}{\partial y},$$

$$\frac{dz}{ds} = \frac{\partial \mathcal{H}}{\partial \delta}, \quad \frac{d\delta}{ds} = -\frac{\partial \mathcal{H}}{\partial z}. \tag{3.91}$$

The *longitudinal* time-like coordinate is a negative deviation in path length from the synchronous particle,

$$z = v(t_0 - t) = s - vt. \tag{3.92}$$

Here t is the time that the particle passes the point s, and t_0 is when the hypothetical synchronous particle would pass the same point. Since we are constructing the linear transfer matrix, we need only keep terms to first order in the canonical coordinates when we invoke Hamilton's equations:

$$\frac{dx}{ds} = \left(1 + \frac{x}{\rho}\right) x' \simeq x',$$

$$\frac{dx'}{ds} = -\left(\frac{1}{\rho} + \frac{x}{\rho^2}\right) + \frac{1}{\rho}(1 + \delta) \simeq \frac{\delta}{\rho} - \frac{x}{\rho^2},$$

$$\frac{dy}{ds} \simeq y',$$

$$\frac{dy'}{ds} = 0,$$

$$\frac{dz}{ds} = -\left(1 + \frac{x}{\rho}\right) + 1 \simeq -\frac{x}{\rho},$$

$$\frac{d\delta}{ds} = 0. \tag{3.93}$$

We can transform the above equations of motion to a set of finite difference equations:

$$\begin{pmatrix} x_1 \\ x_1' \\ y_1 \\ y_1' \\ z_1 \\ \delta_1 \end{pmatrix} = \begin{pmatrix} 1 & ds & 0 & 0 & 0 & 0 \\ \frac{-ds}{\rho^2} & 1 & 0 & 0 & 0 & \frac{ds}{\rho} \\ 0 & 0 & 1 & ds & 0 & 0 \\ 0 & 0 & 0 & 1 & 0 & 0 \\ -\frac{ds}{\rho} & 0 & 0 & 0 & 1 & 0 \\ 0 & 0 & 0 & 0 & 0 & 1 \end{pmatrix} \begin{pmatrix} x_0 \\ x_0' \\ y_0 \\ y_0' \\ z_0 \\ \delta_0 \end{pmatrix}. \tag{3.94}$$

Subtracting 1 from this matrix yields $\mathbf{G}\, ds$,

$$\mathbf{G}\, ds = \begin{pmatrix} 0 & \rho & 0 & 0 & 0 & 0 \\ -\frac{1}{\rho} & 0 & 0 & 0 & 0 & 1 \\ 0 & 0 & 0 & \rho & 0 & 0 \\ 0 & 0 & 0 & 0 & 0 & 0 \\ -1 & 0 & 0 & 0 & 0 & 0 \\ 0 & 0 & 0 & 0 & 0 & 0 \end{pmatrix} \frac{ds}{\rho} = \mathbf{K}\frac{ds}{\rho}. \tag{3.95}$$

We may write $ds = \rho\theta/n$ and let n go to infinity. The linear transformation matrix can then be written as

$$\mathbf{M}(\theta) = \lim_{n \to \infty} \left(\mathbf{I} + \mathbf{K}\frac{\theta}{n} \right)^n = e^{\mathbf{K}\theta}$$

$$= \mathbf{I} + \mathbf{K}\theta + \frac{(\mathbf{K}\theta)^2}{2!} + \frac{(\mathbf{K}\theta)^3}{3!} + \dots . \qquad (3.96)$$

Raising $\mathbf{K}$ to increasing powers produces a repetitive sequence:

$$\mathbf{K}^2 = \begin{pmatrix} -1 & 0 & 0 & 0 & 0 & \rho \\ 0 & -1 & 0 & 0 & 0 & 0 \\ 0 & 0 & 0 & 0 & 0 & 0 \\ 0 & 0 & 0 & 0 & 0 & 0 \\ 0 & -\rho & 0 & 0 & 0 & 0 \\ 0 & 0 & 0 & 0 & 0 & 0 \end{pmatrix}, \qquad (3.97)$$

$$\mathbf{K}^3 = \begin{pmatrix} 0 & -\rho & 0 & 0 & 0 & 0 \\ \frac{1}{\rho} & 0 & 0 & 0 & 0 & -1 \\ 0 & 0 & 0 & 0 & 0 & 0 \\ 0 & 0 & 0 & 0 & 0 & 0 \\ 1 & 0 & 0 & 0 & 0 & -\rho \\ 0 & 0 & 0 & 0 & 0 & 0 \end{pmatrix}, \qquad (3.98)$$

$$\mathbf{K}^4 = \begin{pmatrix} 1 & 0 & 0 & 0 & 0 & -\rho \\ 0 & 1 & 0 & 0 & 0 & 0 \\ 0 & 0 & 0 & 0 & 0 & 0 \\ 0 & 0 & 0 & 0 & 0 & 0 \\ 0 & \rho & 0 & 0 & 0 & 0 \\ 0 & 0 & 0 & 0 & 0 & 0 \end{pmatrix} = -\mathbf{K}^2. \qquad (3.99)$$

Only a finite number of powers of $\mathbf{K}$ must be evaluated, since every $2n \times 2n$ square matrix satisfies its own characteristic function by the Cayley-Hamilton theorem.[12] The eigenvalues λ of $\mathbf{K}$ satisfy the characteristic equation

$$0 = \sum_{j=0}^{2n} A_j \lambda^j \quad \text{with} \quad A_{2n} = 1, \qquad (3.100)$$

so the theorem gives us

$$\mathbf{K}^{2n} = - \sum_{j=0}^{2n-1} A_j \mathbf{K}^j. \qquad (3.101)$$

We can rearrange Eq. (3.96) to give

$$\mathbf{M} = \mathbf{I} + \mathbf{K}\theta + \mathbf{K}^2\left(\frac{\theta^2}{2!} - \frac{\theta^4}{4!} + \dots\right) + \mathbf{K}^3\left(\frac{\theta^3}{3!} - \frac{\theta^5}{5!} + \dots\right)$$

$$= \mathbf{I} + \mathbf{K}\theta + \mathbf{K}^2(1 - \cos\theta) + \mathbf{K}^3(\theta - \sin\theta)$$

$$= \begin{pmatrix} \cos\theta & \rho\sin\theta & 0 & 0 & 0 & \rho(1 - \cos\theta) \\ -\frac{1}{\rho}\sin\theta & \cos\theta & 0 & 0 & 0 & \sin\theta \\ 0 & 0 & 1 & \rho\theta & 0 & 0 \\ 0 & 0 & 0 & 1 & 0 & 0 \\ -\sin\theta & -\rho(1 - \cos\theta) & 0 & 0 & 1 & -\rho(\theta - \sin\theta) \\ 0 & 0 & 0 & 0 & 0 & 1 \end{pmatrix}. \quad (3.102)$$

It is worthwhile to check this result from a purely geometrical argument. This method can also be used to find transfer matrices for other elements.

If higher order terms are kept, then a nonlinear transformation can be obtained. A warning must be given, concerning a composition of two or more nonlinear transformations. When a linear transformation is made of another linear transformation, the result is also linear (i. e., the product of two matrices is another matrix.) When two nonlinear functions with terms of order n are composed, the resultant composition is of order $2n$. This composite function will be accurate only to order n. The truncation involved in such a composition will produce effects which violate the symplectic behavior of the true physics for particles traveling on a trajectory slightly removed from the design trajectory. This results in a seeming violation of Liouville's theorem, but is actually an artifact of the truncation process. A technique has been developed[13] in which the successive transformations are combined to preserve the symplectic nature to some specified order. This method effectively involves the exponentiation of Poisson bracket operators (See Ref. 13.) A good review of Lie algebraic methods used in accelerator physics is given in Ref. 14.

The time derivative a function of coordinates and momenta $f(\vec{x}, \vec{P}; t)$, called the Liouville equation[1] is easily derived using the chain rule:

$$\frac{df}{dt} = \sum_j\left(\frac{\partial f}{\partial x_j}\frac{dx_j}{dt} + \frac{\partial f}{\partial P_j}\frac{dP_j}{dt}\right) + \frac{\partial f}{\partial t}$$

$$= \sum_j\left(\frac{\partial f}{\partial x_j}\frac{\partial H}{\partial P_j} - \frac{\partial f}{\partial P_j}\frac{\partial H}{\partial x_j}\right) + \frac{\partial f}{\partial t}$$

$$= [f, H]_P + \frac{\partial f}{\partial t} \quad (3.103)$$

where for any two functions $f(\vec{x}, \vec{P}; t)$, and $g(\vec{x}, \vec{P}; t)$, the Poisson bracket is defined as

$$[f, g]_P = \frac{\partial f}{\partial x_j} \frac{\partial g}{\partial P_j} - \frac{\partial f}{\partial P_j} \frac{\partial g}{\partial x_j}. \tag{3.104}$$

3.8 Symplectic integrators

In the neighborhood of a reference trajectory $\vec{X} = \widehat{X}(s)$, we can expand the equation of motion Eq. (3.53) about $\widehat{X}(s)$. For the reference trajectory, Eq. (3.53) becomes

$$\frac{d\widehat{X}_i}{ds} = \sum_{j=1}^{6} S_{ij} \frac{\partial H}{\partial X_j} (\widehat{X}). \tag{3.105}$$

Expanding both sides in Taylor series yields

$$\frac{d}{ds} (\widehat{X}_i + \Delta X_i) = \sum_{j=1}^{6} S_{ij} \frac{\partial H}{\partial X_j} (\widehat{X} + \Delta X)$$

$$= \sum_{j=1}^{6} S_{ij} \left[\frac{\partial H}{\partial X_j} (\widehat{X}) + \sum_{k=1}^{6} \frac{\partial^2 H}{\partial X_j \partial X_k} (\widehat{X}) \Delta X_k + \cdots \right]. \tag{3.106}$$

Then after subtracting Eq. (3.105), we have

$$\frac{d\Delta X_i}{ds} = \sum_{j=1}^{6} \sum_{k=1}^{6} S_{ij} \frac{\partial^2 H}{\partial X_j \partial X_k} (\widehat{X}) \Delta X_k + \cdots. \tag{3.107}$$

The matrix of second derivatives

$$C_{jk} = \frac{\partial^2 H}{\partial X_j \partial X_k} (\widehat{X}) \tag{3.108}$$

is symmetric $C_{jk} = C_{kj}$. By comparing this with the previous section (Eq. (3.85)), we find that the group generator for an infinitesimal step ds must be $\mathbf{SC}$. In the case of a constant matrix $\mathbf{C}$ (as for the dipole in the linear approximation of the previous section), we can again solve the linearized Eq. (3.107) via exponentiation to obtain the linearized transport matrix

$$\mathbf{M}(s) = e^{\mathbf{SC}\,s}. \tag{3.109}$$

If $\mathbf{C}$ is not a constant matrix, but is rather a matrix which depends on the time-like coordinate, then we can think of approximating $\mathbf{M}(s)$ by a product of many exponentials of almost infinitesimal steps where $\mathbf{C}(s)$ is taken as constant through each step so

$$\mathbf{M}(s) = \lim_{ds \to 0} e^{\mathbf{SC}(s-ds)ds} \cdots e^{\mathbf{SC}(2ds)ds} e^{\mathbf{SC}(ds)ds} e^{\mathbf{SC}(0)ds}. \tag{3.110}$$

A general $2n \times 2n$-symmetric matrix is described by

$$\frac{(2n)^2 - 2n}{2} + 2n = (2n + 1)n \tag{3.111}$$

degrees of freedom. Since any $2n \times 2n$ symplectic matrix can be written as the exponentiation of $\mathbf{S}$ times a symmetric matrix, the symplectic matrices have the same number $(2n+1)n$ of free parameters. For $n = 1$, 2, and 3 there are 3, 10, and 21 free parameters, respectively.

For example, any 4×4 symmetric real matrix can be written as

$$\mathbf{C} = \sum_{j=1}^{10} \alpha_j \mathbf{c}_j, \tag{3.112}$$

where the α_j are real coefficients of a set of 10 independent symmetric matrices which form a basis for the set of 4×4 symmetric matrices. The α_j are the components of a vector in a 10-dimensional vector space with the basis vectors $\mathbf{c}_j$. One choice for this basis is

$$\mathbf{c}_j: \begin{pmatrix} 0 & 0 & 0 & 0 \\ 0 & 1 & 0 & 0 \\ 0 & 0 & 0 & 0 \\ 0 & 0 & 0 & 1 \end{pmatrix}, \begin{pmatrix} 0 & 0 & 0 & 0 \\ 0 & 1 & 0 & 0 \\ 0 & 0 & 0 & 0 \\ 0 & 0 & 0 & -1 \end{pmatrix}, \begin{pmatrix} 1 & 0 & 0 & 0 \\ 0 & 0 & 0 & 0 \\ 0 & 0 & 1 & 0 \\ 0 & 0 & 0 & 0 \end{pmatrix}, \begin{pmatrix} 1 & 0 & 0 & 0 \\ 0 & 0 & 0 & 0 \\ 0 & 0 & -1 & 0 \\ 0 & 0 & 0 & 0 \end{pmatrix},$$

$$\begin{pmatrix} 0 & 1 & 0 & 0 \\ 1 & 0 & 0 & 0 \\ 0 & 0 & 0 & 0 \\ 0 & 0 & 0 & 0 \end{pmatrix}, \begin{pmatrix} 0 & 0 & 1 & 0 \\ 0 & 0 & 0 & 0 \\ 1 & 0 & 0 & 0 \\ 0 & 0 & 0 & 0 \end{pmatrix}, \begin{pmatrix} 0 & 0 & 0 & 1 \\ 0 & 0 & 0 & 0 \\ 0 & 0 & 0 & 0 \\ 1 & 0 & 0 & 0 \end{pmatrix}, \begin{pmatrix} 0 & 0 & 0 & 0 \\ 0 & 0 & 1 & 0 \\ 0 & 1 & 0 & 0 \\ 0 & 0 & 0 & 0 \end{pmatrix},$$

$$\begin{pmatrix} 0 & 0 & 0 & 0 \\ 0 & 0 & 0 & 1 \\ 0 & 0 & 0 & 0 \\ 0 & 1 & 0 & 0 \end{pmatrix}, \begin{pmatrix} 0 & 0 & 0 & 0 \\ 0 & 0 & 0 & 0 \\ 0 & 0 & 0 & 1 \\ 0 & 0 & 1 & 0 \end{pmatrix}, \text{ in the order of } j \text{ from 1 to 10.} \tag{3.113}$$

The ten products $\mathbf{G}_j = \mathbf{S}\mathbf{c}_j$ form a set of generators for group of 4×4 symplectic matrices, and any 4×4 symplectic matrix may be written in the form

$$\exp\left(\sum_{j=1}^{10} \mathbf{G}_j \alpha_j \right). \tag{3.114}$$

Mechanics of Trajectories

It is easy to see how to extend this to a similar basis set of generators for any order symplectic group.

The product of two exponentials of different matrices can be combined into a single exponential of a third matrix:

$$e^X e^Y = e^Z. \tag{3.115}$$

If $[\mathbf{X}, \mathbf{Y}] = 0$, then we have simply

$$e^{(X+Y)} = e^X e^Y. \tag{3.116}$$

However if the matrices $\mathbf{X}$ and $\mathbf{Y}$ do not commute, then $\mathbf{Z}$ can be calculated from the the Baker-Campbell-Hausdorff (BCH) formula[17,18]:

$$
\begin{aligned}
\mathbf{Z} &= \log\left(e^X e^Y\right) \\
&= \mathbf{X} + \mathbf{Y} + \frac{1}{2}[\mathbf{X}, \mathbf{Y}] + \frac{1}{12}([\mathbf{X}, \mathbf{X}, \mathbf{Y}] + [\mathbf{Y}, \mathbf{Y}, \mathbf{X}]) + \frac{1}{24}[\mathbf{X}, \mathbf{Y}, \mathbf{Y}, \mathbf{X}] \\
&\quad - \frac{1}{720}([\mathbf{Y}, \mathbf{Y}, \mathbf{Y}, \mathbf{Y}, \mathbf{X}] + [\mathbf{X}, \mathbf{X}, \mathbf{X}, \mathbf{X}, \mathbf{Y}]) \\
&\quad + \frac{1}{360}([\mathbf{Y}, \mathbf{X}, \mathbf{X}, \mathbf{X}, \mathbf{Y}] + [\mathbf{X}, \mathbf{Y}, \mathbf{Y}, \mathbf{Y}, \mathbf{X}]) \\
&\quad + \frac{1}{120}([\mathbf{X}, \mathbf{X}, \mathbf{Y}, \mathbf{Y}, \mathbf{X}] + [\mathbf{Y}, \mathbf{Y}, \mathbf{X}, \mathbf{X}, \mathbf{Y}]) + \mathcal{O}(6),
\end{aligned} \tag{3.117}
$$

where the extended commutator notation indicates multiple commutators nested to the right:

$$[a, b, c] = [a, [b, c]], \qquad [a, b, c, d] = [a, [b, [c, d]]], \tag{3.118}$$

and so forth.

The Zassenhaus formula (sort of like the reverse process of the BCH formula) splits a single exponentiation of the sum of two matrices $\mathbf{A}$ and $\mathbf{B}$ into a product of two exponentials of the two matrices times higher order exponentials of commutators:[19]

$$e^{(A+B)h} = e^{Ah} e^{Bh} e^{-[A,B]h^2/2} e^{(2[B,A,B]+[A,A,B])h^3/6} e^{\mathcal{O}(h^4)\cdots} \cdots, \tag{3.119}$$

where the parameter h is a small integration step. To second order in h, this can be written as (see Problem 3–10)

$$e^{(A+B)h} = e^{Ah/2} e^{Bh} e^{Ah/2} + \mathcal{O}(h^3). \tag{3.120}$$

To second order an integration step $e^{\mathbf{SC}h}$ can be expanded into

$$\exp\left(\sum_{j=1}^{n} \alpha_j \mathbf{G}_j h\right) = \left(\prod_{j=1}^{n} \exp(\mathbf{G}_j h/2)\right)\left(\prod_{j=n}^{1} \exp(\mathbf{G}_j h/2)\right), \qquad (3.121)$$

by successive applications of Eq. (3.120). If we partition the sum in Eq. (3.112) into $\mathbf{A} = \alpha_1 \mathbf{G}_1$ and $\mathbf{B} = \sum_{n=2}^{10} \alpha_n \mathbf{G}_n$, then the factor

$$e^{\mathbf{A}h/2} = \begin{pmatrix} 1 & \alpha_1 h/2 & 0 & 0 \\ 0 & 1 & 0 & 0 \\ 0 & 0 & 1 & \alpha_1 h/2 \\ 0 & 0 & 0 & 1 \end{pmatrix} + \mathcal{O}(h^3), \qquad (3.122)$$

produces a drift matrix of length $\alpha_1 h/2$ up to second order in h, and the factor $e^{\mathbf{B}h}$ corresponds to a thin element kick.

If we apply the BCH formula twice to the time-symmetric* product

$$e^{\mathbf{W}} = e^{\mathbf{X}h}\, e^{\mathbf{Y}h}\, e^{\mathbf{X}h}, \qquad (3.123)$$

then it must be that

$$\mathbf{W} = (2\mathbf{X} + \mathbf{Y})h + \frac{1}{6}\left([\mathbf{Y},\mathbf{Y},\mathbf{X}] - [\mathbf{X},\mathbf{X},\mathbf{Y}]\right)h^3 + \mathcal{O}(h^5). \qquad (3.124)$$

If an integrator formula

$$I(h) = e^{\mathbf{g}_1 h + \mathbf{g}_2 h^2 + \mathbf{g}_3 h^3 + \cdots}, \qquad (3.125)$$

with matrices $\mathbf{g}_j$ is "time reversible", then we must have $I(h)I(-h) = 1$. The BCH formula gives to lowest order

$$I(h)I(-h) = e^{2\mathbf{g}_2 h^2 + \mathcal{O}(h^3)}, \qquad (3.126)$$

so we find we must have $\mathbf{g}_2 = 0$. Repeating this with $\mathbf{g}_2 = 0$, now the lowest term would require that $\mathbf{g}_4 = 0$, and by induction we see that all the even powers of h in the exponent in $I(h)$ must have zero coefficients: $\mathbf{g}_{2j} = 0$. To state this another way, $\log(I(h))$ must be an odd function of h, if $I(h)$ is a time-reversible integrator function.

---

* This symmetry is typically referred to as time-symmetric even when the integration variable may be the s-coordinate rather than time, since s is the independent time-like parameter of the Hamiltonian.

Higher order integrators may be constructed from the second order integration function, Eq. (3.120):

$$I_2(h) = e^{Ah/2} \, e^{Bh} \, e^{Ah/2} = e^{g_1 h + g_3 h^3 + \cdots}. \tag{3.127}$$

For example,[17] a fourth order integrator may be constructed by a time-symmetric product of second order integrators:

$$I_4(h) = I_2(ah) \, I_2(bh) \, I_2(ah), \tag{3.128}$$

where a and b are parameters to be determined. Applying Eq. (3.124) gives

$$
\begin{aligned}
I_4(h) &= e^{(2a+b)g_1 h + (2a^3 + b^3)g_3 h^3 + \cdots}, \\
&= e^{(A+B)h + \mathcal{O}(h^5)}.
\end{aligned}
\tag{3.129}
$$

To fourth order this requires that

$$1 = 2a + b, \tag{3.130}$$
$$0 = 2a^3 + b^3, \tag{3.131}$$

which, when solved, yield

$$a = \frac{1}{2 - 2^{1/3}}, \tag{3.132}$$
$$b = \frac{2^{1/3}}{2 - 2^{1/3}}. \tag{3.133}$$

This procedure can be extended[17] to higher order and gives explicit values for 6th and 8th order integrators. These Yoshida-style integrators are more general than just for the symplectic groups, and may be applied to any of the quadratic Lie groups discussed in the next section.

3.8.1 Lie groups and algebras

A Lie algebra is a vector space over a field (for our case either the real numbers $\mathbb{R}$ or complex numbers $\mathbb{C}$) with an additional binary operation $[\cdot, \cdot]$ called the Lie bracket or commutator. The Lie bracket operator satisfies the following properties for any elements x, y, z in the Lie algebra and a, b in the field:

1. Bilinearity: $[ax + by, z] = a[y, z]$, $[z, ax + by] = a[z, x] + b[z, y]$;
2. Anticommutativity: $[x, y] = -[y, x]$;

3. Jacobi Identity: $[x,[y,z]] + [y,[z,x]] + [z,[x,y]] = 0$, with 0 here being the identity element in the vector space.

An example of this operator is the commutator of matrices used in the previous section. The Poisson bracket of Eq. (3.104) is another example of a Lie bracket operator.

A quadratic Lie group G_c may be defined by a group representation of $n \times n$ nonsingular (i. e., the inverse must exist) complex matrices:

$$G_c = \{\mathbf{M} \in GL_n(\mathbb{C}) : \mathbf{MJM}^\dagger = \mathbf{J}\}, \tag{3.134}$$

where $GL_n(\mathbb{C})$ is the general linear group of $n \times n$ complex matrices consisting of all nonsingular $n \times n$ complex matrices, and where $\mathbf{J}$ is any particular matrix in $GL_n(\mathbb{C})$. (See Problem 3–3 for the mathematical definition of a group.) Here, as in quantum mechanics, the dagger indicates the Hermitian conjugate:

$$\mathbf{M}^\dagger = (\mathbf{M}^*)^T. \tag{3.135}$$

The Lie algebra corresponding to G_c may be represented by the collection of matrices:

$$\mathfrak{g}_c = \{\mathbf{A} \in \mathbb{C}^{n \times n} : \mathbf{AJ} + \mathbf{JA}^\dagger = \mathbf{0}\}, \tag{3.136}$$

where $\mathbb{C}^{n \times n}$ is the set of all $n \times n$ complex matrices, and $\mathbf{0}$ is the $n \times n$ matrix filled with zeros.

If on the other hand we consider only real matrices then we have

$$G_r = \{\mathbf{M} \in GL_n(\mathbb{R}) : \mathbf{MJM}^T = \mathbf{J}\}, \quad \text{and} \tag{3.137}$$

$$\mathfrak{g}_r = \{\mathbf{A} \in \mathbb{R}^{n \times n} : \mathbf{AJ} + \mathbf{JA}^T = \mathbf{0}\}, \tag{3.138}$$

where $GL_n(\mathbb{R})$ is the general linear group of $n \times n$ nonsingular real matrices and $\mathbf{J}$ is a particular matrix in $GL_n(\mathbb{R})$.

Some examples of these groups used in physics are:

- The unitary group: $U(n) = G_c$ when $\mathbf{J} = \mathbf{I}$ the identity matrix. When restricted to matrices of determinant $+1$, we have the special unitary group $SU(n)$. (In keeping with the general physics convention, we frequently write the shortened form which assumes the complex number field.) The corresponding Lie algebra is the same for the unitary and special unitary groups and is given the name $\mathfrak{su}(n)$.

- The orthogonal group $O(n)$ of real $n \times n$ matrices is obtained from G_r with $\mathbf{J} = \mathbf{I}$. This is the group of rotations in n dimensions and consists of both proper ($|\mathbf{M}| = 1$) and improper ($|\mathbf{M}| = -1$) rotations. The subgroup of proper rotations is called SO(n). Both O(n) and SO(n) have the same algebra $\mathbf{so}(n)$.

- The Lorentz group SO(3,1) (over real numbers) is obtained from G_r when

$$\mathbf{J} = \begin{pmatrix} 1 & 0 & 0 & 0 \\ 0 & 1 & 0 & 0 \\ 0 & 0 & 1 & 0 \\ 0 & 0 & 0 & -1 \end{pmatrix}, \tag{3.139}$$

where we restrict the matrices to be in the special linear group SL(4,$\mathbb{R}$) of 4×4 matrices with determinant of +1.

- For the $2n \times 2n$ symplectic matrices in our particular representation we take $\mathbf{J} = \mathbf{S}$ so that*

$$\text{Sp}(2n, \mathbb{R}; \mathbf{S}) = \{\mathbf{M} \in \text{GL}_{2n}(\mathbb{R}) : \mathbf{M}\mathbf{S}\mathbf{M}^T = \mathbf{S}\}. \tag{3.140}$$

As we have seen in § 3.5 for a Hamiltonian system, in the neighborhood of a trajectory $\widehat{\mathbf{X}}_0(t)$, other nearby trajectories must obey a Taylor series expansion about $\widehat{\mathbf{X}}_0(t)$ where the first order Jacobian matrix must be symplectic. If the Hamiltonian $H(\vec{X}; t)$ is conserved, then a constant value of the Hamiltonian will describe a surface or manifold in the six dimensional phase space. Any trajectory with that value for $H(\vec{X}; t)$ must lie in that surface. The Lie group Sp($2n$,$\mathbb{R}$;$\mathbf{S}$) describes the symplectic geometry of trajectories in flowing in this surface. A tangent plane at a point on the surface is called a *tangent space*, and collectively the tangent spaces to all points on the surface are called the *tangent bundle*. The corresponding Lie algebra $\mathbf{sp}(2n$,$\mathbb{R}$;$\mathbf{S}$) describes the geometry of the tangent spaces to points in the surface.

---

* Here we add the $\mathbf{S}$ in the name of the group Sp($2n, \mathbb{R}; \mathbf{S}$) to emphasize that this is a particular matrix representation of the group depending on the selection of the form of $\mathbf{J}$. In fact any similarity transformation of the matrix $\mathbf{S}$ would give a particular representation of the symplectic group. Another common representation is given by a matrix of the form

$$\mathbf{J} = \begin{pmatrix} \mathbf{0}_n & \mathbf{I}_n \\ -\mathbf{I}_n & \mathbf{0}_n \end{pmatrix},$$

where $\mathbf{I}_n$ and $\mathbf{0}_n$ are the $n \times n$ dimensional identity and zero matrices, respectively.

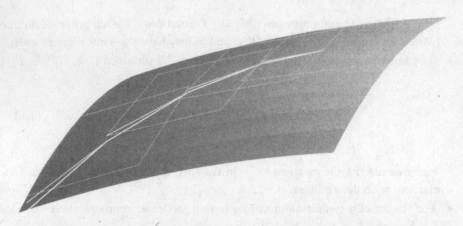

Figure. 3.4 Four integration steps for a trajectory with four integration steps in tangent planes and the resulting integrated trajectory projected back onto the manifold of the Hamiltonian by Φ.

For each time step t_j, a tangent plane to the surface is taken at the point $\mathbf{X}_j$. We can calculate the differential step in the tangent plane for a small time slice, and the resulting point is then mapped from the the tangent plane to the manifold by some function Φ as shown in Fig. 3.4. The function Φ is said to "lift" a straight line through the origin in the tangent plane onto a geodesic in the group (on the manifold). An integration step is made in the tangent plane and then lowered back onto the manifold by Φ. (The obvious choice for the lift Φ is an exponential; however as we will see, the exp function is not the only possibility.)

3.8.2 Cayley transforms and symplectification

Arthur Cayley showed that an orthogonal matrix $\mathbf{Q}$ could be factored[3] as

$$\mathbf{Q} = (\mathbf{I} - \mathbf{A})(\mathbf{I} + \mathbf{A})^{-1}, \tag{3.141}$$

where $\mathbf{A} = -\mathbf{A}^{\mathrm{T}}$ is an antisymmetric matrix. Eq. (3.141) can be solved for the antisymmetric matrix

$$\mathbf{A} = (\mathbf{I} - \mathbf{Q})(\mathbf{I} + \mathbf{Q})^{-1}. \tag{3.142}$$

Similar factorizations have been made for matrices in some of the other groups[3], and these are frequently referred to as Cayley transforms as well.

If we are given a matrix $\mathbf{Q}_0$ which is supposed to be orthogonal, but due to computational errors it is not quite orthogonal, then it is possible to tweak some of the elements of $\mathbf{Q}_0$ to produce a new matrix $\mathbf{Q}_1$ which is exactly orthogonal, i. e.

so that $\mathbf{Q}_1^T \mathbf{Q}_1 = \mathbf{I}$. Given Cayley's factorization, a simple method is to calculate the matrix

$$\mathbf{A}_0 = (\mathbf{I} - \mathbf{Q}_0)(\mathbf{I} + \mathbf{Q}_0)^{-1}, \tag{3.143}$$

which is not quite be antisymmetric $(\mathbf{A}_0^T \neq -\mathbf{A}_0)$. The matrix

$$\mathbf{A}_1 = \frac{\mathbf{A}_0 - \mathbf{A}_0^T}{2} \tag{3.144}$$

is guaranteed to be symmetric, so we can invert the transform to obtain

$$\mathbf{Q}_1 = (\mathbf{I} - \mathbf{A}_1)(\mathbf{I} + \mathbf{A}_1)^{-1}, \tag{3.145}$$

which now must be orthogonal.

For the symplectic matrices, we need to do a bit more work to develop formulae for the corresponding transform and its inverse.[20,21] Given a $2n \times 2n$-matrix $\mathbf{W}$ and recalling that $\mathbf{S}^2 = -\mathbf{I}$, we have

$$(\mathbf{I} - \mathbf{WS})\mathbf{S}(\mathbf{I} + \mathbf{SW}) = (\mathbf{S} + \mathbf{W})(\mathbf{I} + \mathbf{SW}) = \mathbf{S} + \mathbf{WSW} = (\mathbf{S} - \mathbf{W})(\mathbf{I} - \mathbf{SW})$$
$$= (\mathbf{I} + \mathbf{WS})\mathbf{S}(\mathbf{I} - \mathbf{SW}). \tag{3.146}$$

Theorem: A symplectic matrix $\mathbf{M}$ may be written in the form

$$\mathbf{M} = (\mathbf{I} + \mathbf{SW})(\mathbf{I} - \mathbf{SW})^{-1}, \tag{3.147}$$

if and only if $\mathbf{W}$ is a symmetric matrix, and where $\mathbf{S}$ is the metric for the selected representation of the symplectic group. This statement must be qualified with the requirement that

$$|\mathbf{I} - \mathbf{SW}| \neq 0, \tag{3.148}$$
$$|\mathbf{I} + \mathbf{M}| \neq 0. \tag{3.149}$$

The first of these caveats is obvious from Eq. (3.147), but the second is equivalent to saying that $\mathbf{M}$ has no eigenvalue equal to $+1$. This will become apparent when we solve Eq. (3.147) for the symmetric matrix $\mathbf{W}$.

Proof:

Show that if $\mathbf{W}$ is symmetric, then $\mathbf{M}$ is symplectic:

$$
\begin{aligned}
\mathbf{M}^T \mathbf{SM} &= (\mathbf{I} - \mathbf{W}^T \mathbf{S}^T)^{-1}(\mathbf{I} + \mathbf{W}^T \mathbf{S}^T)\mathbf{S}(\mathbf{I} + \mathbf{SW})(\mathbf{I} - \mathbf{SW})^{-1} \\
&= (\mathbf{I} + \mathbf{WS})^{-1}(\mathbf{I} - \mathbf{WS})\mathbf{S}(\mathbf{I} + \mathbf{SW})(\mathbf{I} - \mathbf{SW})^{-1} \\
&= (\mathbf{I} + \mathbf{WS})^{-1}(\mathbf{I} + \mathbf{WS})\mathbf{S}(\mathbf{I} - \mathbf{SW})(\mathbf{I} - \mathbf{SW})^{-1} \\
&= \mathbf{S}.
\end{aligned}
\tag{3.150}
$$

Therefore $\mathbf{M}$ is symplectic if $\mathbf{W}$ is symmetric.

Now let us assume that $\mathbf{W}$ is not symmetric, so it can be written as the sum of symmetric and antisymmetric matrices:

$$
\mathbf{W} = \mathbf{P} + \mathbf{Q},
\tag{3.151}
$$

where $\mathbf{P} = \mathbf{P}^T$ and $\mathbf{Q} = -\mathbf{Q}^T$.

$$
\begin{aligned}
\mathbf{M}^T \mathbf{SM} &= \mathbf{S} \\
&= (\mathbf{I} - \mathbf{W}^T \mathbf{S}^T)^{-1}(\mathbf{I} + \mathbf{W}^T \mathbf{S}^T)\mathbf{S}(\mathbf{I} + \mathbf{SW})(\mathbf{I} - \mathbf{SW})^{-1} \\
&= (\mathbf{I} - \mathbf{W}^T \mathbf{S}^T)^{-1}(\mathbf{S} + \mathbf{P} - \mathbf{Q} - \mathbf{P} - \mathbf{Q} + \mathbf{W}^T \mathbf{SW})(\mathbf{I} - \mathbf{SW})^{-1} \\
&= (\mathbf{I} - \mathbf{W}^T \mathbf{S}^T)^{-1}[(\mathbf{S} - \mathbf{W}^T)(\mathbf{I} - \mathbf{SW}) - 4\mathbf{Q}](\mathbf{I} - \mathbf{SW})^{-1} \\
&= \mathbf{S} - 4(\mathbf{I} - \mathbf{W}^T \mathbf{S}^T)^{-1}\mathbf{Q}(\mathbf{I} - \mathbf{SW})^{-1}.
\end{aligned}
\tag{3.152}
$$

So assuming that the inverses in the last line exist, then $\mathbf{Q} = 0$. (Actually the inverse $(\mathbf{I} - \mathbf{W}^T \mathbf{S}^T)^{-1}$ must exist if its transpose $(\mathbf{I} - \mathbf{SW})^{-1}$ exists.)

This proves the theorem.

Given the symplectic matrix $\mathbf{M}$, the symmetric matrix $\mathbf{W}$ is then obtained by rearranging Eq. (3.147):

$$
\mathbf{W} = \mathbf{S}(\mathbf{I} + \mathbf{M})^{-1}(\mathbf{I} - \mathbf{M}).
\tag{3.153}
$$

So an almost symplectic matrix $\mathbf{M}_0$ can be symplectified by calculating the almost symmetric matrix

$$
\mathbf{W}_0 = \mathbf{S}(\mathbf{I} + \mathbf{M}_0)^{-1}(\mathbf{I} - \mathbf{M}_0),
\tag{3.154}
$$

which can then be symmetrized to

$$
\mathbf{W}_1 = \frac{1}{2}(\mathbf{W}_0 + \mathbf{W}_0^T).
\tag{3.155}
$$

The new symplectified matrix is then

$$M_1 = (I + SW)(I - SW)^{-1}. \tag{3.156}$$

Ref. 21 explores this algorithm and its deficiencies in more detail.

It is worth noting that Eqs. (3.147 and 3.153) provide another pair of functions (other than exp and log) to map between the manifold of the trajectories and the local tangent spaces since W and the C of Eq. (3.108) are both symmetric matrices.

3.8.3 Integration by Cayley transform[22,23]

In general we can integrate a multidimensional first order differential equation of any quadratic group

$$\dot{X}(t) = A(X(t), t) X(t), \tag{3.157}$$

with an exponential integrator as described in § 3.8 for the symplectic group, i. e. Eq. (3.105) with $A = SC$. (Here we have written the integration parameter as t and the dot over the function indicates a derivative with respect to t.) In Chapter 13 we will integrate the spin precession equations for $SU(2,\mathbb{C})$ using the exponential function.

The inverse of an exponential function will map a symplectic matrix M into the matrix $\log(M) = SC$ where C is a symmetric matrix. In the previous section, we saw that the inverse of a Cayley transform can map a symplectic matrix M into the matrix

$$SW = (I + M)^{-1}(M - I), \tag{3.158}$$

where W is also a symmetric matrix. This suggests that a Cayley transform rather than the exponential function can be used for integration, since both SC and SW must be elements of the same Lie algebra $\mathfrak{sp}(2n,\mathbb{R})$.

A generalized Cayley transform is defined as

$$cay(z, \alpha) = \frac{1 + \alpha z}{1 - \alpha z}, \tag{3.159}$$

where $\alpha \neq 0$ is some constant. In terms of matrices, we can define

$$\Phi(A(t)) = cay(A(t)) = cay\left(A(t), \tfrac{1}{2}\right) = \left[I - \tfrac{1}{2}A(t)\right]^{-1}\left[I + \tfrac{1}{2}A(t)\right], \tag{3.160}$$

where we have taken* $\alpha = \tfrac{1}{2}$.

---

* It is perhaps worth noting that the first Padé approximant of e^z is $cay(z, 1/2)$, i. e., $cay(z, 1/2) = 1 + z + z^2/2 + z^3/4 + \cdots = e^z + \mathcal{O}(z^3)$.

For the exponential function, we would have

$$\mathbf{X}(t) = e^{\mathbf{\Omega}(t)}\mathbf{X}_0, \quad \text{with} \quad \mathbf{\Omega}(0) = \mathbf{0}, \tag{3.161}$$

for some matrix function $\mathbf{\Omega}(t) \in \mathfrak{sp}(2n,\mathbb{R})$. For the Cayley transform, we want to find a matrix function $\mathbf{\Omega}(t)$ with $\mathbf{\Omega}(t_0) = \mathbf{0}$, so that

$$\mathbf{X}(t) = \text{cay}(\mathbf{\Omega}(t))\mathbf{X}_0 = \left(\mathbf{I} - \tfrac{1}{2}\mathbf{\Omega}\right)^{-1}\left(\mathbf{I} + \tfrac{1}{2}\mathbf{\Omega}\right)\mathbf{X}_0. \tag{3.162}$$

Taking the derivative of $\mathbf{X}$ we obtain

$$\begin{aligned}
\dot{\mathbf{X}} &= \left[\tfrac{1}{2}\left(\mathbf{I} - \tfrac{1}{2}\mathbf{\Omega}\right)^{-1}\dot{\mathbf{\Omega}}\left(\mathbf{I} - \tfrac{1}{2}\mathbf{\Omega}\right)^{-1}\left(\mathbf{I} + \tfrac{1}{2}\mathbf{\Omega}\right) + \tfrac{1}{2}\left(\mathbf{I} - \tfrac{1}{2}\mathbf{\Omega}\right)^{-1}\dot{\mathbf{\Omega}}\right]\mathbf{X}_0 \\
&= \tfrac{1}{2}\left(\mathbf{I} - \tfrac{1}{2}\mathbf{\Omega}\right)^{-1}\dot{\mathbf{\Omega}}\left(\mathbf{I} - \tfrac{1}{2}\mathbf{\Omega}\right)^{-1}\left[\left(\mathbf{I} + \tfrac{1}{2}\mathbf{\Omega}\right) + \left(\mathbf{I} - \tfrac{1}{2}\mathbf{\Omega}\right)\right]\mathbf{X}_0 \\
&= \left(\mathbf{I} - \tfrac{1}{2}\mathbf{\Omega}\right)^{-1}\dot{\mathbf{\Omega}}\ \left(\mathbf{I} - \tfrac{1}{2}\mathbf{\Omega}\right)^{-1}\mathbf{X}_0, \tag{3.163}
\end{aligned}$$

Solving Eq. (3.162) for $\mathbf{X}_0$:

$$\mathbf{X}_0 = \left(\mathbf{I} + \tfrac{1}{2}\mathbf{\Omega}\right)^{-1}\left(\mathbf{I} - \tfrac{1}{2}\mathbf{\Omega}\right)\mathbf{X}(t), \tag{3.164}$$

and inserting into the previous equation gives us

$$\begin{aligned}
\dot{\mathbf{X}} &= \mathbf{A}(\mathbf{X}, t)\mathbf{X} \\
&= \left(\mathbf{I} - \tfrac{1}{2}\mathbf{\Omega}\right)^{-1}\dot{\mathbf{\Omega}}\left(\mathbf{I} - \tfrac{1}{2}\mathbf{\Omega}\right)^{-1}\left(\mathbf{I} + \tfrac{1}{2}\mathbf{\Omega}\right)^{-1}\left(\mathbf{I} - \tfrac{1}{2}\mathbf{\Omega}\right)\mathbf{X}, \\
&= \left(\mathbf{I} - \tfrac{1}{2}\mathbf{\Omega}\right)^{-1}\dot{\mathbf{\Omega}}\left(\mathbf{I} + \tfrac{1}{2}\mathbf{\Omega}\right)^{-1}\mathbf{X}, \tag{3.165}
\end{aligned}$$

so we would like to find a function $\mathbf{\Omega}(t)$ which satisfies

$$\mathbf{A}(\mathbf{X}, t) = \left(\mathbf{I} - \tfrac{1}{2}\mathbf{\Omega}\right)^{-1}\dot{\mathbf{\Omega}}\left(\mathbf{I} - \tfrac{1}{2}\mathbf{\Omega}\right)^{-1}\left(\mathbf{I} + \tfrac{1}{2}\mathbf{\Omega}\right)^{-1}\left(\mathbf{I} - \tfrac{1}{2}\mathbf{\Omega}\right), \tag{3.166}$$

and solving for $\dot{\mathbf{\Omega}}$ produces

$$\begin{aligned}
\dot{\mathbf{\Omega}} &= \left(\mathbf{I} - \tfrac{1}{2}\mathbf{\Omega}\right)\mathbf{A}\left(\mathbf{I} + \tfrac{1}{2}\mathbf{\Omega}\right) \\
&= \mathbf{A} - \tfrac{1}{2}[\mathbf{\Omega}, \mathbf{A}] - \tfrac{1}{4}\mathbf{\Omega}\mathbf{A}\mathbf{\Omega}. \tag{3.167}
\end{aligned}$$

It is obvious that the first and second term are both elements of the Lie algebra; however it is not so obvious that the symmetric product $\mathbf{\Omega}\mathbf{A}\mathbf{\Omega}$ is. Since we are restricting ourselves to Lie algebras of the form in Eq. (3.136) corresponding to quadratic Lie groups, it turns out that the symmetric product $\mathbf{\Omega}\mathbf{A}\mathbf{\Omega}$ is in fact a member of the Lie algebra. This follows from

$$\mathbf{J}(\mathbf{\Omega}^T\mathbf{A}^T\mathbf{\Omega}^T) = (-\mathbf{\Omega}\mathbf{J})\mathbf{A}^T\mathbf{\Omega}^T = -\mathbf{\Omega}(-\mathbf{A}\mathbf{J})\mathbf{\Omega}^T = -(\mathbf{\Omega}\mathbf{A}\mathbf{\Omega})\mathbf{J}. \tag{3.168}$$

It is interesting to contrast Eq. (3.167) having only a few terms with the infinite BCH series of Eq. (3.117) using the exponential lift.

Table I. Generic Butcher table.

c_1	a_{11}	a_{12}	$\ldots$	a_{1p}
c_2	a_{21}	a_{22}	$\ldots$	a_{2p}
$\vdots$	$\vdots$	$\vdots$	$\ddots$	$\vdots$
c_p	a_{p1}	a_{p2}	$\ldots$	a_{pp}
	b_1	b_2	$\ldots$	b_p

Table II. Butcher table for a common 4th order method.

0	0	0	0	0
$\frac{1}{2}$	$\frac{1}{2}$	0	0	0
$\frac{1}{2}$	0	$\frac{1}{2}$	0	0
1	0	0	1	0
	$\frac{1}{6}$	$\frac{1}{3}$	$\frac{1}{3}$	$\frac{1}{6}$

3.8.4 Runge-Kutta-Munthe-Kas integration

Numerical integration by the traditional Runge-Kutta method[22] is done with a set of difference equations of the general form

$$\dot{y} = f(t, y) \tag{3.169}$$

$$k_i = hf\left(t_n + c_i h, y_n + \sum_{j=1}^{\nu} a_{ij} k_j\right) \tag{3.170}$$

$$y_{n+1} = \sum_{j=1}^{\nu} b_j k_j, \tag{3.171}$$

where h is the integration step size. Note that y, k, and f may be vectors. The coefficients and weights for a particular method are frequently given in something called a Butcher table with the general form given in Table I. For some tables when some steps depend on earlier steps, evaluation may need to be iterated, and the method is called an implicit Runge-Kutta integrator.

The Butcher table for the popular fourth order Runge-Kutta integrator[24] is shown in Table II and is called explicit. Since the values of the a_{ij} are zero on an above the diagonal, evaluation is straight forward with each step depending only on previous evaluations of the function.

By itself this standard Runge-Kutta method is not a symplectic integrator. However a Lie algebraic version of the Runge-Kutta method was developed by Munthe-Kas[23] and combines integrating in the tangent planes with lifts back to the trajectory manifold. The Runge-Kutta-Munthe-Kas (RKMK) algorithm will integrate the equation

$$\dot{\mathbf{y}} = \mathbf{A}(t, \mathbf{y})\,\mathbf{y}, \tag{3.172}$$

using a lift function $\Phi(\Omega(t))$ to get

$$\mathbf{y}(t) = \boldsymbol{\Phi}\left(\Omega(t)\right)\mathbf{y}_0, \tag{3.173}$$

where the matrix function $\Omega(t)$ is defined by the differential equation

$$\dot{\Omega} = \mathcal{F}(t, \Omega(t), \mathbf{A}(t)). \tag{3.174}$$

for some function $\mathcal{F}$. For the Cayley transform we found that $\mathcal{F}$ was given by Eq. (3.167) at the end of the previous section:

$$\Phi(t) = \mathrm{cay}(\Omega(t)), \tag{3.175}$$

$$\mathcal{F} = \dot{\Omega} = \mathbf{A} - \tfrac{1}{2}[\Omega, \mathbf{A}] - \tfrac{1}{4}\Omega\mathbf{A}\Omega. \tag{3.176}$$

For the exponential function may be derived from the BCH formula[23]:

$$\Phi(t) = e^{\Omega(t)}, \tag{3.177}$$

$$\mathcal{F} = \mathbf{A} - \tfrac{1}{2}[\Omega, \mathbf{A}] + \tfrac{1}{12}[\Omega, \Omega, \mathbf{A}] + \cdots. \tag{3.178}$$

The RKMK algorithm then becomes

$$\omega_i = \sum_{j=1}^{\nu} a_{ij}\mathbf{F}_j, \tag{3.179}$$

$$\mathbf{A}_i = h\mathcal{F}\left(t_n + c_i h, \Phi(\omega_i)\mathbf{y}_n\right), \tag{3.180}$$

$$\mathbf{F}_i = \mathcal{F}(t_n + c_i h, \omega_i, \mathbf{A}_i), \tag{3.181}$$

$$\Omega_{n+1} = \sum_{j=1}^{\nu} b_j \omega_j, \tag{3.182}$$

$$\mathbf{y}_{n+1} = \boldsymbol{\Phi}(\Omega_{n+1})\mathbf{y}_n. \tag{3.183}$$

Problems

3–1   The transformation of coordinates from the fixed system to the local system moving along the design orbit is given by the equations (See Fig. 3.2.)

$$x = \sqrt{\xi^2 + \eta^2} - \rho, \quad \text{and}$$

$$s = \rho \tan^{-1}\left(\frac{\eta}{\xi}\right).$$

Show that the momenta given by

$$p_x = p_r = \vec{p} \cdot \hat{x} \quad \text{and}$$

$$p_s = \left(1 + \frac{x}{\rho}\right) \vec{p} \cdot \hat{s}$$

are canonically conjugate to the coordinates x and s. Hint: Construct a generating function $F_3(p_\xi, p_\eta, x, s)$ as described in Appendix C.

3–2 Prove that the inverse of a symplectic matrix $\mathbf{M}$ is given by

$$\mathbf{M}^{-1} = -\mathbf{S}\mathbf{M}^T\mathbf{S}.$$

If we write a symplectic $\mathbf{M}$ in terms of 2×2-blocks $\mathbf{A}$, $\mathbf{B}$, $\mathbf{C}$, and $\mathbf{D}$ as

$$\mathbf{M} = \begin{pmatrix} \mathbf{A} & \mathbf{B} \\ \mathbf{C} & \mathbf{D} \end{pmatrix},$$

show that

$$\mathbf{M}^{-1} = \begin{pmatrix} \tilde{\mathbf{A}} & \tilde{\mathbf{C}} \\ \tilde{\mathbf{B}} & \tilde{\mathbf{D}} \end{pmatrix}.$$

Find a similar formula for a 6×6 symplectic matrix.

3–3 Demonstrate that the collection of real $2n$-dimensional symplectic matrices, $\mathrm{Sp}(2n, \mathbb{R}; \mathbf{S})$, forms a group. The properties of a group[78] are:

   i. for any two elements $a, b \in G$, then $ab \in G$;

   ii. if a, b, $c \in G$, then $a(bc) = (ab)c$;

   iii. there is a unique element $e \in G$ such that $ea = a = ae$ for any element $a \in G$;

   iv. for each $a \in G$ there is an element $a^{-1} \in G$ such that $a^{-1}a = e = aa^{-1}$.

3–4 The usual representation for the Pauli spin matrices of SU(2) is:

$$\sigma_x = \begin{pmatrix} 0 & 1 \\ 1 & 0 \end{pmatrix}, \quad \sigma_y = \begin{pmatrix} 0 & -i \\ i & 0 \end{pmatrix}, \quad \sigma_z = \begin{pmatrix} 1 & 0 \\ 0 & -1 \end{pmatrix}.$$

For one degree of freedom, $\mathbf{S} = i\sigma_y$. This suggests a connection between SU(2) and Sp(2n).

a) Given a matrix $\mathbf{M}$ for some beam line, show that the transformation defined by

$$\check{\mathbf{M}} = \sigma_z \mathbf{M}^{-1} \sigma_z,$$

is the transfer matrix for the mirror image of the beam line. (For example, if

$$M = M_q M_b M_d,$$

for a drift followed by a bend and then a quadrupole, then

$$\check{M} = M_d M_b M_q.)$$

Note that if the independent variable used is time, then σ_z is the time reversal operator.

b) Show that the following relations hold:

$$M^T = \sigma_x \check{M} \sigma_x,$$
$$M^{-1} = \sigma_y M^T \sigma_y.$$

3–5   Use the symplectic generator method to find the transfer matrix for a drift of length L.

3–6   For general static magnetic elements with no xy-coupling and no rf electric fields the transport matrix for (x, x', z, δ) has the general form:

$$\begin{pmatrix} C & S & 0 & D \\ C' & S' & 0 & D' \\ E & F & 1 & G \\ 0 & 0 & 0 & 1 \end{pmatrix}$$

a) using the symplectic condition $M^T S M = S$ express E and F in terms of the other elements and show that there is no constraint on the value of G.

b) Show that matrices of this form make a subgroup of $Sp(4)$ the real symplectic group of 4×4 matrices.

3–7   Use the symplectic generator method to find the transfer matrix for a quadrupole magnet of length L with $A_x = A_y = 0$ and

$$A_s = \frac{g}{2}(x^2 - y^2).$$

3–8   a) Find the equations of motion for a particle in a uniform magnetic field along the s-axis, with the vector potential

$$\vec{A} = \begin{cases} \left(-\frac{B_0}{2}y, \frac{B_0}{2}x, 0\right), & \text{if } 0 < s < l; \\ 0, & \text{otherwise.} \end{cases}$$

Be sure to use a canonical system of coordinates inside the magnet.

b) Find the generator **G** and use it to obtain the linear transformation matrix,

$$
\mathbf{M} = \begin{pmatrix}
\frac{1+\cos\phi}{2} & r\sin\phi & \frac{\sin\phi}{2} & r(1-\cos\phi) \\
-\frac{\sin\phi}{4r} & \frac{1+\cos\phi}{2} & -\frac{1-\cos\phi}{4r} & \frac{\sin\phi}{2} \\
-\frac{\sin\phi}{2} & -r(1-\cos\phi) & \frac{1+\cos\phi}{2} & r\sin\phi \\
\frac{1-\cos\phi}{4r} & -\frac{\sin\phi}{2} & -\frac{\sin\phi}{4r} & \frac{1+\cos\phi}{2}
\end{pmatrix},
$$

for the transformation through a solenoid magnet in the hard-edge approximation, with $r = p_s/(qB_0)$, and $\phi = l/r$.

3–9 Repeat the previous problem with the vector potential

$$
\vec{A} = \begin{cases} (0, B_0 x, 0), & \text{if } 0 < s < l; \\ 0, & \text{otherwise.} \end{cases}
$$

Note that this give a constant field along the axis of the magnet just like the previous problem, but the resulting transfer matrix is quite different. Explain this difference. What does the magnet look like?

3–10 Verify Eq. (3.120) to second order in h.

3–11 Show that

$$
e^{\mathbf{SC}} = [\mathbf{I} + \tanh(\mathbf{SC}/2)]\,[\mathbf{I} - \tanh(\mathbf{SC}/2)]^{-1},
$$

where **C** is defined by Eq. (3.108). Compare this with Eq. (3.147).

References for Chapter 3

[1] J. Liouville, "Sur la Théorie de la Variation des constantes arbitraires", *Journal de Math. Pures et Appl.*, p.342, 3 (1838).

[2] R. C. Tolman, *The Principles of Statistical Mechanics*, Oxford (1938).

[3] Hermann Weyl, *The Classical Groups*, Princeton NJ (1946).

[4] V. I. Arnold, *Mathematical Methods of Classical Mechanics*, Springer–Verlag, New York (1978).

[5] K. G. Steffen, *High Energy Beam Optics*, John Wiley and Sons, New York, (1965).

[6] H. Goldstein, *Classical Mechanics*, Addison–Wesley, Reading, Massachusetts (1980).

[7] A. Clark, *Elements of Abstract Algebra*, Wadsworth Publ. Co., Belmont (1971).

[8] R. Gilmore, *Lie Groups, Lie Algebras, and Some of Their Applications*, John Wiley & Sons, New York (1974).

[9] L. D. Landau and E. M. Lifshitz *Mechanics*, Pergamon, Oxford (1976).

[10] R. D. Ruth, "Single Particle Dynamics and Nonlinear Resonances in Circular Accelerators", SLAC-PUB-3836 (1985).

[11] A. A. Kolomensky and A. N. Lebedev, *Theory of Cyclic Accelerators*, John Wiley and Sons, New York (1962).

[12] Robert A. Horn and Charles R. Johnson, *Matrix Analysis*, Cambridge (1985).

[13] A. J. Dragt, "Lectures on Nonlinear Orbit Dynamics", *Physics of High Energy Accelerators* (Fermilab Summer School, 1981), A. I. P. Conf. Proc. 87, R. A. Carrigan, F. R. Huson, and M. Month, editors (1982)

[14] Étienne Forest, J. Phys. A: Math. Gen. **39**, 5321 (2006).

[15] L. C. Teng, "Concerning n-Dimensional Coupled Motions", FN-229 0100, National Accelerator Laboratory, Batavia (1971).

[16] Ronald D. Ruth, IEEE Trans. on Nucl. Sci. **NS-30**, 2669 (1983).

[17] Haruo Yoshida, Phys. Lett. **A 150**, 262 (1990).

[18] E. B. Dynkin, Dokl. Akad. Nauk SSSR (N. S.) **57**, 323 (1947).

[19] Wilhelm Magnus, Communications on Pure and Applied Mathematics, **7**, 649 (1954).

[20] L. M. Healy, "Lie Algebraic Methods for Treating Parameter Errors in Particle Accelerators", Doctoral Thesis, U. of Maryland (1986).

[21] W. W. MacKay, Proc. of EPAC 2006, Edinburgh Scotland, 2281 (2006).

[22] Arieh Iserles, "On Cayley-transform methods for the discretization of Lie-group equations", NA1999/04, Dept. of Applied Mathematics and Theoretical Physics, Univ. of Cambridge (1999).
[23] Arieh Iserles et al., "Lie-group methods", NA2000/03, Dept. of Applied Mathematics and Theoretical Physics, Univ. of Cambridge (2000).
[24] Milton Abramowitz and Irene A. Stegun, *Handbook of Mathematical Functions*, Dover Pub. (1970).

4

Optical Elements with Static Magnetic Fields

In this chapter we discuss the fields of several different types of magnets used in accelerators. Ignoring end effects, the typical magnets either have fields transverse to the motion of the reference particle or along the particle's direction of motion, i. e., a longitudinal field.

Magnets are used to change the direction of motion of a particle by exerting a force perpendicular to both the motion of the particle and the magnetic field, via the Lorentz force:

$$\vec{F} = q\vec{v} \times \vec{B}. \tag{4.1}$$

A transverse field which is perpendicular to the particle's motion will provide a much larger force than a field of the same magnitude which is almost in the direction of the particle's trajectory. Because of this, magnets with transverse fields make up the backbone of accelerators and beam transport systems.

A longitudinal field can be produced by a solenoid whose axis is collinear with the design trajectory. Such a magnet is frequently found in the experimental detector in a colliding beam accelerator or as a lens for a low energy beam. The lithium lens invented at Novosibirsk[1] produces a toroidal field by passing a large current through a light metal (lithium) with a low nuclear cross section. The particle beam is sent through the block of lithium along the direction of the current. Lithium lenses have been used for collection of antiprotons in an antiproton source at Fermilab[2].

Static magnetic fields for accelerators are produced either by electric currents or permanent magnet material[3] such as alloys of samarium–cobalt or neodymium–boron–iron. Iron is frequently placed around the currents or permanent magnet material to decrease the magnetic reluctance and shape the field distribution.

Electrically powered magnets can be made with warm conductors such as copper or aluminum that are sometimes cooled by water, or superconductors—the most popular being an alloy of niobium–titanium embedded in a copper stabilizer. Excellent discussions of some of the practical considerations of these magnets can be found in an article by G. E. Fischer[4], and the book by M. N. Wilson[5].

The magnetic field must obey Maxwell's equations,

$$\nabla \cdot \vec{B} = 0 \tag{4.2}$$

$$\nabla \times \vec{H} = \vec{j} + \epsilon_0 \frac{\partial \vec{E}}{\partial t}, \tag{4.3}$$

with $\vec{B} = \mu \vec{H}$. The vector potential, $\vec{A}$ with $\vec{B} = \nabla \times \vec{A}$, is extremely useful for calculations, particularly in problems which may be reduced to a plane, such as the case of transverse fields in long magnets, where the end effects become negligible. For static fields, $\partial \vec{E} / \partial t = 0$.

4.1 Transverse fields and multipoles

A magnetic field which is transverse to the direction of the beam obeys a symmetry which mathematically requires a magnet of infinite extent along the z or s axis. This translates to a practical rule of thumb that the two dimensional model is a good approximation in region of the magnet which is more than one and a half gap heights from the end of the magnet. For a preliminary design, it is usually alright to ignore the end effects; however, they must be taken into account when actually building an accelerator. This frequently requires either modifying the magnet ends, or installing small correction magnets. End effects will be discussed in a later section.

In the two dimensional model Eqs. (4.2, and 4.3) become

$$\frac{\partial B_x}{\partial x} + \frac{\partial B_y}{\partial y} = 0 \tag{4.4}$$

$$\frac{\partial B_x}{\partial y} = \frac{\partial B_y}{\partial x}. \tag{4.5}$$

The field of a bending magnet can be expanded in a series of multipoles[6],

$$B_y = B_0 \sum_{n=0}^{\infty} \left(\frac{r}{a}\right)^n (b_n \cos n\theta - a_n \sin n\theta), \tag{4.6}$$

$$B_x = B_0 \sum_{n=0}^{\infty} \left(\frac{r}{a}\right)^n (a_n \cos n\theta + b_n \sin n\theta), \tag{4.7}$$

where $x = r \cos \theta$, $y = r \sin \theta$, the central field is given by B_0, and a is a reference radius of the expansion which should be the same order of magnitude as the aperture of the magnet (usually $\approx \frac{2}{3}$ the radius of the beam pipe.) Eqs. (4.6 and 4.7) can be rewritten in complex form using Cartesian notation:

$$B_x - iB_y = B_0 \sum_{n=0}^{\infty} (a_n - ib_n) \left(\frac{x + iy}{a} \right)^n. \qquad (4.8)$$

These multipoles are different from the distributions of electric multipoles given by a symmetric distribution of point charges. For example, the lowest order multipole describes a uniform magnetic field, which could be produced by two infinite parallel planes of north and south magnetic poles. This is, of course, mathematically identical to the infinite parallel plate capacitor. These *dipole* terms are given by the $n = 0$ terms in the above expansion. A magnet with $b_0 = 1$ and all other b_n and a_n equal to zero is called a *normal dipole*, since it requires two pole faces to produce this field and bends a particle in the "normal" horizontal direction. The b_n coefficients are called *normal* multipoles, and the a_n coefficients are called *skew* multipoles because they are identical to the normal multipoles except for being skewed about the z–axis.

The coefficient, b_n, gives the relative strength of the $2(n + 1)$–pole. These *normal* multipoles have a symmetric gap between pole faces along the x–axis, and they satisfy the symmetry relations:

$$B_x(x, y) = -(-1)^n B_x(-x, y),$$
$$B_x(x, y) = -B_x(x, -y),$$
$$B_y(x, y) = (-1)^n B_y(-x, y),$$
$$B_y(x, y) = B_y(x, -y). \qquad (4.9)$$

Similarly, the *skew* multipoles have a gap between pole faces along the y–axis, no currents on the x–axis, and obey the symmetry relations:

$$B_x(x, y) = (-1)^n B_x(-x, y),$$
$$B_x(x, y) = B_x(x, -y),$$
$$B_y(x, y) = -(-1)^n B_y(-x, y),$$
$$B_y(x, y) = -B_y(x, -y). \qquad (4.10)$$

These two categories of multipoles are frequently divided into subfamilies of even and odd n:

i) normal multipoles with even n which have the current symmetries $I(x, y) = -I(-x, y) = I(x, -y)$,

ii) normal multipoles with odd n which have the current symmetries $I(x, y) = I(-x, y) = I(x, -y)$,

iii) skew multipoles with even n which have the current symmetries $I(x, y) = I(-x, y) = -I(x, -y)$, and

iv) skew multipoles with odd n which have the current symmetries $I(x, y) = -I(-x, y) = -I(x, -y)$.

When a magnet is being designed, use of these symmetries removes many of the unwanted multipoles in a very simple way; however, when a real magnet is built, it is impossible to place the conductors and iron exactly where the design calls for them. The multipoles which are not forbidden by the symmetry of the design are called *allowed* multipoles, and the multipoles which are not allowed by the symmetry are called *forbidden* multipoles. In a real magnet, construction errors can break the symmetry and result in small amounts of the forbidden multipoles.

4.2 Equipotential surfaces and pole face contours

Inside the aperture, where there is no magnetic material or currents (ignoring the beam), the magnetic field can be expressed as the gradient of the magnetostatic potential, i. e.,

$$\vec{H} = \vec{\nabla}\Psi. \tag{4.11}$$

In cylindrical coordinates, (r, θ, z), one can obtain the following relation for the potential:

$$\Psi = \sum_{n=0}^{\infty} \frac{a}{n+1} \left(\frac{r}{a}\right)^{n+1} [F_n \cos\left((n+1)\theta\right) + G_n \sin\left((n+1)\theta\right)], \tag{4.12}$$

where we have let $G_n = B_0 b_n / \mu_0$, and $F_n = B_0 a_n / \mu_0$ be the respective normal and skew strengths for the $2(n+1)$th multipoles. Similarly to electric fields, the magnetic field must be perpendicular to the equipotential lines of Ψ. If we assume that our magnet were made of iron with an infinite permeability, then for a pure normal $2(n+1)$–multipole, we would expect the optimum shape of the pole face to be the same as the equipotential of the term proportional to G_n:

$$\left(\frac{r}{a}\right)^{n+1} \sin\left[(n+1)\theta\right] = 1. \tag{4.13}$$

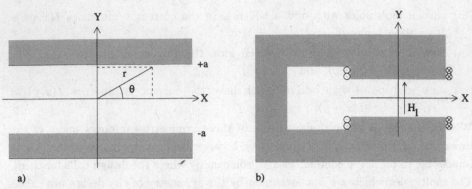

Figure. 4.1 a) Pole profile for a "dipole" magnet.   b) Dipole magnet with coils and dipole field.

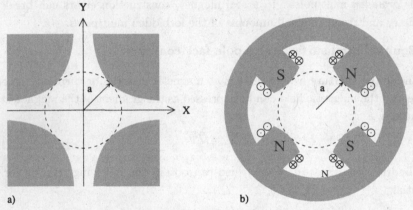

Figure. 4.2 a) Pole–face profile of equation $xy = a^2/2$. b) Quadrupole magnet with coils.

Fig. 4.1a illustrates the pole profile for a dipole magnet with equipotential surfaces given by

$$r \sin \theta = y = \text{constant.} \qquad (4.14)$$

Fig. 4.1b shows how such a magnet might be realized with a gap of $2a$ and N turns of current I wound about each pole. We may use Ampere's law to estimate the field in the gap, using the approximation $\mu_{\text{iron}} = \infty$:

$$2NI = \oint \vec{H} \cdot d\vec{l} = 2aH, \qquad (4.15)$$

giving $H = NI/a$.

For a normal quadrupole magnet the optimum pole profile is hyperbolic, since the profile equation is

$$r^2 \sin 2\theta = 2xy = a^2, \tag{4.16}$$

with a constant. For coils of N turns of wire about each pole face (see Fig. 4.2b). Applying Ampere's law again gives

$$2NI = \int_{\text{north}}^{\text{south}} \vec{H} \cdot d\vec{l} = \Psi_{\text{south}} - \Psi_{\text{north}} \tag{4.17}$$

$$= G_1 \frac{a}{2}(1 - (-1)) = G_1 a, \tag{4.18}$$

or

$$G_1 = \frac{2NI}{a}. \tag{4.19}$$

Evaluating the potential gives

$$\Psi = \frac{2NI}{a^2} xy \tag{4.20}$$

to produce the field

$$\vec{H} = \frac{2NI}{a^2}(y\hat{x} + x\hat{y}). \tag{4.21}$$

In general for a $2(n + 1)$–pole with N turns of current I about each pole, we may estimate

$$G_n = (n + 1)\frac{NI}{a}, \tag{4.22}$$

to get a potential of

$$\Psi_n = NI \left(\frac{r}{a}\right)^{n+1} \sin((n + 1)\theta), \tag{4.23}$$

and calculating the magnet field, we have

$$H_x = (n + 1)\frac{NI}{a} \left(\frac{r}{a}\right)^n \sin n\theta, \tag{4.24}$$

$$H_y = (n + 1)\frac{NI}{a} \left(\frac{r}{a}\right)^n \cos n\theta. \tag{4.25}$$

The explicit form for a sextupole ($n = 2$) is given by

$$\vec{H} = \frac{3NI}{a^3}[2xy\hat{x} + (x^2 - y^2)\hat{y}]. \tag{4.26}$$

4.3 Quadrupole lenses

A quadrupole lens is a magnet with four poles, arranged as in Fig. 4.2, expanded along the z-axis by a length l which is considerably longer than a.

In the absence of electric fields, the motion equation is

$$\gamma m \frac{d\vec{v}}{dt} = q\vec{v} \times \vec{B} = q\mu_0 \vec{v} \times \vec{H},$$
(4.27)

which in the paraxial approximation yields

$$\frac{d^2 x}{ds^2} + kx = 0$$
(4.28)

$$\frac{d^2 y}{ds^2} - ky = 0$$
(4.29)

where $s = \beta ct$ and

$$k = \frac{\mu_0 q G_1}{pa} = \frac{q}{p} \frac{\partial B_y}{\partial x}\bigg|_{x=0, y=x}.$$
(4.30)

$$G_1 = 2\frac{NI}{a}.$$
(4.31)

Solutions of Eqs. (4.28 and 4.29) for a magnet of length l are respectively

$$\begin{pmatrix} x \\ x' \end{pmatrix} = \begin{pmatrix} \cos\sqrt{k}l & \frac{1}{\sqrt{k}}\sin\sqrt{k}l \\ -\sqrt{k}\sin\sqrt{k}l & \cos\sqrt{k}l \end{pmatrix} \begin{pmatrix} x_0 \\ x'_0 \end{pmatrix} = \mathbf{M_F} \begin{pmatrix} x_0 \\ x'_0 \end{pmatrix}, \quad \text{and}$$
(4.32)

$$\begin{pmatrix} y \\ y' \end{pmatrix} = \begin{pmatrix} \cosh\sqrt{k}l & \frac{1}{\sqrt{k}}\sinh\sqrt{k}l \\ \sqrt{k}\sinh\sqrt{k}l & \cosh\sqrt{k}l \end{pmatrix} \begin{pmatrix} y_0 \\ y'_0 \end{pmatrix} = \mathbf{M_D} \begin{pmatrix} y_0 \\ y'_0 \end{pmatrix}.$$
(4.33)

When horizontal motion is focused, vertical motion is defocused; the opposite happens if the field is reversed.

For $\sqrt{k}l \ll 1$ (thin lens approximation with $l \to 0$ as kl remains constant) the matrices reduce to

$$\mathbf{M_{F,D}} = \begin{pmatrix} 1 & 0 \\ \mp kl & 1 \end{pmatrix} = \begin{pmatrix} 1 & 0 \\ \mp\frac{1}{f} & 1 \end{pmatrix},$$
(4.34)

where f is the *focal length* of the lens.

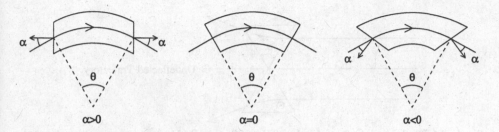

Figure. 4.3 The three most common situations regarding the angle α formed by the beam direction and the normal $\vec{n}$ to the magnet yokes.

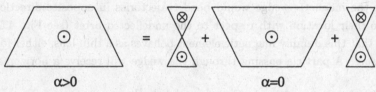

Figure. 4.4 A magnet with positively slanted ends, $\alpha > 0$, can be considered as the combination of a normal bending magnet and two magnetic wedges.

4.4 End field effects in a dipole

In all the bending magnets met so far, we have implicitly assumed that the design trajectory is orthogonal to both ends of the magnet. Indeed, this cannot be true in all circumstances and there will be cases where the side of the magnet is tilted. For practical reasons, we shall always consider rotations of the yoke-ends around vertical axes by an angle α, obtaining the two possible situations, illustrated in Fig. 4.3 together with the normal case, $\alpha = 0$.

Evidently, particles inside the magnets will experience different focusing and defocusing forces according to whether α is positive, negative, or zero. In fact for $\alpha > 0$, particles beyond the reference orbit travel a shorter path in the magnet than for $\alpha = 0$ and $\alpha < 0$. Of course similar arguments can be conceived for particles traveling closer to the center of curvature.

In order to evaluate these focusing effects, let the $\alpha > 0$ case be considered. The $\alpha < 0$ case can be found by just changing the sign of α. Such a magnet can be described as a normal one encompassed by two magnetic wedges, as shown in Fig. 4.4.

The most common bending magnet used in the present day is a dipole with

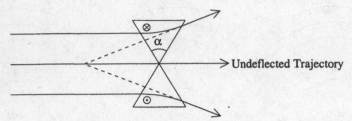

Figure. 4.5 The defocusing effect of a thin wedge with $\alpha > 0$, showing an increase in path length with $B_y = -B_0$ for $x > 0$, and $B_y = +B_0$ for $x < 0$.

flat pole-faces and no field index, $\partial B_y / \partial x = 0$.

Since the magnetic wedge would bend trajectories in opposite directions according to their location with respect to the undeflected orbit (see Fig. 4.5), one can deem that this dummy magnetic element behaves as a thin lens, either focusing or defocusing. A particle passing through the wedge will receive a horizontal kick given by

$$\Delta x' = -\frac{q}{p} \int_{-x\tan\alpha}^{0} B_y \, ds \simeq \frac{qB_0}{p} \tan\alpha \, x. \tag{4.35}$$

The horizontal displacement for a thin wedge is negligible, $\Delta x \simeq 0$, and the effect of the kick may be represented by the matrix

$$\mathbf{M_W} = \begin{pmatrix} 1 & 0 \\ \frac{\tan\alpha}{\rho} & 1 \end{pmatrix}, \tag{4.36}$$

or

$$\mathbf{M_W} = \begin{pmatrix} 1 & 0 & 0 \\ \frac{\tan\alpha}{\rho} & 1 & 0 \\ 0 & 0 & 1 \end{pmatrix}, \tag{4.37}$$

if dispersion is taken into account.

It is trivial to write down the matrix characterizing the sum of magnetic elements of Fig. 4.4;

$$\mathbf{M_H} = \mathbf{M_W}\mathbf{M_H}(\theta)\mathbf{M_W}, \tag{4.38}$$

where $\mathbf{M_W}$ is given by Eq. (4.37), and $\mathbf{M_H}(\theta)$ by the (x, x', δ) components of the matrix in Eq. (3.102).

Bearing in mind that

$$\cos\theta + \tan\alpha\sin\theta = \frac{\cos(\theta - \alpha)}{\cos\alpha}, \tag{4.39}$$

$$\sin\theta - \tan\alpha\cos\theta = \frac{\sin(\theta - \alpha)}{\cos\alpha}, \tag{4.40}$$

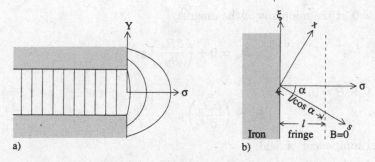

Figure. 4.6 a) Projection of the end of a dipole magnet showing the fringe field at the end of the magnet. b) Horizontal projection of the end of a dipole with a fringe field. The coordinates σ and ξ are the horizontal coordinates that are perpendicular and parallel to the iron, respectively. The coordinates s and x are the usual longitudinal and radial coordinates of the design trajectory.

$$\frac{\tan\alpha}{\rho}\frac{\cos(\theta-\alpha)}{\cos\alpha} - \frac{1}{\rho}\frac{\sin(\theta-\alpha)}{\cos\alpha} = -\frac{\sin(\theta-2\alpha)}{\rho\cos^2\alpha}, \tag{4.41}$$

and

$$(1-\cos\theta)\tan\alpha + \sin\theta = \frac{\sin(\theta-\alpha)+\sin\alpha}{\cos\alpha}, \tag{4.42}$$

the horizontal matrix can be written as

$$\mathbf{M_H} = \begin{pmatrix} \frac{\cos(\theta-\alpha)}{\cos\alpha} & \rho\sin\theta & \rho(1-\cos\theta) \\ -\frac{\sin(\theta-2\alpha)}{\rho\cos^2\alpha} & \frac{\cos(\theta-\alpha)}{\cos\alpha} & \frac{\sin(\theta-\alpha)+\sin\alpha}{\cos\alpha} \\ 0 & 0 & 1 \end{pmatrix}. \tag{4.43}$$

As far as the vertical motion is concerned, the procedure is somewhat subtler[9]. For a simplified understanding, we will assume that the magnetic field is constant and vertical inside the iron of the magnet and then drops off linearly in some small region of thickness l beyond the iron, as shown in Fig. 4.6. The vertical kick from the fringe field may be obtained from

$$\Delta y' = \frac{q}{p}\int_{\text{fringe}} B_x\, ds. \tag{4.44}$$

Using the coordinates (ξ, y, σ) as defined in Fig. 4.6, we have

$$B_y = B_0\left(1 - \frac{\sigma}{l}\right) \quad \text{for} \quad 0 < \sigma < l, \tag{4.45}$$

and assuming that the poles are wide,

$$B_\xi = 0. \tag{4.46}$$

Taking $y = 0$ at the midplane of the magnet,

$$B_\sigma \simeq 0 + \left(\frac{\partial B_\sigma}{\partial y}\right) y, \tag{4.47}$$

or, using $\nabla \times \vec{B} = 0$,

$$B_\sigma \simeq \left(\frac{\partial B_y}{\partial \sigma}\right) y = -\frac{B_0}{l} y. \tag{4.48}$$

The radial component of field,

$$B_x = B_\xi \cos\alpha + B_\sigma \sin\alpha = -\frac{B_0 \sin\alpha}{l} y, \tag{4.49}$$

may be inserted into Eq. (4.44) and integrated along s, from 0 to $(l/\cos\alpha)$, giving

$$\Delta y' = \frac{q}{p} \int_0^{\frac{l}{\cos\alpha}} \left(-\frac{B_0 \sin\alpha}{l} y\right) ds = -\left(\frac{\tan\alpha}{\rho}\right) y, \tag{4.50}$$

for the size of the vertical kick, which may be noted to be identical to Eq. (4.36) except for the sign of the focusing term. This linear thin-lens approximation of the fringe field is frequently very good; however, higher order corrections may be required for large fringe fields since the particle trajectory actually curves in the fringe field and B_y does not really drop off linearly at the end of the magnet. Such higher order corrections require detailed knowledge of the actual fringe field which is usually obtained by mapping the field with a probe such as a Hall probe or rotating coils.

As before with the horizontal case, we may modify the matrix for the vertical components for a sector magnet to obtain the matrix for a bend magnet with canted ends:

$$
\begin{aligned}
\mathbf{M_V} &= \begin{pmatrix} 1 & 0 \\ -\frac{\tan\alpha}{\rho} & 1 \end{pmatrix} \begin{pmatrix} 1 & \rho\theta \\ 0 & 1 \end{pmatrix} \begin{pmatrix} 1 & 0 \\ -\frac{\tan\alpha}{\rho} & 1 \end{pmatrix} \\
&= \begin{pmatrix} 1 - \theta\tan\alpha & \rho\theta \\ -\frac{\tan\alpha}{\rho}(2 - \theta\tan\alpha) & 1 - \theta\tan\alpha \end{pmatrix}.
\end{aligned} \tag{4.51}
$$

The most common and interesting case occurs when the magnet's poles have a rectangular plan, as shown in Fig. 4.7. Then for $\alpha = \theta/2$,

$$\mathbf{M_H} = \begin{pmatrix} 1 & \rho\sin\theta & \rho(1 - \cos\theta) \\ 0 & 1 & 2\tan\frac{\theta}{2} \\ 0 & 0 & 1 \end{pmatrix}, \tag{4.52}$$

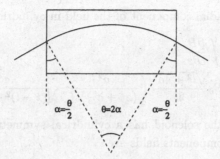

Figure. 4.7 A rectangular magnet.

$$M_V = \begin{pmatrix} 1 - \theta \tan\frac{\theta}{2} & \rho\theta \\ -\frac{1}{\rho}\tan\frac{\theta}{2}\left(2 - \theta\tan\frac{\theta}{2}\right) & 1 - \theta\tan\frac{\theta}{2} \end{pmatrix}. \tag{4.53}$$

For $\theta \ll 1$, keeping quadratic terms in θ, the horizontal matrix reduces to

$$M_H = \begin{pmatrix} 1 & \rho\theta & \frac{1}{2}\rho\theta^2 \\ 0 & 1 & \theta \\ 0 & 0 & 1 \end{pmatrix}, \tag{4.54}$$

and the vertical matrix becomes

$$M_V \simeq \begin{pmatrix} 1 - \frac{\theta^2}{2} & \rho\theta \\ -\frac{\theta}{\rho} & 1 - \frac{\theta^2}{2} \end{pmatrix} \simeq \begin{pmatrix} \cos\theta & \rho\sin\theta \\ -\frac{\sin\theta}{\rho} & \cos\theta \end{pmatrix}, \tag{4.55}$$

which means that what is lost in horizontal focusing is gained in the vertical one. Nevertheless, for $\theta \ll 1$, the 2×2 nondispersive part of the horizontal matrix is almost equal to the vertical matrix:

$$M_H \simeq M_V \simeq \begin{pmatrix} 1 & \rho\theta \\ 0 & 1 \end{pmatrix}, \tag{4.56}$$

and the vertical and horizontal are almost the same as a simple drift of length $\rho\theta$.

4.5 Ideal solenoid with thin fringes[10]

The magnetic field on the axis ($r = 0$) of a solenoid of length l and radius a can be written as

$$B(s) = B_0 \hat{s} \frac{\sqrt{l^2 + 4a^2}}{2l}\left[\frac{z}{\sqrt{z^2 + a^2}} + \frac{l - z}{\sqrt{(z - l)^2 + a^2}}\right]. \tag{4.57}$$

Here we have placed the upstream end of the solenoid at $s = 0$. Recalling Maxwell's equation

$$\nabla \cdot \vec{B} = 0, \tag{4.58}$$

we may solve for the radial component of the field in cylindrical coordinates as

$$B_r(r,s) = -\int \frac{\partial B_s}{\partial s} dr$$

$$= -B_0 \frac{\sqrt{l^2 + 4a^2}}{2l} \left(\frac{a^2}{(z^2 + a^2)^{3/2}} - \frac{a^2}{[(z-l)^2 + a^2]^{3/2}} \right) r \qquad (4.59)$$

near the axis. Since the solenoid has a cylindrical symmetry $B_\phi = 0$, and the transverse Cartesian components fields are

$$B_x = B_r \frac{x}{r}, \quad \text{and} \quad B_y = B_r \frac{y}{r} \qquad (4.60)$$

where $r = \sqrt{x^2 + y^2}$.

Integrating the trajectories through the field of Eqs. 4.57 and 4.59 is nontrivial. For solenoids of small radius with $a \ll l$, it is convenient to linearize the field components with thin fringes. The longitudinal component may be written to first order as

$$B_s(s) = \begin{cases} 0, & s \in (-\infty, -a), \\ B_0 \left(\frac{s+a}{2a} \right), & s \in [-a, a), \\ B_0, & s \in [a, l-a], \\ B_0 \left(\frac{l+a-s}{2a} \right), & s \in (l-a, l+a], \\ 0, & s \in (l+a, \infty), \end{cases} \qquad (4.61)$$

and the radial component as

$$B_r(s) = \begin{cases} 0, & s \in (-\infty, -a), \\ -\frac{B_0 r}{2a}, & s \in [-a, a), \\ 0, & s \in [a, l-a], \\ \frac{B_0 r}{2a}, & s \in (l-a, l+a], \\ 0, & s \in (l+a, \infty). \end{cases} \qquad (4.62)$$

Linearizing the differential equations from Problem 1–3 we have

$$x'' \simeq \frac{q}{p}(y'B_s - B_y), \qquad (4.63)$$

$$y'' \simeq -\frac{q}{p}(x'B_s - B_x). \qquad (4.63)$$

In the body of the magnet $\vec{B} = B_0 \hat{s}$ so these reduce to

$$\frac{d}{ds} \begin{pmatrix} x' \\ y' \end{pmatrix} = \frac{qB_0}{p} \begin{pmatrix} 0 & 1 \\ -1 & 0 \end{pmatrix} \begin{pmatrix} x' \\ y' \end{pmatrix}, \qquad (4.64)$$

which may be integrated from 0 to s to yield the slopes

$$\begin{pmatrix} x'(s) \\ y'(s) \end{pmatrix} = \exp\left[\begin{pmatrix} 0 & 1 \\ -1 & 0 \end{pmatrix} ks\right] \begin{pmatrix} x'_0 \\ y'_0 \end{pmatrix}$$

$$= \begin{pmatrix} \cos ks & \sin ks \\ -\sin ks & \cos ks \end{pmatrix} \begin{pmatrix} x'_0 \\ y'_0 \end{pmatrix} \tag{4.65}$$

where

$$k = \frac{qB_0}{p}. \tag{4.66}$$

Integrating a second time gives

$$\begin{pmatrix} x(s) \\ y(s) \end{pmatrix} = \begin{pmatrix} x_0 \\ y_0 \end{pmatrix} + \frac{1}{k} \begin{pmatrix} \sin ks & (1 - \cos ks) \\ -(1 - \cos ks) & \sin ks \end{pmatrix} \begin{pmatrix} x'_0 \\ y'_0 \end{pmatrix}. \tag{4.67}$$

Now it is easy to write the transfer matrix through the body of the solenoid of length l in the thin fringe limit of $a = 0$ as

$$\mathbf{M}_{\text{body}} = \begin{pmatrix} 1 & \frac{\sin kl}{k} & 0 & \frac{1-\cos kl}{k} \\ 0 & \cos kl & 0 & \sin kl \\ 0 & -\frac{1-\cos kl}{k} & 1 & \frac{\sin kl}{k} \\ 0 & -\sin kl & 0 & \cos kl \end{pmatrix}. \tag{4.68}$$

For the transfer matrix through the first fringe field we find the differential equations for transverse motion:

$$x'' = \frac{qB_0}{2pa}[(s+a)y' + y], \tag{4.69}$$

$$y'' = -\frac{qB_0}{2pa}[(s+a)x' + x]. \tag{4.70}$$

When integrated over a thin fringe from $s = -a$ to 0, the terms with $(s+a)x'$ and $(s+a)y'$ will be of order a^2 or higher so they become negligible as a tends toward zero. For the thin lens approximation, then we are left with

$$x'' = \frac{k}{2a}y, \tag{4.71}$$

$$y'' = -\frac{k}{2a}x, \tag{4.72}$$

or after integration

$$x' = x'_0 + \frac{k}{2}y_0, \tag{4.73}$$

$$y' = y'_0 - \frac{k}{2}x_0. \tag{4.74}$$

In the thin lens limit, the trajectories must still be continuous, so the upstream fringe matrix may be written as

$$
M_{up} = \begin{pmatrix} 1 & 0 & 0 & 0 \\ 0 & 1 & \frac{k}{2} & 0 \\ 0 & 0 & 1 & 0 \\ -\frac{k}{2} & 0 & 0 & 1 \end{pmatrix}.
\tag{4.75}
$$

The downstream fringe will be the same, but with the slope of the field reversed, so

$$
M_{dn} = \begin{pmatrix} 1 & 0 & 0 & 0 \\ 0 & 1 & -\frac{k}{2} & 0 \\ 0 & 0 & 1 & 0 \\ \frac{k}{2} & 0 & 0 & 1 \end{pmatrix}.
\tag{4.76}
$$

Multiplying the matrices for the the body and fringes then gives a matrix for the whole solenoid:

$$
\begin{aligned}
M_{sol} &= M_{dn} M_{body} M_{up} \\
&= \begin{pmatrix} 1 & 0 & 0 & 0 \\ 0 & 1 & -\frac{k}{2} & 0 \\ 0 & 0 & 1 & 0 \\ \frac{k}{2} & 0 & 0 & 1 \end{pmatrix} \begin{pmatrix} 1 & \frac{\sin kl}{k} & 0 & \frac{1-\cos kl}{k} \\ 0 & \cos kl & 0 & \sin kl \\ 0 & -\frac{1-\cos kl}{k} & 1 & \frac{\sin kl}{k} \\ 0 & -\sin kl & 0 & \cos kl \end{pmatrix} \begin{pmatrix} 1 & 0 & 0 & 0 \\ 0 & 1 & \frac{k}{2} & 0 \\ 0 & 0 & 1 & 0 \\ -\frac{k}{2} & 0 & 0 & 1 \end{pmatrix} \\
&= \begin{pmatrix} \frac{1+\cos kl}{2} & \frac{\sin kl}{k} & \frac{\sin kl}{2} & \frac{1-\cos kl}{k} \\ -\frac{k \sin kl}{4} & \frac{1+\cos kl}{2} & -k\frac{1-\cos kl}{4} & \frac{\sin kl}{2} \\ -\frac{\sin kl}{2} & -\frac{1-\cos kl}{k} & \frac{1+\cos kl}{2} & \frac{\sin kl}{k} \\ k\frac{1-\cos kl}{4} & -\frac{\sin kl}{2} & -\frac{k \sin kl}{4} & \frac{1+\cos kl}{2} \end{pmatrix},
\end{aligned}
\tag{4.77}
$$

which agrees with the result of Problem 3–8 if we make the substitutions $r = 1/k$ and $\phi = kl$.

Another interesting way of writing the matrix for the solenoid is as a combination of a rotation and and a symmetric focusing element:

$$
M_{sol} = RM_f = M_f R,
\tag{4.78}
$$

where

$$
R = \begin{pmatrix} I \cos \frac{kl}{2} & I \sin \frac{kl}{2} \\ -I \sin \frac{kl}{2} & I \cos \frac{kl}{2} \end{pmatrix},
\tag{4.79}
$$

$$
M_f = \begin{pmatrix} F & 0 \\ 0 & F \end{pmatrix},
\tag{4.80}
$$

$$
F = \begin{pmatrix} \cos \frac{kl}{2} & \frac{2}{k} \sin \frac{kl}{2} \\ -\frac{k}{2} \sin \frac{kl}{2} & \cos \frac{kl}{2} \end{pmatrix},
\tag{4.81}
$$

and I is the 2×2 identity matrix.

We leave it to the reader to verify that M_{sol} is symplectic whereas the individual pieces M_{up}, M_{body}, and M_{dn} are not symplectic. The reason for this nonsymplectic behavior is that the transverse vector potential is not zero inside the magnet. So in the transverse direction, the canonical momentum coordinates differ from the kinetic momentum coordinates used to calculate the slopes x' and y' in the paraxial approximation. When transfer maps are calculated from integration of symplectic generators with a vector potential as in Chapter 3, it is important to select an appropriate vector potential which matches the fringe fields of the electromagnetic element. For magnets this means picking vector potentials which flow along with the currents. Outside the magnet in the field-free region, the canonical momentum should revert back to the kinetic momentum with $\vec{A} \to 0$. A suitable vector potential in the Coulomb gauge is

$$\vec{A}(\vec{r}) = \frac{\mu_0}{4\pi} \int \frac{\vec{J}(\vec{r}')}{|\vec{r} - \vec{r}'|} \, d^3r, \qquad (4.82)$$

which clearly flows with the currents and drops to zero away from the magnet. If there is any chance that an unsuitable vector potential was used to generate a map, it is recommended that the map be compared with a direct integration using the Lorentz force and magnetic field.

Problems

4-1   A lithium lens of length, l, and radius, a, has a current, I, flowing through it with a uniform current density. Consider a beam of antiprotons with momentum, p. What is the focal length of this lens?

4-2   Show that Eqs. (4.11 and 4.12) yield field components of the forms given in Eqs. (4.6 and 4.7), which in turn satisfy Maxwell's equations.

4-3   a) Assuming an infinite plane separates the boundary between two magnetic materials of permeability μ_1 and μ_2 respectively, use Maxwell's equations to calculate the change in both the normal and tangential components of both $\vec{H}$ and $\vec{B}$ across the boundary. b) Using these boundary conditions, estimate the magnetic field in the air gap of an iron dipole magnet like that illustrated in Fig. 4.1b, with an approximate path through the iron of $L = 30$ cm and a gap $g = 2$ cm. There are a total of 100 turns of wire carrying 100 A. Assume that $\mu/\mu_0 = 1000$. What

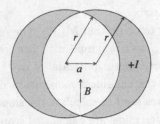

Figure. 4.8 The ideal cross section of a $\cos\theta$ style magnet. The central intersection of two identical circles will have a uniform magnetic field, if equal and opposite currents flow in the two outer lunes.

is the magnitude of B? What is the radius of curvature of a proton with a kinetic energy of 600 MeV? c) How would this change if $\mu/\mu_0 = 20$?

4–4   Using the coordinates x, $x' = dx/ds$, y, $y' = dy/ds$, $z = -v\,\Delta t$, and δ, (Δt is the deviation in the time that a particle passes the location s from the time that the design particle passes the same location. In other words, z is the path length difference of the particle in question from the design particle.) Find an exact transformation function for a trajectory through a drift space of length l. Note that the z-component is nonlinear.

4–5   Find a formula for the excess path length of a trajectory passing through a thick quadrupole lens element.

4–6   Ignoring fringe field effects, find exact expressions for the transformation of a particle through a sector magnet with ends perpendicular to the design particle trajectory.

4–7   a) Assuming that $x/\rho \ll 1$, find an expression for the vector potential $\vec{A}$ which yields the field given by Eqs. (4.6 and 4.7).
b) Find a first order correction in x/ρ to this.

4–8   A magnet is constructed from conductors with a transverse cross section as shown in Fig. 4.8. Show that there is a uniform magnetic field in the region of intersection of the two circles.

4–9   Show that current distributed in a thin cylindrical shell with a strength $I(\theta) = I_0 \cos(n\theta)$, will produce a pure $2n$-multipole distribution inside the cylinder.

4–10   Show that current distributed in a thin cylindrical shell with a strength $I(\theta) = \frac{I_0}{n\pi} \cos(n\theta)$, will produce a pure $2n$-multipole distribution inside the cylinder.

4–11   A beam must be bent by an angle $\theta = 1$ mrad by an element of length $\ell = 0.5$ m. Compare the required fields needed for both an electrostatic (parallel plate) and magnetic (dipole magnet) deflectors for proton beams of kinetic energy

10 MeV, 100 MeV, 1 GeV, 10 GeV, and 100 GeV. What voltages would be required if the parallel plates were separated by 5 cm?

4–12   Consider a 0.5 m long window-frame dipole with the cross section shown below.

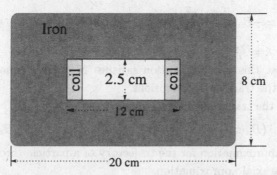

a) Estimate the number of ampere-turns needed to achieve a 0.6 T field in the gap. Assume that the iron is not saturated (i.e., $\mu_r \gtrsim 5000$).

b) Air-cooled copper coils can carry as much as 1.5 A/mm^2; whereas water-cooled copper can carry almost 10 times as much (averaged over conductor and water channels). For the given dimensions of the magnet, would you recommend air-cooled or water-cooled coils. How wide would the gap between the coils be?

c) If the magnet is to be powered by a supply with a maximum current of 1000 A, how many turns should be in the coil.

d) What is the stored energy in the gap.

e) Assuming a constant field in the iron, estimate the additional energy stored in the iron yoke.

f) Estimate the inductance of the magnet.

4–13   An iron dominated dipole magnet has tilted pole faces. If we assume that the iron is not saturated, find a relation between the field index and gap height as a function of radial distance (x) from the design trajectory.

4–14   Consider a helical dipole magnet twisted through a full 360° in a length

λ with approximate transverse field components given by

$$B_x \simeq -B_0 \left\{ \left[1 + \frac{k^2}{8}(3x^2 + y^2) \right] \sin kz - \frac{k^2}{4} xy \cos kz \right\}$$

$$B_y \simeq B_0 \left\{ \left[1 + \frac{k^2}{8}(x^2 + 3y^2) \right] \cos kz - \frac{k^2}{4} xy \sin kz \right\}$$

$$B_z \simeq -B_0 k \left[1 + \frac{k^2}{8}(x^2 + y^2) \right] (x \cos kz + y \sin kz)$$

where the pitch is $k = \frac{2\pi}{\lambda}$.

a) Show that this parametrization satisfies Maxwell's equations to second order in the transverse dimensions.

b) Ignoring the transverse dependence of the field

$$(B_x \simeq -B_0 \sin kz, \quad B_y \simeq B_0 \cos kz, \quad \text{and} \quad B_z \simeq 0)$$

find an approximation for the trajectory of a particle through the magnet in the paraxial approximation.

4–15

a) Using Eqs. (4.6 & 4.7), show that the field in the polar coordinate representation is

$$B_r = B_0 \sum_{n=0}^{\infty} \left(\frac{r}{a} \right)^n [a_n \cos((n+1)\theta) + b_n \sin((n+1)\theta)]$$

$$B_\theta = B_0 \sum_{n=0}^{\infty} \left(\frac{r}{a} \right)^n [b_n \cos((n+1)\theta) - a_n \sin((n+1)\theta)].$$

b) Show that the vector potential

$$A_r = 0$$

$$A_\theta = 0$$

$$A_z = B_0 \sum_{n=0}^{\infty} \frac{a}{n+1} \left(\frac{r}{a} \right)^{n+1} [a_n \sin((n+1)\theta) - b_n \sin((n+1)\theta)].$$

leads to the components given in part (a).

4–16 Verify that $\mathbf{M}_{body}$ is not symplectic, but that $\mathbf{M}_{sol}$ is symplectic.

4–17 Verify Eq. (4.78).

4–18 Show that the focusing matrix $\mathbf{F}$ in Eq. 4.81 may be written as the product of a drift followed by a thin lens and then a second drift of the same length as the first.

References for Chapter 4

[1] B. F. Bayanov et al., "Proceedings of the 12th International Conference on High–Energy Accelerators", Fermilab, 1983.

[2] G. Biallas et al., "Proceedings of the 12th International Conference on High–Energy Accelerators", Fermilab, 1983.

[3] K. Halbach, "Specialty Magnets", AIP Conference Proceedings No. 153 Vol. 2, New York, 1278 (1987).

[4] G. E. Fischer, "Iron Dominated Magnets", AIP Conference Proceedings No. 153 Vol. 2, New York, 1120 (1987).

[5] M. N. Wilson, *Superconducting Magnets*, Clarendon Press, Oxford 1983.

[6] J. Herrera et al., "Magnetic Field Measurements of Superconducting Magnets for the Colliding Beam Accelerator", *12th International Conference on High Energy Accelerators*, p. 563, FNAL (1983).

[7] E. Durand, Magnétostatique, Masson Ed., Paris 1968, p.563.

[8] C. Audoin, Revue de Physique Appliquée, 1(1966)2.

[9] K. G. Steffen, *High Energy Beam Optics*, J. Wiley, New York (1965).

[10] Rudy Larsen, "Transport matrices for a magnetic solenoid", SPEAR-107 (1971).

5

Strong Focusing

To keep the beam dimensions within certain limits can require increasing the betatron tunes Q_H, Q_V so that they are greater than one, as hinted at the end of Chapter 2. Unfortunately gradient magnets cannot simultaneously focus horizontal and vertical oscillations unless $0 < n < 1$ (see Eqs. (2.36 and 2.37)), which leads to tunes less than one. A quadrupole lens also cannot focus in both planes (see Eqs. (4.32 and 4.33)).

Using geometrical optics, it is possible to show that a pair of thin quadrupole lenses, one converging with focal length $f_F > 0$, and the other diverging with focal length $f_D < 0$, separated by a distance l can act as a focusing system (see Fig. 5.1) of focal length

$$f = \frac{f_F f_D}{f_F + f_D - l}. \tag{5.1}$$

This system is called a *doublet*. The principle of strong focusing is just based upon the doublet idea.

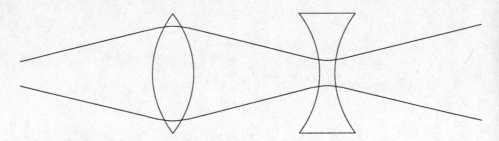

Figure. 5.1 System made of a pair of converging-diverging lenses (doublet).

In the beginning, the doublet consisted of two gradient magnets, separated by a short straight section, with field indices of opposite sign (alternating gradient). The structure or *lattice* of such synchrotrons (AGS, CERN PS, Serpukhov) is thus defined as the *combined-function* type. Later on, synchrotrons and, above all, storage rings began to have magnets, with $n = 0$, that bend and quadrupole-doublets that focus, i. e., *separated-function* lattices. Ref. 6 contains a rather exhaustive list of the high-energy accelerators up to 1989.

5.1 Transfer matrix approach and stability of the linear system

For a periodic cell, one can build up a transfer matrix $M(s)$ by multiplying the matrices referring to drifts, focusing and defocusing gradient magnets in the case of a combined-function lattice, or to drifts, bending magnets, and quadrupoles in the example of a separated-function lattice. Then a generic particle, described at s_0 by a displacement z_0 (which can either be x_0 or y_0) and a slope z'_0, is characterized at s by a vector $\vec{z}(s) \equiv (z(s), z'(s))$ obtained by the following matrix multiplication:

$$\vec{z}(s) = M(s)\vec{z}_0 \quad \text{or} \quad \begin{pmatrix} z(s) \\ z'(s) \end{pmatrix} = \begin{pmatrix} a(s) & b(s) \\ c(s) & d(s) \end{pmatrix} \begin{pmatrix} z_0 \\ z'_0 \end{pmatrix}. \quad (5.2)$$

Recalling the definition of periodic cell given in Chapter 2, and combining quadrupoles, bending magnets, and straight sections, one has

$$M = \begin{pmatrix} a & b \\ c & d \end{pmatrix}, \quad (5.3)$$

for the transfer matrix of a unit cell, where a, b, c, and d are constants, built up by the multiplication of the matrices of the items forming the periodic cell.

By examining the eigenvalues of the transfer matrix, it is possible to make some general statements about the stability of a beam while ignoring effects of nonlinear resonances. The eigenvalues for a 2×2 matrix,

$$M = \begin{pmatrix} a & b \\ c & d \end{pmatrix}, \quad (5.4)$$

can be obtained from the characteristic equation

$$\lambda^2 - (a+d)\lambda + 1 = 0. \quad (5.5)$$

Since $\det(M) = 1$ and M is real, the two eigenvalues must be reciprocals of each other,

$$\lambda_1 = \frac{1}{\lambda_2}, \quad (5.6)$$

and, if not purely real, must also be complex conjugates of each other,

$$\lambda_1 = \lambda_2^*. \quad (5.7)$$

From these last two conditions it is easy to see that either both roots are real or both roots are complex with absolute values of one. The roots of the characteristic equation are

$$\lambda_1 = \frac{1}{2}\mathrm{tr}(\mathbf{M}) + \sqrt{\left(\frac{1}{2}\mathrm{tr}(\mathbf{M})\right)^2 - 1}, \tag{5.8}$$

and

$$\lambda_2 = \frac{1}{2}\mathrm{tr}(\mathbf{M}) - \sqrt{\left(\frac{1}{2}\mathrm{tr}(\mathbf{M})\right)^2 - 1}. \tag{5.9}$$

If $|\mathrm{tr}(\mathbf{M})| < 2$, it will be shown that the focusing is such that the beam can be stable (barring any resonant effects). But if $|\mathrm{tr}(\mathbf{M})| > 2$ the beam diverges to larger and larger extent after many turns. As the number of turns tends to infinity, the slope grows without bound.

To quantify this, we make a mathematical definition of stability. Here we use a definition due to Liapunov[1]. A point $\vec{x}_0$ is called a fixed point of the map $\mathbf{M}$, if $\mathbf{M}\vec{x}_0 = \vec{x}_0$. A system is said to be *Liapunov stable* about a fixed point, $\vec{x}_0$, if for each $\epsilon > 0$, there exists a $\delta > 0$ such that if $\|\vec{x} - \vec{x}_0\| < \delta$, then $\|\mathbf{M}^n\vec{x} - \mathbf{M}^n\vec{x}_0\| < \epsilon$ for all $0 < n < \infty$. That is, there is some neighborhood about the fixed point which remains bounded for all time.

For the example of the linear transfer matrix,

$$\vec{x}_0 = \begin{pmatrix} 0 \\ 0 \end{pmatrix} \tag{5.10}$$

is a fixed point, i. e., a point on the closed orbit.

If the eigenvalues are not degenerate and $b \neq 0$, corresponding eigenvectors are

$$\vec{v}_1 = \begin{pmatrix} b \\ \lambda_1 - a \end{pmatrix}, \quad \text{and} \quad \vec{v}_2 = \begin{pmatrix} b \\ \lambda_2 - a \end{pmatrix}. \tag{5.11}$$

Note that if $b = c = 0$, and the eigenvalues $\lambda_1 = a \neq \lambda_2 = d$, then

$$\begin{pmatrix} 1 \\ 0 \end{pmatrix} \quad \text{and} \quad \begin{pmatrix} 0 \\ 1 \end{pmatrix} \tag{5.12}$$

are independent eigenvectors. Then either $|a| > 1$ or $|d| > 1$, and successive applications of the matrix $\mathbf{M}$ will cause either the position or the slope to diverge to infinity. If $b = 0$ and $c \neq 0$, then the eigenvectors

$$\vec{u}_1 = \begin{pmatrix} \lambda_1 - d \\ c \end{pmatrix}, \quad \text{and} \quad \vec{u}_2 = \begin{pmatrix} \lambda_2 - d \\ c \end{pmatrix}, \tag{5.13}$$

may be used instead of $\vec{v}_1$ and $\vec{v}_2$ in the following arguments.

A point $\vec{x}$ in the neighborhood of $\vec{x}_0$ may be expanded in terms of the eigenvectors:

$$\vec{x} = A\vec{v}_1 + B\vec{v}_2, \tag{5.14}$$

where

$$A = \frac{(a - \lambda_2)x + bx'}{b(\lambda_1 - \lambda_2)}, \tag{5.15}$$

and

$$B = \frac{(a - \lambda_1)x + bx'}{b(\lambda_2 - \lambda_1)}. \tag{5.16}$$

The deviation of the trajectory from the closed orbit after n turns is

$$D_n = \|A\lambda_1^n \vec{v}_1 + B\lambda_2^n \vec{v}_2\| < |\lambda_1^n| |A| \|\vec{v}_1\| + |\lambda_2^n| |B| \|\vec{v}_2\|. \tag{5.17}$$

If $|\text{tr}(\mathbf{M})| < 2$, then $\text{tr}(\mathbf{M})$ may be written as $2\cos\mu$ for some real angle μ. In this case the eigenvalues may be written as $e^{j\mu}$ and $e^{-j\mu}$, with $j = \sqrt{-1}$. The deviation of the trajectory from the closed orbit

$$D_n < |A| \|\vec{v}_1\| + |B| \|\vec{v}_2\|, \tag{5.18}$$

which is bounded for all n, and thus the closed orbit is stable.

If $\text{tr}(\mathbf{M}) > 2$, then $\text{tr}(\mathbf{M})$ may be written as $2\cosh\mu$ for some real $\mu > 0$ and the eigenvalues are $\lambda_1 = e^{\mu}$ and $\lambda_2 = e^{-\mu}$. For this case,

$$D_n \sim (e^{\mu})^n |A| \|\vec{v}_1\|, \tag{5.19}$$

which goes to infinity as $n \to \infty$.

If $\text{tr}(\mathbf{M}) < -2$, then $\text{tr}(\mathbf{M})$ may be written as $-2\cosh\mu$ for some $\mu > 0$, and the eigenvalues become $\lambda_1 = -e^{\mu}$ and $\lambda_2 = -e^{-\mu}$. Again D_n blows up as $n \to \infty$.

For the case when the eigenvalues are degenerate, either $\lambda_1 = \lambda_2 = +1$ or $\lambda_1 = \lambda_2 = -1$. Mathematically this may or may not be bounded depending on the off diagonal terms, but small perturbations will make this of solution unstable, as will be demonstrated in Chapter 10.

For the stable region, $|\text{tr}(\mathbf{M})| < 2$, the matrix $\mathbf{M}$ may be parametrized as follows. The diagonal elements may be written as $a = \cos\mu + k_1$, and $d = \cos\mu - k_1$, for some real k_1. Now $ad = \cos^2\mu - k_1^2$, and $\det(\mathbf{M}) = ad - bc = 1$ yield $bc = -k_1^2 - \sin^2\mu$. Since $|\cos\mu| < 1$, $\sin\mu \neq 0$, and k_1 may be replaced by $\alpha\sin\mu$ for some real α, giving $bc = -(1 + \alpha^2)\sin^2\mu$. By writing $b = \beta\sin\mu$, for some real β, we must have $c = -\gamma\sin\mu$, where

$$\gamma = \frac{1 + \alpha^2}{\beta}, \quad \text{or} \quad \beta\gamma - \alpha^2 = 1. \tag{5.20}$$

Thus the transfer matrix can be written as

$$\mathbf{M} = \begin{pmatrix} \cos\mu + \alpha\sin\mu & \beta\sin\mu \\ -\gamma\sin\mu & \cos\mu - \alpha\sin\mu \end{pmatrix}$$

$$= \begin{pmatrix} 1 & 0 \\ 0 & 1 \end{pmatrix}\cos\mu + \begin{pmatrix} \alpha & \beta \\ -\gamma & -\alpha \end{pmatrix}\sin\mu, \tag{5.21}$$

which may also be written as

$$\mathbf{M} = \mathbf{I}\cos\mu + \mathbf{J}\sin\mu = e^{\mathbf{J}\mu}, \tag{5.22}$$

since $\mathbf{J}^2 = -\mathbf{I}$. There is a sign ambiguity left in the phase μ and the matrix $\mathbf{J}$. This is removed by requiring that $\beta > 0$, which also implies $\gamma > 0$.

The parameters β, α, and γ are sometimes called the Twiss parameters, and are not to be confused with the momentum compaction factor and the relativistic parameters! Every time some ambiguity will arise, these quantities will be written as $\alpha_H(\alpha_V)$, $\beta_H(\beta_V)$, $\gamma_H(\gamma_V)$.

5.2 Analytical approach

Using the Hamiltonian of Eq. (3.83), the equations of motion may be found to be

$$x'' + k_x(s)x = \frac{\delta}{\rho(s)}, \quad \text{and} \tag{5.23}$$

$$y'' + k_y(s)y = 0, \tag{5.24}$$

where

$$k_x(s) = \frac{1}{\rho^2} + \frac{q}{p_0}\frac{\partial B_y}{\partial x}, \quad \text{and} \tag{5.25}$$

$$k_y(s) = -\frac{q}{p_0}\frac{\partial B_y}{\partial x}, \tag{5.26}$$

since $\partial B_y/\partial x = \partial B_x/\partial y$.

Ignoring the inhomogeneous term in the horizontal equation, we may generically write both equations as

$$\frac{d^2z}{ds^2} + k(s)z = 0, \tag{5.27}$$

where z represents either x or y, and k represents the respective function k_x or k_y .

The function, $k(s)$, is a periodic function of period L, describing the variation of the focusing strengths along the orbit s, and Eq. (5.27) is called Hill's equation. Floquet's theorem states that Hill's equation has the quasi-periodic solution

$$z(s) = z_A(s) + z_B(s) \tag{5.28}$$

with

$$z_A(s) = Aw(s)\cos\Psi(s) \tag{5.29}$$

and

$$z_B(s) = Bw(s)\sin\Psi(s), \tag{5.30}$$

where A and B are constants determined by initial conditions. The amplitude function $w(s)$ is periodic with the period L of the unit cell, and $\Psi(s)$ is a non-periodic phase of the oscillations. Since both $z_A(s)$ and $z_B(s)$ separately satisfy Hill's equation, substitution of z_A into Eq. (5.27) yields

$$A[(w'' - w\Psi'^2 + kw)\cos\Psi - (2w'\Psi' + w\Psi'')\sin\Psi] = 0. \tag{5.31}$$

This last equation must hold for all values of Ψ, requiring that the coefficients of the trigonometric functions must be identically zero:

$$w'' - w\Psi'^2 + kw = 0, \tag{5.32}$$

and

$$2w'\Psi' + w\Psi'' = 0. \tag{5.33}$$

The latter equation may easily be integrated giving $w^2\Psi' = C$, where C is an arbitrary constant. This arbitrary constant may be taken equal to one, since C may be absorbed into w, if we rescale the two constants A and B. Thus by defining the normalization of w, such that $C = 1$, we have

$$\Psi' = \frac{1}{w^2}, \tag{5.34}$$

and

$$w'' + k(s)w - \frac{1}{w^3} = 0. \tag{5.35}$$

Therefore the most general solution is

$$z(s) = Aw \cos \Psi + Bw \sin \Psi, \tag{5.36}$$

and the slope is

$$z'(s) = A\left(w' \cos \Psi - \frac{\sin \Psi}{w}\right) + B\left(w' \sin \Psi + \frac{\cos \Psi}{w}\right). \tag{5.37}$$

At $s = s_0$, the initial conditions $z(s_0) = z_0$, $z'(s_0) = z'_0$, $\Psi(s_0) = \Psi_0$, and $w(s_0) = w_0$ can be combined to give

$$z_0 = Aw_0 \cos \Psi_0 + Bw_0 \sin \Psi_0, \tag{5.38}$$

and

$$z'_0 = A\left(w'_0 \cos \Psi_o - \frac{\sin \Psi_0}{w_0}\right) + B\left(w'_0 \sin \Psi_o + \frac{\cos \Psi_0}{w_0}\right). \tag{5.39}$$

Solving for A and B yields

$$A = \left(w'_0 \sin \Psi_0 + \frac{\cos \Psi_0}{w_0}\right) z_0 - (w_0 \sin \Psi_0)\, z'_0, \tag{5.40}$$

and

$$B = -\left(w'_0 \cos \Psi_0 - \frac{\sin \Psi_0}{w_0}\right) z_0 + (w_0 \cos \Psi_0)\, z'_0, \tag{5.41}$$

which substituted into Eqs. (5.36 and 5.37), give the linear relation between $z(s)$, $z'(s)$ and z_0, z'_0, i. e., the elements of the matrix $\mathbf{M}(s)$ in Eq. (5.2)

$$a(s) = \frac{w(s)}{w_0} \cos\left[\Psi(s) - \Psi_0\right] - w(s)w'_0 \sin\left[\Psi(s) - \Psi_0\right], \tag{5.42}$$

$$b(s) = w(s)w_0 \sin\left[\Psi(s) - \Psi_0\right], \tag{5.43}$$

$$c(s) = -\frac{1 + w(s)w_0 w'(s)w'_0}{w(s)w_0} \sin\left[\Psi(s) - \Psi_0\right]$$
$$- \left[\frac{w'_0}{w(s)} - \frac{w'(s)}{w_o}\right] \cos\left[\Psi(s) - \Psi_0\right], \tag{5.44}$$

and

$$d(s) = \frac{w_0}{w(s)} \cos\left[\Psi(s) - \Psi_0\right] + w_0 w'(s) \sin\left[\Psi(s) - \Psi_0\right]. \tag{5.45}$$

Defining

$$\beta(s) = w^2(s), \tag{5.46}$$

$$\alpha(s) = -\frac{1}{2}\beta' = -w(s)w'(s), \quad \text{and} \tag{5.47}$$

$$\mu(s) = \Psi(s) - \Psi_0, \tag{5.48}$$

one has

$$-w_0'w(s) = -\frac{w(s)}{w_0}w_0w_0' = \sqrt{\frac{\beta(s)}{\beta_0}}\alpha_0, \tag{5.49}$$

$$\frac{w_0'}{w(s)} - \frac{w'(s)}{w_0} = \frac{w_0w_0'}{w_0w(s)} - \frac{w(s)w'(s)}{w_0w(s)} = \frac{\alpha(s) - \alpha_0}{\sqrt{\beta_0\beta(s)}}, \tag{5.50}$$

and

$$\frac{w(s)}{w_0} = \sqrt{\frac{\beta(s)}{\beta_0}}. \tag{5.51}$$

The function $\beta(s)$ is called the *betatron function*, $\mu(s)$ is called the *betatron phase*, and $\alpha(s)$ is sometimes referred to as a *correlation function* since it is proportional to the covariance of a beam with a Gaussian distribution (see Problem 5-4.) Since $w(s)$ is a periodic function, then $\beta(s)$, and $\alpha(s)$ must also be periodic with the same period as $k(s)$. The transfer matrix can then be written in the following way:

$$\mathbf{M}(s) = \begin{pmatrix} \sqrt{\frac{\beta(s)}{\beta_0}}[\cos\mu(s) + \alpha_0\sin\mu(s)] & \sqrt{\beta_0\beta(s)}\sin\mu(s) \\ -\frac{[\alpha(s) - \alpha_0]\cos\mu(s) + [1 + \alpha_0\alpha(s)]\sin\mu(s)}{\sqrt{\beta_0\beta(s)}} & \sqrt{\frac{\beta_0}{\beta(s)}}[\cos\mu(s) - \alpha(s)\sin\mu(s)] \end{pmatrix}. \tag{5.52}$$

Due to Eqs. (5.34, 5.46, and 5.48) $\mu'(s) = 1/\beta(s)$, then $\mathbf{M}(s)$ of Eq. (5.52) can be written as:

$$\mathbf{M}(s) = \begin{pmatrix} C & S \\ C' & S' \end{pmatrix}, \tag{5.53}$$

where the second row is the derivative with respect to s of the first row. Since for $s = s_0 + L$, one has $\mu(s) = \mu$, $z(s_0) = z_0$, $z'(s_0) = z_0'$, $w(s_0) = w_0$, and $w'(s_0) = w_0'$, the matrix of Eq. (5.52) coincides with the matrix of Eq. (5.21).

From the above definition one can write the phase advance per cell as

$$\mu = \int_s^{s+L} \Psi' ds = \int_s^{s+L} \frac{ds}{w^2} = \int_s^{s+L} \frac{ds}{\beta(s)} = Q\frac{2\pi}{N}, \tag{5.54}$$

$$Q_{\mathrm{H,V}} = \frac{N}{2\pi} \int_s^{s+L} \frac{ds}{\beta_{\mathrm{H,V}}}, \tag{5.55}$$

where N equals the number of cells, and Q_H and Q_V are the horizontal and vertical betatron tunes.

5.3 Emittances

Let the matrix (5.52) be inserted into Eq. (5.2), to result in

$$z(s) = \sqrt{\frac{\beta(s)}{\beta_0}} \left[\cos\mu(s) + \alpha_0 \sin\mu(s)\right] z_0 + \sqrt{\beta_0 \beta(s)} \sin\mu(s) z_0', \tag{5.56}$$

and

$$z'(s) = -\frac{[\alpha(s) - \alpha_0]\cos\mu(s) + [1 + \alpha_0\alpha(s)]\sin\mu(s)}{\sqrt{\beta_0\beta(s)}} z_0$$
$$+ \sqrt{\frac{\beta_0}{\beta(s)}} \left[\cos\mu(s) - \alpha(s)\sin\mu(s)\right] z_0'. \tag{5.57}$$

The equation for $z(s)$ can be rearranged to give

$$z(s) = \beta(s)^{\frac{1}{2}} \left[\frac{z_0}{\sqrt{\beta_0}}\cos\mu(s) + \left(\frac{z_0}{\sqrt{\beta_0}}\alpha_0 + \sqrt{\beta_0}z_0'\right)\sin\mu(s)\right]. \tag{5.58}$$

Notice that $z(s)$ is of the form $\sqrt{\beta(s)}(A\cos\mu(s) + B\sin\mu(s))$; therefore, we may write

$$z(s) = \sqrt{\mathcal{W}\beta(s)}\cos\left[\mu(s) + \mu_0\right], \tag{5.59}$$

if we require that

$$\frac{z_0}{\sqrt{\beta_0}} = \sqrt{\mathcal{W}}\cos\mu_0, \quad \text{and} \tag{5.60}$$

$$-\left(\frac{z_0}{\sqrt{\beta_0}}\alpha_0 + \sqrt{\beta_0}z_0'\right) = \sqrt{\mathcal{W}}\sin\mu_0, \tag{5.61}$$

where $\mathcal{W}$ and μ_0 are constants which depend on the initial conditions at s_0. Similarly, for $z'(s)$, we find that

$$z'(s) = -\sqrt{\frac{\mathcal{W}}{\beta(s)}} \left\{\alpha(s)\cos\left[\mu(s) + \mu_0\right] + \sin\left[\mu(s) + \mu_0\right]\right\}. \tag{5.62}$$

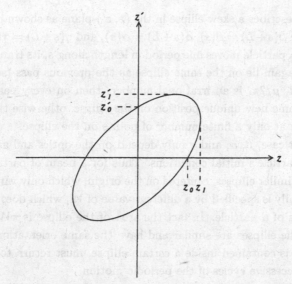

Figure. 5.2 The ellipse of Eq. (5.67) in terms of the parameters α, β, γ, and W. The indicated intercepts are given by $z_0 = \sqrt{W/\gamma}$ and $z_0' = \sqrt{W/\beta}$, and the maxima are given by $z_1 = \sqrt{W\beta}$ and $z_1' = \sqrt{W\gamma}$. The area of the ellipse is πW.

The constant W may be obtained by squaring and adding Eqs. (5.60 and 5.61):

$$W = \gamma_0 z_0^2 + 2\alpha_0 z_0 z_0' + \beta_0 z_0'^2$$
$$= \frac{1}{\beta_0}\left[z_0^2 + (\alpha_0 z_0 + \beta_0 z_0')^2\right]. \tag{5.63}$$

Indeed, this constant has a deeper meaning than just an initial condition, as will be shown immediately. Let Eqs. (5.59 and 5.62) be squared and multiplied by each other:

$$z^2 = W\beta \cos^2\left[\mu(s) + \mu_0\right], \tag{5.64}$$

$$z'^2 = \frac{W}{\beta}\left\{\alpha^2 \cos^2\left[\mu(s) + \mu_0\right] + 2\alpha \sin\left[\mu(s) + \mu_0\right]\cos\left[\mu(s) + \mu_0\right]\right.$$
$$\left. + \sin^2\left[\mu(s) + \mu_0\right]\right\}, \tag{5.65}$$

$$zz' = -W\left[\alpha \cos^2\left[\mu(s) + \mu_0\right] + \sin\left[\mu(s) + \mu_0\right]\cos\left[\mu(s) + \mu_0\right]\right], \tag{5.66}$$

which, when multiplied respectively by γ, β, 2α, sum up to

$$\gamma z^2 + 2\alpha zz' + \beta z'^2 = W\left[(\beta\gamma - \alpha^2)\cos^2\left[\mu(s) + \mu_0\right] + \sin^2\left[\mu(s) + \mu_0\right]\right] = W. \tag{5.67}$$

Thus an invariant of motion has been found, which is called the Courant-Snyder invariant.

Eq. (5.67) describes a skew ellipse in the (z, z')-plane as shown in Fig. 5.2. For a periodic lattice $\beta(s+L) = \beta(s)$, $\alpha(s+L) = \alpha(s)$, and $\gamma(s+L) = \gamma(s)$. This means that every time a particle moves one period in length along s, its transverse position and slope must again lie on the same ellipse as the previous pass (see Fig. 5.2). If the tune per cell, $\mu/2\pi$, is an irrational number, then on every pass, the particle will appear at some new unique position on the ellipse, otherwise the particle will appear cyclically at only a finite number of points on the ellipse.

In the linear case, β, α, and γ only depend on the optics and are independent of any specific particle's initial conditions. Thus for a beam of particles, there is a whole family of similar ellipses, centered on the origin, which only vary in size. Each ellipse in the family is specified by a different value of $\mathcal{W}$, which does depend on the initial conditions of a particle. In fact the area of the ellipse is $\pi\mathcal{W}$ (see Problem 5-3). Since all the ellipses are similar and have the same orientation, we find that a particle which is contained inside a certain ellipse, must return to a point inside the ellipse on successive cycles of the periodic motion.

In an actual accelerator, there is some region of acceptance in phase space in which particles will stay in motion in the accelerator. This region can be limited by physical obstructions, such as the walls of the beam pipe, and by the effects of nonlinear fields. The actual shape of this region can be quite complicated, but it is frequently approximated by an ellipse. The beam must fit inside this acceptance region. The region of phase space in which the fields may be approximated by linear behavior is called the linear aperture.

For the linear approximation, it is convenient to consider some ellipse given by

$$\gamma z^2 + 2\alpha z z' + \beta z'^2 = \epsilon, \tag{5.68}$$

which contains some fraction of the beam. This ellipse then will always contain the same fraction of the beam on successive cycles. This result is just another statement of Liouville's theorem.

There are several different conventions of specifying the fraction of the beam and the constant ϵ. (See Appendix A for a discussion of different conventions.) In this chapter, we will define the *emittance* to be the area of the ellipse, $\pi\epsilon$, which contains 90% of the particles. This is a convention which is frequently used for proton accelerators. Using this definition, when we give an actual value, we will always write the emittance with the factor of π explicitly written out (e. g. $\pi\epsilon = 6\pi \times 10^{-6}$ m.) (In Chapter 8, when we deal with synchrotron radiation in electron rings, we will use a different definition of emittance.)

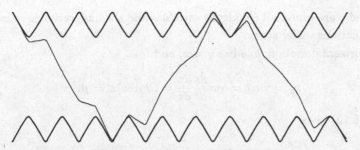

Figure. 5.3 This figure shows a plot of the trajectory $\sqrt{\mathcal{W}\beta(s)}\cos\phi(s)$ through ten identical FODO cells (focus, drift, defocus, drift). The heavy lines are the curves, $z = \pm\sqrt{\mathcal{W}\beta(s)}$. In this case the betatron oscillation is seen to go through one complete cycle after about 6.5 cells.

The trajectory of a particle may be written as

$$z(s) = \sqrt{\mathcal{W}\beta(s)}\cos(\mu(s) + \mu_0), \tag{5.69}$$

which means that the maximum transverse position is

$$z_{\max} = \sqrt{\mathcal{W}\beta(s)}. \tag{5.70}$$

Since each particle in the beam has a different phase, μ_0, it is easy to see that the size of the envelope containing 90% of the beam is

$$z_{90\%} = \sqrt{\epsilon\beta(s)}. \tag{5.71}$$

Therefore the square root of the betatron function may be considered as an envelope function of the beam, specifying the transverse size of the beam as a function of s. This behavior is illustrated in Figure 5.3.

Of course, there are both a horizontal emittance (ϵ_H) and a vertical emittance (ϵ_V).

5.4 Adiabatic invariants

In an accelerator, particles are accelerated. A question arises whether an invariant exists, which can describe the acceleration cycle. Since the energy increase is extremely slower than the betatron oscillations, the so-called *adiabatic approximation* applies, letting the use of the Poincaré-Cartan theorem:

$$I = \oint p\,dq \approx \text{constant}, \tag{5.72}$$

where p and q are canonical conjugate variables and the integration is over a period of the oscillations under study.

For horizontal motion, one has $q = x$, and

$$p_x = \gamma m \dot{x} = \gamma m \frac{ds}{dt}\frac{dx}{ds} = (\beta \gamma m c)x' = p x', \tag{5.73}$$

then Eq. (5.72) gives

$$I_{\mathrm{H}} = p \oint x'\, dx = p(\mathrm{Area}) = p\pi \epsilon_{\mathrm{H}} = \pi m c \beta \gamma \epsilon_{\mathrm{H}} = \pi m c \epsilon_{\mathrm{H}}^* \tag{5.74}$$

similarly

$$I_{\mathrm{V}} = \pi m c \beta \gamma \epsilon_{\mathrm{V}} = \pi m c \epsilon_{\mathrm{V}}^*, \tag{5.75}$$

having defined as *normalized emittances*

$$\pi \epsilon_{\mathrm{H}}^* = \beta \gamma \pi \epsilon_{\mathrm{H}}, \quad \text{and} \quad \pi \epsilon_{\mathrm{V}}^* = \beta \gamma \pi \epsilon_{\mathrm{V}}. \tag{5.76}$$

Note that here β and γ are the relativistic factors and not the envelope functions.

If the energy is adiabatically increased, the area of the ellipse in the xp_x-plane will remain constant, but the area of the ellipse in the xx'-plane will shrink. This effect of shrinking area in the xx'-plane is called *adiabatic damping*.

5.5 Dispersion

For deviations in momentum, we must keep the inhomogeneous term in the horizontal equation of motion:

$$\frac{d^2 x}{ds^2} + k(s)x = \frac{1}{\rho(s)}\delta, \tag{5.77}$$

with $\delta = \Delta p/p_0$. The solution to this equation may be written as

$$x(s) = C(s)\,x_0 + S(s)\,x_0' + D(s)\,\delta_0, \tag{5.78}$$

where the functions C, S, and D have the initial conditions:

$$C(0) = S'(0) = 1, \tag{5.79}$$

$$C'(0) = S(0) = 0, \tag{5.80}$$

and

$$D(0) = D'(0) = 0. \tag{5.81}$$

The functions C and S are solutions to the homogeneous equation, and D is a particular solution to the inhomogeneous equation with $\delta = 1$. The slope of the trajectory is then clearly given by

$$x'(s) = C'(s)\, x_0 + S'(s)\, x_0' + D'(s)\, \delta_0. \tag{5.82}$$

The trajectory equations for x, x', and δ may be written in matrix form as

$$\begin{pmatrix} x \\ x' \\ \delta \end{pmatrix} = \begin{pmatrix} C(s) & S(s) & D(s) \\ C'(s) & S'(s) & D'(s) \\ 0 & 0 & 1 \end{pmatrix} \begin{pmatrix} x_0 \\ x_0' \\ \delta_0 \end{pmatrix}. \tag{5.83}$$

One remarkable thing about this transfer matrix is that one of its eigenvalues is $+1$, and the other two eigenvalues are determined only by the 2×2-matrix

$$\begin{pmatrix} C(s) & S(s) \\ C'(s) & S'(s) \end{pmatrix}, \tag{5.84}$$

as in section 5.1. For a periodic cell, the eigenvector corresponding to the $\lambda_3 = +1$ eigenvalue may be written

$$\begin{pmatrix} \eta\,\delta \\ \eta'\,\delta \\ \delta \end{pmatrix} = \begin{pmatrix} \eta \\ \eta' \\ 1 \end{pmatrix} \delta. \tag{5.85}$$

The periodic function η is called the *dispersion function*. The matrix equation for this eigenvector is then

$$\begin{pmatrix} \eta \\ \eta' \\ 1 \end{pmatrix} = \begin{pmatrix} C(s) & S(s) & D(s) \\ C'(s) & S'(s) & D'(s) \\ 0 & 0 & 1 \end{pmatrix} \begin{pmatrix} \eta \\ \eta' \\ 1 \end{pmatrix}, \tag{5.86}$$

where the matrix elements are evaluated for the periodic cell starting at position s. Solving this for η and η' yields

$$\eta(s) = \frac{[1 - S'(s)]D(s) + S(s)D'(s)}{2(1 - \cos\mu)} \tag{5.87}$$

$$\eta'(s) = \frac{[1 - C(s)]D'(s) + C'(s)D(s)}{2(1 - \cos\mu)}. \tag{5.88}$$

It is important to realize that the periodic dispersion function $\eta(s) = dx/d\delta$ is different from the matrix element $D(s) = \partial x/\partial\delta$ (see Appendix A). It can be seen that the trajectory $x(s)$ is composed of two parts: a part due to betatron oscillations, $x_\beta(s)$, and a part due to dispersion, $x_p(s) = \eta(s)\,\delta$, i. e.,

$$x(s) = x_\beta(s) + x_p(s).$$ (5.89)

The various orbits of the particles will have independent components of motion due to the betatron oscillations and the momentum spread of the beam. Statistically we must add the beam size, σ_β, due to betatron oscillations in quadrature with the size due to dispersion:

$$\sigma_{\text{tot}} = \sqrt{\sigma_\beta^2 + \left(\eta\frac{\sigma_p}{p}\right)^2},$$ (5.90)

for the rms momentum spread, σ_p.

5.6 Momentum compaction

Recalling the definition of momentum compaction, we have

$$\alpha_p = \frac{\Delta L}{L} \bigg/ \frac{\Delta p}{p}$$ (5.91)

with the circumference being given by

$$L = \oint ds$$ (5.92)

and

$$L + \Delta L = \oint d\sigma,$$ (5.93)

where $d\sigma$ is an infinitesimal arc of trajectory referred to an off–momentum particle, while ds is referring to a particle with the design momentum. Since both ds and $d\sigma$ cover the same infinitesimal azimuthal angle

$$d\theta = \frac{d\sigma}{\rho + x_p} = \frac{ds}{\rho},$$ (5.94)

we have

$$d\sigma = \left(1 + \frac{x_p}{\rho}\right) ds,$$ (5.95)

which inserted into Eq. (5.93), yields

$$\Delta L = \oint \frac{x_p}{\rho}\, ds.$$ (5.96)

Using the relation, $x_p = \eta\,\delta$, produces

$$\Delta L = \frac{\Delta p}{p} \oint \frac{\eta(s)}{\rho(s)}\,ds, \tag{5.97}$$

and gives the momentum compaction as

$$\alpha_p = \frac{1}{L} \oint \frac{\eta(s)}{\rho(s)}\,ds. \tag{5.98}$$

Problems

5–1   Using Eq. (5.21), evaluate $\mathbf{M}^k$, $\mathbf{M}^{-1}$, $\mathbf{MM}^{-1}$, and $\mathbf{M_1 M_2}$, where $\mathbf{M_1}$ and $\mathbf{M_2}$ have different phase advances but the same Twiss parameters.

5–2   Show that Eqs. (5.23 and 5.24) may be obtained from the Hamiltonian of Eq. (3.83).

5–3   a) Show that the equation

$$\gamma x^2 + 2\alpha xy + \beta y^2 = \mathcal{W},$$

with $\beta\gamma - \alpha^2 = 1$ is the equation of an ellipse, centered on the origin.
b) Show that the area of this ellipse is given by $A = \pi\mathcal{W}$.
c) Show that

i) $x_{\max} = \sqrt{\beta\mathcal{W}}$ when $y = -\alpha\sqrt{\mathcal{W}/\beta}$
ii) $y_{\max} = \sqrt{\gamma\mathcal{W}}$ when $x = -\alpha\sqrt{\mathcal{W}/\gamma}$
iii) $y_0 = \sqrt{\mathcal{W}/\beta}$ when $x = 0$,
and iv) $x_0 = \sqrt{\mathcal{W}/\gamma}$ when $y = 0$.

Note that the area $A = \pi x_0 y_{\max} = \pi y_0 x_{\max}$.

5–4   a) Show that the transformation

$$\begin{pmatrix} \xi \\ \zeta \end{pmatrix} = \begin{pmatrix} \beta^{-\frac{1}{2}} & 0 \\ \alpha\beta^{-\frac{1}{2}} & \beta^{\frac{1}{2}} \end{pmatrix} \begin{pmatrix} z \\ z' \end{pmatrix},$$

transforms the transfer matrix,

$$\mathbf{M} = e^{\mathbf{J}\mu},$$

into the matrix,

$$\mathbf{N} = \begin{pmatrix} \cos\mu & \sin\mu \\ -\sin\mu & \cos\mu \end{pmatrix}.$$

These new coordinates (ξ, ζ) are sometimes referred to as Floquet or Courant-Snyder coordinates. Note that the ellipse of the Courant-Snyder invariant has been transformed to a circle. Show that the invariant remains unchanged by this transformation.

b) Consider a Gaussian distribution of particles in the new coordinates,

$$f = \frac{N}{2\pi\epsilon} \exp\left(-\frac{\xi^2 + \zeta^2}{2\epsilon}\right).$$

Find the distribution in the old coordinates (z, z'). Evaluate the variances $\sigma_z^2 = \langle(z - \langle z\rangle)^2\rangle$, and $\sigma_{z'}^2 = \langle(z' - \langle z'\rangle)^2\rangle$, and the covariance $\sigma_{zz'}^2 = \langle(z - \langle z\rangle)(z' - \langle z'\rangle)\rangle$.

5-5   Using the invariant

$$\mathcal{W} = \gamma z^2 + 2\alpha z z' + \beta z'^2,$$

show that the Twiss parameters transform from s_1 to s_2 by the matrix transformation

$$\begin{pmatrix} \beta_2 \\ \alpha_2 \\ \gamma_2 \end{pmatrix} = \begin{pmatrix} M_{11}^2 & -2M_{11}M_{12} & M_{12}^2 \\ -M_{11}M_{21} & M_{11}M_{22} + M_{12}M_{21} & -M_{12}M_{22} \\ M_{21}^2 & -2M_{21}M_{22} & M_{22}^2 \end{pmatrix} \begin{pmatrix} \beta_1 \\ \alpha_1 \\ \gamma_1 \end{pmatrix}, \quad \text{if}$$

$$\begin{pmatrix} z_2 \\ z_2' \end{pmatrix} = \begin{pmatrix} M_{11} & M_{12} \\ M_{21} & M_{22} \end{pmatrix} \begin{pmatrix} z_1 \\ z_1' \end{pmatrix}.$$

5-6   a) Show that the transfer matrix for the Twiss parameters, given in the Problem 5-5, has a determinant of 1. What does this say about the eigenvalues?

b) Find the eigenvalues and compare the requirements for a stable beam with the results of §5.2.

5-7   a) Show that a differential equation for the Twiss parameters may be written in matrix form by

$$\begin{pmatrix} \beta' \\ \alpha' \\ \gamma' \end{pmatrix} = \begin{pmatrix} 0 & -2 & 0 \\ k & 0 & -1 \\ 0 & 2k & 0 \end{pmatrix} \begin{pmatrix} \beta \\ \alpha \\ \gamma \end{pmatrix}.$$

b) Solve this equation for constant $k \neq 0$. Compare the result with Twiss matrix of Problem (5-5) evaluated for a quadrupole.

c) Solve this equation for $k = 0$. Notice that β is quadratic in free space.

d) Show that $\beta''' + 4k\beta' + 2k'\beta = 0$.

5-8   a) Show that the particular solution of Eq. (5.77) may be written as

$$D(s) = S(s) \int_0^s \frac{C(\tau)}{\rho(\tau)} \, d\tau - C(s) \int_0^s \frac{S(\tau)}{\rho(\tau)} \, d\tau.$$

b) Show that the periodic dispersion function may be written as

$$\eta(s) = \frac{\sqrt{\beta(s)}}{2 \sin(\pi Q_{\mathrm{H}})} \int_s^{s+L} \frac{\sqrt{\beta(\tau)}}{\rho(\tau)} \cos\left[\phi(\tau) - \phi(s) - \pi Q_{\mathrm{H}}\right] d\tau.$$

5-9   Given the transfer matrix for a series of elements and the initial Twiss parameters, what is the betatron phase advance through this set of elements?

5-10   Show that the conversion from rms to 90% and 95% emittances are approximately $\epsilon_{90\%} \simeq 4.605\epsilon_{\mathrm{rms}}$, and $\epsilon_{90\%} \simeq 5.991\epsilon_{\mathrm{rms}}$, for a Gaussian distribution.

5-11   A gold ($^{197}\mathrm{Au}^{+79}$) beam passes through a 1 mm thick $\mathrm{Al_2O_3}$ flag, at a location in the beam line where the Twiss parameters are $\beta_{\mathrm{H}} = \beta_{\mathrm{V}} = 6$ m and $\alpha_{\mathrm{H}} = \alpha_{\mathrm{V}} = 0$. Multiple Coulomb scattering takes place in the flag adding to the angular divergence of the beam. A good approximation[9] of the deflection is a Gaussian distribution with an rms angle given by

$$\bar{\theta} = \sqrt{\langle \theta^2 \rangle} \simeq z \frac{20\,\mathrm{MeV/c}}{p\beta} \sqrt{\frac{x}{L_{\mathrm{rad}}}} \left(1 + \frac{1}{9} \log_{10} \frac{x}{L_{\mathrm{rad}}}\right),$$

where z is the beam particle's charge, x is the thickness of the flag expressed in $[\mathrm{g \cdot cm^{-2}}]$. Assume[10] $L_{\mathrm{rad}} = 24\,\mathrm{g \cdot cm^{-2}}$ and $\rho_{\mathrm{Al_2O_3}} = 3.7\,\mathrm{g \cdot cm^{-3}}$. The gold beam has a total energy of 10 GeV/nucleon with $M_{\mathrm{Au}}/A = 0.93113$ GeV/nucleon.

   a) Evaluate $\bar{\theta}$.

   b) The beam has a normalized emittance $\epsilon_{95\%}^N = 10\ \mu$m just before the flag. Estimate the blowup in emittance from the flag.

References for Chapter 5

[1] V. I. Arnold, *Mathematical Methods of Classical Mechanics*, Springer-Verlag, New York (1978).

[2] M. V. Klein, *Optics*, John Wiley and Sons, London-New York 1970 (Chapter 3).

[3] N. C. Christofilos: Unpublished report (1950).

[4] E. D. Courant, M. S. Livingston and H. S. Snyder, Phys. Rev., 88, 1190 (1952).

[5] E. D. Courant and H. S. Snyder, *Ann. Physics*, 3, 1 (1958)

[6] *Catalogue of High-Energy Accelerators*, S. Kurokawa ed., Tsukuba, Japan (1989).

[7] K. G. Steffen: High Energy Beam Optics, John Wiley and Sons, London-New York 1965.

[8] K. G. Steffen, " Basic course on accelerator optics", *Proc. of the CERN Accelerator School, General Accelerator Physics, Gif-sur-Yvette*, CERN 85-19, Geneva (1985).

[9] William R. Leo, *Techniques for Nuclear and Particle Physics Experiments*, Springer-Verlag, Berlin (1987).

[10] Yung-Su Tsai, Rev. Mod. Phys., 46, 815 (1974).

6

Lattice Exercises

6.1 The FODO lattice

The most common magnetic structure of modern accelerators and storage rings consists of a pair of F-D quadrupoles, interspaced by a bending magnet, whose effects on the optics are minor compared to the focusing effects of the quadrupoles. Instead the bending magnet plays the fundamental and unique role in giving rise to dispersion.

Bearing in mind the substantial optical identity between horizontal and vertical motions, one can define a FODO cell as the simplest block of items spanning the machine by repeating itself. (The O's in the FODO notation designate the space between quadrupoles.) Clearly in the vertical plane the corresponding pattern of the cell would be DOFO. For a total cell length of l, with equivalent spacing and bends between thin lens quadrupoles, one obtains

$$
\mathbf{M}_{\text{FODO}} = \begin{pmatrix} 1 & \frac{l}{2} \\ 0 & 1 \end{pmatrix} \begin{pmatrix} 1 & 0 \\ \frac{1}{f} & 1 \end{pmatrix} \begin{pmatrix} 1 & \frac{l}{2} \\ 0 & 1 \end{pmatrix} \begin{pmatrix} 1 & 0 \\ -\frac{1}{f} & 1 \end{pmatrix}
$$
$$
= \begin{pmatrix} 1 - \frac{l}{2f} - \frac{l^2}{4f^2} & l + \frac{l^2}{4f} \\ -\frac{l}{2f^2} & 1 + \frac{l}{2f} \end{pmatrix}. \tag{6.1}
$$

Here we have assumed that the two thin lens quadrupoles have identical strengths, $1/f$. In the next section, quadrupoles of different focal lengths will be considered.

Setting this matrix equal to the most general strong-focusing matrix we have that

$$
\begin{pmatrix} 1 - \frac{l}{2f} - \frac{l^2}{4f^2} & l + \frac{l^2}{4f} \\ -\frac{l}{2f^2} & 1 + \frac{l}{2f} \end{pmatrix} = \begin{pmatrix} \cos\mu + \alpha\sin\mu & \beta\sin\mu \\ -\gamma\sin\mu & \cos\mu - \alpha\sin\mu \end{pmatrix}. \tag{6.2}
$$

Taking the trace yields

$$
1 - \frac{l^2}{8f^2} = \cos\mu = 1 - 2\sin^2\frac{\mu}{2}, \tag{6.3}
$$

or

$$
\sin\frac{\mu}{2} = \pm\frac{l}{4f}. \tag{6.4}
$$

115

The extremes of the beta function are then

$$\beta_{\max} = \frac{l}{\sin\mu}\left(1 + \sin\frac{\mu}{2}\right) = 2f\sqrt{\frac{1+\sin\frac{\mu}{2}}{1-\sin\frac{\mu}{2}}}, \quad \text{and} \tag{6.5}$$

$$\beta_{\min} = \frac{l}{\sin\mu}\left(1 - \sin\frac{\mu}{2}\right) = 2f\sqrt{\frac{1-\sin\frac{\mu}{2}}{1+\sin\frac{\mu}{2}}}. \tag{6.6}$$

Knowing the phase-advance per cell given by Eq. (6.4), one can obtain as betatron tunes

$$Q_{\mathrm{h,v}} = N_c\frac{\mu_{\mathrm{h,v}}}{2\pi}, \tag{6.7}$$

where $N_c = L/l$ is the total number of cells, L is the machine circumference, and l is again the length of the FODO cell.

6.2 Stability diagrams

Let the quantities, used to obtain Eq. (6.1), be slightly modified as follows:

$$-\frac{l}{2f_F} = -F; \quad \frac{l}{2f_D} = D. \tag{6.8}$$

Then one has for the horizontal motion,

$$\mathbf{M_{FODO}} = \begin{pmatrix} 1 & l/2 \\ 0 & 1 \end{pmatrix}\begin{pmatrix} 1 & 0 \\ \frac{2D}{l} & 1 \end{pmatrix}\begin{pmatrix} 1 & l/2 \\ 0 & 1 \end{pmatrix}\begin{pmatrix} 1 & 0 \\ -\frac{2F}{l} & 1 \end{pmatrix}. \tag{6.9}$$

Calculating the trace now yields

$$\cos\mu = 1 + D - F - \frac{FD}{2}, \quad \text{and}$$

$$\sin^2\frac{\mu}{2} = \frac{FD}{4} + \frac{F-D}{2}. \tag{6.10}$$

For stability we must have $-1 < \cos\mu < 1$. This gives two equations for the boundary of a region of stability:

$$\sin^2\frac{\mu}{2} = 0, \quad \text{and} \tag{6.11}$$

$$\sin^2\frac{\mu}{2} = 1. \tag{6.12}$$

For the first case, $\sin^2\frac{\mu}{2} = 0$, and $0 = FD + 2F - 2D$, or $F = \frac{2D}{2+D}$, which gives

$$F = 0 \quad \text{for} \quad D = 0, \quad \text{and} \tag{6.13}$$

$$F = 1 \quad \text{for} \quad D = 2. \tag{6.14}$$

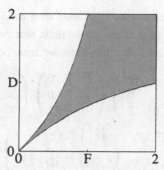

Figure. 6.1 Stability or "necktie" diagram for an alternate focusing lattice. The shaded area is the region of stability.

The second case, $\sin^2 \frac{\mu}{2} = 1$, yields the limit $F = 2$. For vertical motion the roles of the focusing and defocusing quadrupoles are reversed, giving a stability requirement which results in a reflection of these two limit boundaries about the line $D = F$. Fig. 6.1 shows the region of stability for the focusing and defocusing quadrupoles on a plot of $l/2f_D$ vs $l/2f_F$.

Having found this stability diagram, it is worth while to mention that a similar figure can be found also in the combined-function structure. Considering, for simplicity, a cell made of two magnets with opposite field indices, of such big absolute values that $1 - n \approx -n$; then the cell-matrix is

$$\mathbf{M} = \begin{pmatrix} \cos(\sqrt{n}\frac{\pi}{N}) & \frac{\rho}{\sqrt{n}}\sin(\sqrt{n}\frac{\pi}{N}) \\ -\frac{\sqrt{n}}{\rho}\sin(\sqrt{n}\frac{\pi}{N}) & \cos(\sqrt{n}\frac{\pi}{N}) \end{pmatrix} \begin{pmatrix} \cosh(\sqrt{n}\frac{\pi}{N}) & \frac{\rho}{\sqrt{n}}\sinh(\sqrt{n}\frac{\pi}{N}) \\ \frac{\sqrt{n}}{\rho}\sinh(\sqrt{n}\frac{\pi}{N}) & \cosh(\sqrt{n}\frac{\pi}{N}) \end{pmatrix}$$

(6.15)

having used Eqs. (2.27 and 2.36), and having considered $2N$ cells.

It is trivial to demonstrate that

$$\cos\mu = \frac{1}{2}\text{tr}(\mathbf{M}) = \cos\left(\sqrt{n}\frac{\pi}{N}\right)\cosh\left(\sqrt{n}\frac{\pi}{N}\right),$$

(6.16)

which gives rise to a necktie diagram similar to the one shown in Fig. 6.1, provided that $-n_-$ replaces $l/2f_D$, and n_+ replaces $l/2f_F$. The central condition $\cos\mu = 0$ yields $\sqrt{n}\,\pi/N = \pi/2$ or $n = N^2/4$.

6.3 FODO cell dispersion

In order to deal with the dispersion problem, the FODO cell is slightly modified into the symmetric cell $\frac{1}{2}$FODO$\frac{1}{2}$F. This change would not be strictly necessary for what will be considered in this section, but it will provide nicer matrices for the *dispersion suppressors* which will be discussed in § 6.6. Here we will assume that the cell length, $l = \rho\theta_c$, and that there are no drift sections in the periodic cell. The total bending angle of the cell is θ_c. The transfer matrix for a single cell is

$$
\mathbf{M} = \begin{pmatrix} 1 & 0 & 0 \\ -\frac{1}{2f} & 1 & 0 \\ 0 & 0 & 1 \end{pmatrix} \begin{pmatrix} 1 & \frac{l}{2} & \frac{l\theta_c}{8} \\ 0 & 1 & \frac{\theta_c}{2} \\ 0 & 0 & 1 \end{pmatrix} \begin{pmatrix} 1 & 0 & 0 \\ \frac{1}{f} & 1 & 0 \\ 0 & 0 & 1 \end{pmatrix} \begin{pmatrix} 1 & \frac{l}{2} & \frac{l\theta_c}{8} \\ 0 & 1 & \frac{\theta_c}{2} \\ 0 & 0 & 1 \end{pmatrix} \begin{pmatrix} 1 & 0 & 0 \\ -\frac{1}{2f} & 1 & 0 \\ 0 & 0 & 1 \end{pmatrix}
$$

$$
= \begin{pmatrix} 1 - \frac{l^2}{8f^2} & l + \frac{l^2}{4f} & \frac{1}{2}l\left(1 + \frac{l}{8f}\right)\theta_c \\ -\frac{l}{4f^2}\left(1 - \frac{l}{4f}\right) & 1 - \frac{l^2}{8f^2} & \left(1 - \frac{l}{8f} - \frac{l^2}{32f^2}\right)\theta_c \\ 0 & 0 & 1 \end{pmatrix}, \tag{6.17}
$$

where we have used the thin lens approximation for the quadrupoles and have assumed that the bend angle $\theta_c \ll 1$. Notice that the top-left 2×2 matrix can be extracted and written as

$$
\mathbf{M}' = \begin{pmatrix} 1 - \frac{l^2}{8f^2} & l\left(1 + \frac{l}{4f}\right) \\ -\frac{l}{4f^2}\left(1 - \frac{l}{4f}\right) & 1 - \frac{l^2}{8f^2} \end{pmatrix} = \begin{pmatrix} \cos\mu & \beta\sin\mu \\ -\frac{\sin\mu}{\beta} & \cos\mu \end{pmatrix}, \tag{6.18}
$$

with $\cos\mu = 1 - \frac{l^2}{8f^2}$ and $\beta = \frac{l}{\sin\mu}\left(1 \pm \sin\frac{\mu}{2}\right)$ exactly as before, since $\alpha = 0$ due to the symmetry. The periodic dispersion function may be obtained as before from the equation,

$$
\begin{pmatrix} \eta \\ \eta' \\ 1 \end{pmatrix} = \mathbf{M}\begin{pmatrix} \eta \\ \eta' \\ 1 \end{pmatrix} = \begin{pmatrix} M_{11} & M_{12} & M_{13} \\ M_{21} & M_{22} & M_{23} \\ M_{31} & M_{32} & M_{33} \end{pmatrix}\begin{pmatrix} \eta \\ \eta' \\ 1 \end{pmatrix}, \tag{6.19}
$$

or

$$
\begin{pmatrix} \eta \\ \eta' \end{pmatrix} = \mathbf{M}'\begin{pmatrix} \eta \\ \eta' \end{pmatrix} + \begin{pmatrix} M_{13} \\ M_{23} \end{pmatrix}. \tag{6.20}
$$

Solving for the dispersion we get

$$
\begin{pmatrix} \eta \\ \eta' \end{pmatrix} = (\mathbf{I} - \mathbf{M}')^{-1}\begin{pmatrix} M_{13} \\ M_{23} \end{pmatrix} \tag{6.21}
$$

with

$$
\mathbf{I} - \mathbf{M}' = \begin{pmatrix} \frac{l^2}{8f^2} & -l\left(1 + \frac{l}{4f}\right) \\ \frac{l}{4f^2}\left(1 - \frac{l}{4f}\right) & \frac{l^2}{8f^2} \end{pmatrix} \tag{6.22}
$$

and

$$
\det(\mathbf{I} - \mathbf{M}') = \left(\frac{l}{2f}\right)^2 \tag{6.23}
$$

which can be combined to give

$$\begin{pmatrix} \eta \\ \eta' \end{pmatrix} = \left(\frac{2f}{l} \right)^2 \begin{pmatrix} \frac{l^2}{8f^2} & l\left(1 + \frac{l}{4f}\right) \\ -\frac{l}{4f^2}\left(1 - \frac{l}{4f}\right) & \frac{l^2}{8f^2} \end{pmatrix} \begin{pmatrix} \frac{1}{2}\left(1 + \frac{l}{8f}\right)\theta_c \\ \left(1 - \frac{l}{8f} - \frac{l^2}{32f^2}\right)\theta_c \end{pmatrix}. \quad (6.24)$$

Trivial but tedious calculations lead to

$$\eta = \eta_+ = \frac{l\theta_c}{4} \frac{1 + \frac{1}{2}\sin\frac{\mu}{2}}{\sin^2\frac{\mu}{2}}, \quad \text{and} \quad (6.25)$$

$$\eta' = 0. \quad (6.26)$$

This value of η_+ is the maximum value of η in the cell. One would find for a mirror-symmetric $\frac{1}{2}$DOFO$\frac{1}{2}$D cell

$$\eta = \eta_- = \frac{l\theta_c}{4} \frac{1 - \frac{1}{2}\sin\frac{\mu}{2}}{\sin^2\frac{\mu}{2}}, \quad (6.27)$$

and again $\eta' = 0$. Here η_- is the minimum value of η in the cell.

Recalling the most general definition (Eq. 5.98) of momentum compaction factor, α_p may be approximated by

$$\alpha_p = \frac{1}{2\pi R} \oint \frac{\eta(s)}{\rho(s)} \, ds = \frac{1}{2\pi R\rho} \oint \eta(s) \, ds \simeq \frac{\eta_+ + \eta_-}{2R}, \quad (6.28)$$

where we have assumed that the bending radius, ρ, is constant inside the bending magnets and is infinite outside the bending magnets. By writing the circumference as $2\pi R$, we allow for additional length of the quadrupoles and some drift spaces. Substituting for η_+ and η_- gives

$$\alpha_p \simeq \frac{4f^2}{\rho R}. \quad (6.29)$$

6.4 A few explicit forms of the Twiss matrix

Let us evaluate some Twiss-matrices for particular cases. The basic matrix is the one given in Problem 5–5. If $\mathbf{M} = \pm\mathbf{I}$ then the Twiss-matrix is

$$\mathbf{T} = \begin{pmatrix} 1 & 0 & 0 \\ 0 & 1 & 0 \\ 0 & 0 & 1 \end{pmatrix}. \quad (6.30)$$

For a 90° periodic cell,

$$\mathbf{M} = \mathbf{J} = \begin{pmatrix} \alpha & \beta \\ -\gamma & -\alpha \end{pmatrix}, \tag{6.31}$$

and

$$\mathbf{T} = \begin{pmatrix} \alpha^2 & -2\alpha\beta & \beta^2 \\ \alpha\gamma & -(\alpha^2 + \beta\gamma) & \alpha\beta \\ \gamma^2 & -2\alpha\gamma & \alpha^2 \end{pmatrix}. \tag{6.32}$$

For a thin lens with focal length f, we have

$$\mathbf{M} = \begin{pmatrix} 1 & 0 \\ \mp\frac{1}{f} & 1 \end{pmatrix}, \quad \text{and}$$

$$\mathbf{T} = \begin{pmatrix} 1 & 0 & 0 \\ \pm\frac{1}{f} & 1 & 0 \\ \frac{1}{f^2} & \pm\frac{2}{f} & 1 \end{pmatrix}. \tag{6.33}$$

Notice that a thin quadrupole does not alter the beta function $\beta_{h,v}$, but it does change the slope -2α.

Of course for a straight section of length l we have

$$\mathbf{T} = \begin{pmatrix} 1 & -2l & l^2 \\ 0 & 1 & -l \\ 0 & 0 & 1 \end{pmatrix}. \tag{6.34}$$

By applying this to the vector $(\beta_0, \alpha_0, \gamma_0)$, we obtain

$$\beta(s) = \beta_0 - 2\alpha_0 s + \gamma_0 s^2, \tag{6.35}$$

whose importance will be seen in the following sections. Notice that this relation can be obtained by integration of the differential equation in Problem 5–7d, since for a straight section $\beta''' = 0$.

Last but not least, the sector dipole magnet of Eq. (2.47) gives rise to

$$\mathbf{T}(\theta) = \begin{pmatrix} \cos^2\theta & -\rho\sin 2\theta & \rho^2\sin^2\theta \\ \frac{1}{2\rho}\sin 2\theta & \cos 2\theta & -\frac{1}{2}\rho\sin 2\theta \\ \frac{1}{\rho^2}\sin^2\theta & \frac{1}{\rho}\sin 2\theta & \cos^2\theta \end{pmatrix} \tag{6.36}$$

in the horizontal plane.

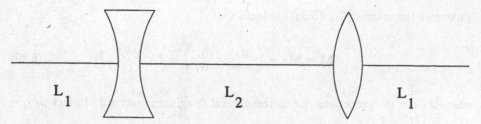

Figure. 6.2 The $\frac{\pi}{2}$-insertion.

6.5 Insertions

In any FODO cell one or both quadrupole magnets can be moved away leaving room for actual straight sections, slightly modifying the lattice dispersion without altering the optics of the lattice, as thoroughly discussed in the previous sections.

Obviously, such free spaces may not be sufficient to host long items like septum-magnets for injection and extraction, chains of rf cavities, etc. Longer straight sections have to be provided, leaving the optics of the machine unaffected.

An elegant solution is the so called $\frac{\pi}{2}$-insertion, consisting of a lattice-interruption, filled by a long straight section l_2, encompassed by two shorter free flights l_1, to which it is connected via two F and D quadrupoles, as shown in Fig. 6.2. The related matrix is

$$\mathbf{M} = \begin{pmatrix} 1 & l_1 \\ 0 & 1 \end{pmatrix} \begin{pmatrix} 1 & 0 \\ -\frac{1}{f} & 1 \end{pmatrix} \begin{pmatrix} 1 & l_2 \\ 0 & 1 \end{pmatrix} \begin{pmatrix} 1 & 0 \\ \frac{1}{f} & 1 \end{pmatrix} \begin{pmatrix} 1 & l_1 \\ 0 & 1 \end{pmatrix}, \tag{6.37}$$

or

$$\mathbf{M} = \begin{pmatrix} 1 - \frac{l_1 l_2}{f^2} + \frac{l_2}{f} & 2l_1 + l_2 - \frac{l_1^2 l_2}{f^2} \\ -\frac{l_2}{f^2} & 1 - \frac{l_1 l_2}{f^2} - \frac{l_2}{f} \end{pmatrix}$$

$$= \begin{pmatrix} \cos\mu + \alpha\sin\mu & \beta\sin\mu \\ -\gamma\sin\mu & \cos\mu - \alpha\sin\mu \end{pmatrix} \tag{6.38}$$

$$\cos\mu = 1 - \frac{l_1 l_2}{f^2}, \quad \beta\sin\mu = \left(2 - \frac{l_1 l_2}{f^2}\right) l_1 + l_2, \quad \gamma\sin\mu = \frac{l_2}{f^2} \tag{6.39}$$

$$l_2 = \alpha f \sin\mu. \tag{6.40}$$

Eq. (6.40) shows that l_2 is a maximum for $\sin\mu = 1$, i. e., for $\mu = \frac{\pi}{2}$. Then from Eq. (6.39), we obtain $\cos\mu = 0$, and

$$f^2 = l_1 l_2, \quad \alpha_I = \frac{l_2}{f}, \quad \gamma_I = \frac{l_2}{f^2}, \quad \beta_I = l_1 + l_2. \tag{6.41}$$

Therefore the matrix (Eq. (6.38)) reduces to

$$\mathbf{M} = \mathbf{M}_I = \begin{pmatrix} \alpha_I & \beta_I \\ -\gamma_I & -\alpha_I \end{pmatrix} = \mathbf{J}_I, \tag{6.42}$$

where $\mathbf{J}_I$ has the same behavior as the general imaginary matrix $\mathbf{J}$. In fact $\beta_I \gamma_I - \alpha_I^2 = l_1 l_2 / f^2 = 1$.

Bearing in mind the mirror symmetry about the center of a low beta insertion, one can expect the lattice parameters α, β, and γ at the interruption A to be identical to those at the interruption B.

$$\begin{pmatrix} \beta \\ \alpha \\ \gamma \end{pmatrix}_A = \begin{pmatrix} \alpha_I^2 & -2\alpha_I\beta_I & \beta_I^2 \\ \alpha_I\gamma_I & -\alpha_I^2 - \beta_I\gamma_I & \alpha_I\beta_I \\ \gamma_I^2 & -2\alpha_I\gamma_I & \alpha_I^2 \end{pmatrix} \begin{pmatrix} \beta_I \\ \alpha_I \\ \gamma_I \end{pmatrix}. \tag{6.43}$$

Notice that $(\beta, \alpha, \gamma)_B = (\beta_I, \alpha_I, \gamma_I)$, and, as stated before, $\mathbf{T}(A \to B) = \mathbf{T}(B \to A)$; then

$$\begin{pmatrix} \beta \\ \alpha \\ \gamma \end{pmatrix}_A = \begin{pmatrix} \beta_I(\beta_I\gamma_I - \alpha_I^2) \\ \alpha_I(\beta_I\gamma_I - \alpha_I^2) \\ \gamma_I(\beta_I\gamma_I - \alpha_I^2) \end{pmatrix} = \begin{pmatrix} \beta_I \\ \alpha_I \\ \gamma_I \end{pmatrix} = \begin{pmatrix} \beta \\ \alpha \\ \gamma \end{pmatrix}_B. \tag{6.44}$$

Although we have shown that $\mathbf{Z} = \mathbf{MZ}$, the matrix $\mathbf{M}$ is not the identity matrix, but is a matrix whose eigenvalues are one, since it must satisfy the equation

$$(\mathbf{M} - \mathbf{I})\mathbf{Z} = 0. \tag{6.45}$$

6.6 Dispersion suppressors

In several straight sections, like the ones hosting rf cavities, extraction systems, etc., it is preferable to have $\eta(s) = \eta'(s) = 0$, i. e., no dispersion. This can be realized by entering this straight section with $\eta = \eta' = 0$, since in the matrix for a drift,

$$\mathbf{M}_0 = \begin{pmatrix} 1 & l & 0 \\ 0 & 1 & 0 \\ 0 & 0 & 1 \end{pmatrix}, \tag{6.46}$$

has the dispersive elements, $M_{13} = M_{23} = 0$.

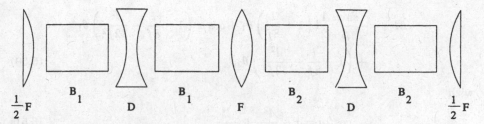

Figure. 6.3 A dispersion suppressor consisting of two FODO cells.

The most common dispersion suppressor consists of two FODO cells of equal length and quadrupole strengths, but with a different bending field in each FODO cell. Fig. 6.3 shows the typical arrangement, with two distinct values of the magnetic field B. This pair of modified cells must be duplicated at the other end of the dispersion free straight section.

The matrix describing the system of Fig. 6.3 may be obtained using Eq. (6.17);

$$\mathbf{M} = \begin{pmatrix} \cos\mu & \beta\sin\mu & \frac{1}{2}l\left(1+\frac{l}{8f}\right)\theta_2 \\ -\frac{\sin\mu}{\beta} & \cos\mu & \left(1-\frac{l}{8f}-\frac{l^2}{32f^2}\right)\theta_2 \\ 0 & 0 & 1 \end{pmatrix}$$

$$\times \begin{pmatrix} \cos\mu & \beta\sin\mu & \frac{1}{2}l\left(1+\frac{l}{8f}\right)\theta_1 \\ -\frac{\sin\mu}{\beta} & \cos\mu & \left(1-\frac{l}{8f}-\frac{l^2}{32f^2}\right)\theta_1 \\ 0 & 0 & 1 \end{pmatrix}. \qquad (6.47)$$

Note how Eq. (6.47) nicely reconfirms that the optics of the lattice is not at all affected by the change of the bending strengths; the only effect is over the dispersive elements M_{13} and M_{23}. Performing the matrix multiplication yields

$$\mathbf{M} = \begin{pmatrix} \cos 2\mu & \beta\sin 2\mu & M_{13} \\ -\frac{\sin 2\mu}{\beta} & \cos 2\mu & M_{23} \\ 0 & 0 & 1 \end{pmatrix}, \qquad (6.48)$$

where

$$M_{13} = \cos\mu \, \frac{1}{2}l\left(1+\frac{l}{8f}\right)\theta_1 + \beta\sin\mu\left(1-\frac{l}{8f}-\frac{l^2}{32f^2}\right)\theta_1$$

$$+ \frac{1}{2}l\left(1+\frac{l}{8f}\right)\theta_2 \qquad (6.49)$$

$$M_{23} = -\frac{\sin\mu}{\beta} \frac{1}{2} l \left(1 + \frac{l}{8f}\right) \theta_1 + \cos\mu \left(1 - \frac{l}{8f} - \frac{l^2}{32f^2}\right) \theta_1$$
$$+ \left(1 - \frac{l}{8f} - \frac{l^2}{32f^2}\right) \theta_2, \tag{6.50}$$

with

$$\cos 2\mu = 1 - \frac{l^2}{2f^2} + \frac{l^4}{32f^4}, \tag{6.51}$$

$$\beta \sin 2\mu = 2l \left(1 - \frac{l^2}{8f^2}\right) \left(1 + \frac{l}{4f}\right), \quad \text{and} \tag{6.52}$$

$$\frac{\sin 2\mu}{\beta} = \frac{l}{2f^2} \left(1 - \frac{l^2}{8f^2}\right) \left(1 - \frac{l}{4f}\right). \tag{6.53}$$

The elements M_{13} and M_{23} may also be written as

$$M_{13} = \frac{1}{2} l \left(1 + \frac{l}{8f}\right) \left[\left(3 - \frac{l^2}{4f^2}\right) \theta_1 + \theta_2\right], \tag{6.54}$$

$$M_{23} = \left(1 - \frac{l}{8f} - \frac{l^2}{32f^2}\right) \left[\left(1 - \frac{l^2}{4f^2}\right) \theta_1 + \theta_2\right]. \tag{6.55}$$

Having demonstrated in §6.3 that in the middle of a focusing quadrupole, the dispersion has its maximum value η_+, the required condition is

$$\begin{pmatrix} 0 \\ 0 \\ 1 \end{pmatrix} = \begin{pmatrix} \cos 2\mu & \beta \sin 2\mu & M_{13} \\ -\frac{\sin 2\mu}{\beta} & \cos 2\mu & M_{23} \\ 0 & 0 & 1 \end{pmatrix} \begin{pmatrix} \eta_+ \\ 0 \\ 1 \end{pmatrix}, \tag{6.56}$$

or

$$\cos 2\mu \, \eta_+ + M_{13} = 0, \quad \text{and} \tag{6.57}$$

$$-\frac{\sin 2\mu}{\beta} \eta_+ + M_{23} = 0. \tag{6.58}$$

With a slight modification of Eq. (6.25), using $\sin(\mu/2) = l/4f$, we get

$$\eta_+ = \frac{4f^2}{l} \left(1 + \frac{l}{8f}\right) \theta, \tag{6.59}$$

where θ must be the total bend in the suppressor, $\theta_1 + \theta_2$. Substituting this into the conditions for dispersion suppression, Eqs. (6.57 and 6.58), results in

$$\left(3 - \frac{l^2}{4f^2}\right) \theta_1 + \theta_2 = \left(4 - \frac{l^2}{4f^2} - \frac{8f^2}{l^2}\right) \theta, \tag{6.60}$$

$$\left(1 - \frac{l^2}{4f^2}\right) \theta_1 + \theta_2 = \left(2 - \frac{l^2}{4f^2}\right) \theta \tag{6.61}$$

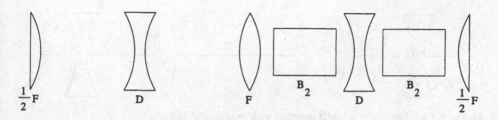

Figure. 6.4 Dispersion suppressor with missing magnets.

or

$$\theta_1 = \left(1 - \frac{1}{4\sin^2\frac{\mu}{2}}\right)\theta, \tag{6.62}$$

$$\theta_2 = \frac{\theta}{4\sin^2\frac{\mu}{2}}, \quad \text{and} \tag{6.63}$$

$$\theta_1 + \theta_2 = \theta. \tag{6.64}$$

For a phase-advance per cell of $\mu = \pi/3$,

$$4\sin^2\frac{\mu}{2} = 1, \tag{6.65}$$

$$\theta_1 = 0 \text{ (missing magnet)} \tag{6.66}$$

and

$$\theta_2 = \theta; \tag{6.67}$$

the situation is sketchily illustrated in Fig. 6.4.

If $\mu = \pi/2$, then $4\sin^2\frac{\mu}{2} = 2$, and $\theta_1 = \theta_2 = \theta/2$ (half-strength bends.) For $\mu < \pi/3$, we must have a reverse bend in one of the magnets. Notice how Eq. (6.64) proves that any dispersion suppressor, however realized, effectively requires a cell length without any bending, thus increasing the circumference by $2l$, since there must be two such suppressors in the ring.

6.7 Low beta insertion

In § 1.1 we saw that the interaction rate in a colliding beam accelerator is roughly proportional to the product of the densities of the two beams at the interaction point. Since the particle density increases as the transverse size of the beam decreases, we should be able to enlarge the data rate by shrinking the betatron

Figure. 6.5 A schematic of a low-β insertion with quadrupole doublets.

functions. A typical collider ring may be made of FODO cells which are placed with a mirror-symmetry about a straight section containing the interaction point. (Actually for beams consisting of single bunches of particles, there will be two such interaction points on opposite sides of the ring, unless some sort of electrostatic separators are used to remove one of the interaction points.) In this straight section, two strong focusing quadrupoles may be symmetrically placed to decrease the betatron function at the collision point. With this symmetry, the betatron function about the collision point is

$$\beta(s) = \beta^* + \frac{s^2}{\beta^*}. \tag{6.68}$$

Actually this only works in one plane, since the focusing quadrupoles are defocusing quadrupoles in the other plane.

In order to remedy this deficiency, two mirror-symmetrically placed doublets may be inserted into the straight section instead of single quadrupoles. This will produce a smaller β^* in one plane than the other. Typically in e^+e^- rings, only doublets are used, since the charge of one beam will act as a nonlinear lens on the other beam, and vice versa. By allowing the beams to spread out in one dimension, this lens effect can greatly be reduced. The effect of the beam-beam interaction is much smaller for protons and antiprotons, since the rest mass of the proton is much greater than an electron. To make a beam crossing with $\beta_x^* = \beta_y^*$, two quadrupole triplets are required. We will discuss the beam-beam effect in Chapter 11.

As with other insertions, we must match the Twiss parameters at the ends of the straight section to the parameters at the ends of the FODO cells. This will keep the rest of the ring unchanged. For the doublet case, the transfer matrix from the beam crossing to the edge of the insertion is (see Fig. 6.5)

$$\mathbf{M} = \begin{pmatrix} 1 & l_3 \\ 0 & 1 \end{pmatrix} \begin{pmatrix} 1 & 0 \\ \frac{1}{f_D} & 1 \end{pmatrix} \begin{pmatrix} 1 & l_2 \\ 0 & 1 \end{pmatrix} \begin{pmatrix} 1 & 0 \\ -\frac{1}{f_F} & 1 \end{pmatrix} \begin{pmatrix} 1 & l_1 \\ 0 & 1 \end{pmatrix}$$

$$= \begin{pmatrix} C_x & S_x \\ C_x' & S_x' \end{pmatrix}. \tag{6.69}$$

There is a similar matrix for the yy'-plane with f_F and f_D replaced by $-f_F$ and $-f_D$, respectively. These yield four equations

$$(C_x\beta_x^*)^2 + S_x^2 = \beta_x^*\beta_{xc}, \tag{6.70}$$

$$(C_y\beta_y^*)^2 + S_y^2 = \beta_y^*\beta_{yc}, \tag{6.71}$$

$$C_x C_x' \beta_x^{*2} + S_x S_x' = -\alpha_{xc}\beta_x^*, \quad \text{and} \tag{6.72}$$

$$C_y C_y' \beta_y^{*2} + S_y S_y' = -\alpha_{yc}\beta_y^*, \tag{6.73}$$

where the subscript c indicates value of β or α at the end of the insertion. The asterisk indicates the minimum value of β at the symmetry point. Using these, we can select appropriate values for l_1, l_2, l_3, f_F, and f_D. In the real application, the high energy physicist will want l_3 to be as large as possible, so that he may install a titanic detector without any interfering accelerator components. These considerations will limit the minimum attainable size of β_x^*, and β_y^*. In practice, the last few FODO cells may be modified to reduce dispersion and decrease the beam waist at the interaction points.

6.8 Coupled motion

So far we have only dealt with the equations of motion which may be treated as separate cases for both x and y motion. If we insert some element into our lattice which couples the horizontal and vertical components of the motion, the solution of the equation of motion becomes more complicated. As a concrete example, one might consider the addition of a skew quadrupole which has a transfer matrix given by

$$
\begin{aligned}
\mathbf{M}_{\text{skew}} = &
\begin{pmatrix}
\cos\psi & 0 & \sin\psi & 0 \\
0 & \cos\psi & 0 & \sin\psi \\
-\sin\psi & 0 & \cos\psi & 0 \\
0 & -\sin\psi & 0 & \cos\psi
\end{pmatrix} \\
\times &
\begin{pmatrix}
\cos\phi & \frac{1}{\sqrt{k}}\sin\phi & 0 & 0 \\
-\sqrt{k}\sin\phi & \cos\phi & 0 & 0 \\
0 & 0 & \cosh\phi & \frac{1}{\sqrt{k}}\sinh\phi \\
0 & 0 & \sqrt{k}\sinh\phi & \cosh\phi
\end{pmatrix} \\
\times &
\begin{pmatrix}
\cos\psi & 0 & -\sin\psi & 0 \\
0 & \cos\psi & 0 & -\sin\psi \\
\sin\psi & 0 & \cos\psi & 0 \\
0 & \sin\psi & 0 & \cos\psi
\end{pmatrix},
\end{aligned} \tag{6.74}
$$

where we have constructed the skew quadrupole from a normal quadrupole by a rotation through $\psi = 45°$.

The 4×4 transfer matrix from (x_1, P_{x1}, y, P_{y1}) to $(x_2, P_{x2}, y_2, P_{y2})$ at s_2 can be written in block form as

$$T = \begin{pmatrix} M & n \\ m & N \end{pmatrix}, \tag{6.75}$$

where M, N, m, and n are 2×2 matrices. If $n = m = 0$, the motion is clearly decoupled; that is a change in either x_1 or P_{x1} does not effect y_2 and P_{y2}, and a change in either y_1 or P_{y1} does not effect x_2 or P_{x2}. If either m or n are nonzero, then there will be such an effect.

It is possible in most cases to find a transformation of coordinates[6,7], from (x, P_x, y, P_y) to new coordinates (u, P_u, v, P_v) for a periodic cell, which will change the matrix T to a new matrix U which is in block diagonal form:

$$U = \begin{pmatrix} A & 0 \\ 0 & B \end{pmatrix}, \tag{6.76}$$

with A and B being 2×2 matrices. This transformation is a kind of *symplectic rotation*, which is in effect just a transformation to the normal mode coordinates for the coupled oscillator. The rotation can be written in block form as

$$R = \begin{pmatrix} I \cos \phi & D^{-1} \sin \phi \\ -D \sin \phi & I \cos \phi \end{pmatrix}, \tag{6.77}$$

where I is the 2×2 identity matrix, and

$$D = \begin{pmatrix} a & b \\ c & d \end{pmatrix}, \tag{6.78}$$

with the restriction that $ad - bc = 1$. The coordinates transform as

$$\begin{pmatrix} x \\ P_x \\ y \\ P_y \end{pmatrix} = R \begin{pmatrix} u \\ P_u \\ v \\ P_v \end{pmatrix}, \tag{6.79}$$

which gives the matrix transformation

$$T = RUR^{-1}. \tag{6.80}$$

By analogy with § 5.1, the block components of U, may be written as

$$A = I \cos \mu_u + J_u \sin \mu_u, \quad \text{and} \tag{6.81}$$

$$B = I \cos \mu_v + J_v \sin \mu_v. \tag{6.82}$$

As before,

$$\mathbf{J}_u = \begin{pmatrix} \alpha_u & \beta_u \\ -\gamma_u & -\alpha_u \end{pmatrix}, \quad \text{and} \quad \mathbf{J}_v = \begin{pmatrix} \alpha_v & \beta_v \\ -\gamma_v & -\alpha_v \end{pmatrix}, \qquad (6.83)$$

are matrices of the respective Twiss parameters. With these definitions, we may attempt to find the rotation matrix $\mathbf{R}$ that transforms the matrix $\mathbf{T}$ into block diagonal form. After some manipulations (see Problem 6-5) the following conditions are obtained:

$$\cos\mu_u - \cos\mu_v = \frac{1}{2}\mathrm{tr}(\mathbf{M} - \mathbf{N})\left[1 + \frac{2\det(\mathbf{m}) + \mathrm{tr}(\mathbf{nm})}{\left[\frac{1}{2}\mathrm{tr}(\mathbf{M} - \mathbf{N})\right]^2}\right]^{\frac{1}{2}}. \qquad (6.84)$$

$$\cos(2\phi) = \frac{\frac{1}{2}\mathrm{tr}(\mathbf{M} - \mathbf{N})}{\cos\mu_u - \cos\mu_v}. \qquad (6.85)$$

$$\mathbf{D} = -\frac{\mathbf{m} + \tilde{\mathbf{n}}}{(\cos\mu_u - \cos\mu_v)\sin(2\phi)}. \qquad (6.86)$$

$$\mathbf{A} = \mathbf{M} - \mathbf{D}^{-1}\mathbf{m}\tan\phi. \qquad (6.87)$$

$$\mathbf{B} = \mathbf{N} + \mathbf{D}\mathbf{n}\tan\phi. \qquad (6.88)$$

Here $\tilde{\mathbf{n}} = \mathbf{S}\mathbf{n}^T\mathbf{S}^T$ with $\mathbf{S} = \begin{pmatrix} 0 & 1 \\ -1 & 0 \end{pmatrix}$.

There are some coupled symplectic matrices for which this transformation will not work. When $\cos\mu_u = \cos\mu_v$, the matrix $\mathbf{D}$ can blow up; however, it turns out that it is impossible to force the two tunes together. Another case arises when the particle motion is unstable, i. e., without both modes of oscillation. An example of this is given by the transformation of x, x', z, and δ through a sector magnet:

$$\begin{pmatrix} \cos\theta & \rho\sin\theta & 0 & \rho(1 - \cos\theta) \\ -\frac{1}{\rho}\sin\theta & \cos\theta & 0 & \sin\theta \\ -\sin\theta & -\rho(1 - \cos\theta) & 1 & -\rho(\theta - \sin\theta) \\ 0 & 0 & 0 & 1 \end{pmatrix} \qquad (6.89)$$

(see Eq. (3.102)). For off-momentum particles, the longitudinal coordinate keeps growing as s increases, in fact the lower-right 2×2 matrix looks like a drift in one of the transverse dimensions. It should be noticed that the trace of this 2×2 matrix is 2, and therefore the longitudinal tune is at the boundary between the stable and unstable regions. In the next chapter, we will see that the addition of a longitudinal oscillating electric field can act like a focusing quadrupole in the z-direction, thus producing longitudinal stability. This method of decoupling the motion may be generalized to a higher dimension; however, this complicates the matter, since more symplectic rotation matrices are required, e. g., in six dimensional phase space three different matrices are required, just as three different Euler angles are required for a general rotation in three dimensions.

6.9 Finding the eigenvalues of a 4×4 transfer matrix

In order to understand the stability criteria and resonance behavior of linear transverse motion in an accelerator ring, it is worth while to examine the eigenvalues of the one-turn matrix. As in the case of one dimensional motion with 2×2-matrices, stable motion will require eigenvalues of unit magnitude, $\lambda_j = e^{i\mu_j}$ with real phases μ_j. We may write a general symplectic transfer matrix for transverse motion as

$$\mathbf{T} = \begin{pmatrix} \mathbf{M} & \mathbf{n} \\ \mathbf{m} & \mathbf{N} \end{pmatrix} = \begin{pmatrix} a & b & c & d \\ e & f & g & h \\ j & k & l & m \\ n & p & q & r \end{pmatrix}. \tag{6.90}$$

The characteristic equation is fourth order:

$$0 = |\mathbf{T} - \lambda\mathbf{I}| = \lambda^4 + A_3\lambda^3 + A_2\lambda^2 + A_1\lambda^1 + A_0. \tag{6.91}$$

Clearly $A_0 = |\mathbf{T}| = 1$, and $A_3 = -\text{tr}(\mathbf{T})$. The equation for the four eigenvectors of $\mathbf{T}$ is

$$\mathbf{T}\mathbf{v}_j = \lambda_j\mathbf{v}_j \tag{6.92}$$

for j from one to four.[†] Multiplying by $\lambda_j^{-1}\mathbf{M}^{-1}$ yields the eigenvector equation for the inverse matrix:

$$\mathbf{T}^{-1}\mathbf{v}_j = \lambda_j^{-1}\mathbf{v}_j. \tag{6.94}$$

By dividing the above characteristic equation by λ^4 we must obtain the same equation again, so we must have $A_1 = A_3 = -\text{tr}(\mathbf{T})$. With a few long lines of straightforward algebra one may also obtain $A_2 = 2 + \text{tr}(\mathbf{M})\text{tr}(\mathbf{N}) - |\mathbf{m} + \tilde{\mathbf{n}}|$.

[†] The degenerate eigenvalues of $+1$ and -1 may not always give a full set of eigenvectors, e. g., the simple drift matrix

$$\begin{pmatrix} 1 & L \\ 0 & 1 \end{pmatrix} \tag{6.93}$$

has only a single eigenvector solution (up to a constant, of course).

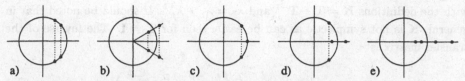

Figure. 6.6 Location of eigenvalues in the complex plane. a)Two pairs of stable eigenvalues with unit radii. Note that $\lambda_j^* = \lambda_j^{-1}$. b)Four unstable eigenvalues with nonzero imaginary components. These four distinct eigenvalues may be written In polar coordinates as $re^{i\mu}$, $re^{-i\mu}$, $r^{-1}e^{i\mu}$, and $r^{-1}e^{-i\mu}$ for some radius $r \neq 1$ and angle $\mu \neq n\pi$ where n is any integer. A matrix with this pattern of eigenvalues cannot be decoupled. c)One pair of stable eigenvalues and two degenerate eigenvalues equal to 1. When there are eigenvalues of ± 1, their corresponding normal mode of motion may or may not be stable. When the motion is stable, these correspond to the integer ($\lambda = 1$) or half integer ($\lambda = -1$) tunes. d)One stable pair of eigenvalues on the real circle, and one unstable pair on the real axis. e)All four eigenvalues real and unstable.

The product of all four eigenvalues must give the determinant, so

$$\lambda_1\lambda_2\lambda_3\lambda_4 = 1. \tag{6.95}$$

Also remember that if λ is an eigenvalue, then so is λ^* since $\mathbf{T}$ is a real matrix. In the 2×2 case, the reciprocal of an eigenvalue must equal its complex conjugate. With four eigenvalues, the reciprocal and complex conjugate may be different eigenvalues. For stable motion, all four eigenvalues must lie on the unit circle in the complex plane. Fig. 6.6 illustrates some of these requirements on the location of eigenvalues in the complex plane.

In order for a 4×4 symplectic matrix (or even more generally a $2n \times 2n$ symplectic matrix) to be decoupled by a similarity transformation, the eigenvalues must be able to be grouped into pairs so that each pair contains contains both the reciprocals and complex conjugates of both eigenvalues within the pair. In other words, the decoupled matrix must consist of 2×2 blocks along the diagonal which are independently symplectic: for each block, either $\lambda_j^* = \lambda_j^{-1}$, or both λ_j are real with $\lambda_1 = \lambda_2^{-1}$. For the cases shown in Fig. 6.7, diagram b is the only one which does not allow decoupling.

Rather than try to solve the fourth order characteristic equation,

$$\lambda^4 - \text{tr}(\mathbf{T})\lambda^3 + [2 + \text{tr}(\mathbf{M})\text{tr}(\mathbf{N}) - |m + \tilde{n}|]\lambda^2 - \text{tr}(\mathbf{T})\lambda + 1 = 0 \tag{6.96}$$

directly, we will use the fact that the eigenvalues come in reciprocal pairs.

Since both $\mathbf{T}$ and $\mathbf{T}^{-1}$ have the same eigenvalue-eigenvector solutions, we may write

$$\mathbf{K}\mathbf{v}_j = (\mathbf{T} + \mathbf{T}^{-1})\mathbf{v}_j = (\lambda_j + \lambda_j^{-1})\mathbf{v}_j = \kappa_j\mathbf{v}_j, \tag{6.97}$$

with the definitions $\mathbf{K} = \mathbf{T} + \mathbf{T}^{-1}$ and $\kappa_j = \lambda_j + \lambda_j^{-1}$. It should be noted that in general, $\mathbf{K}$ is not symplectic as can be easily seen for $\mathbf{T} = \mathbf{I}$. The inverse of the transfer matrix is

$$\mathbf{T}^{-1} = \begin{pmatrix} \tilde{\mathbf{M}} & \tilde{\mathbf{m}} \\ \tilde{\mathbf{n}} & \tilde{\mathbf{N}} \end{pmatrix} = \begin{pmatrix} f & -b & p & -k \\ -c & a & -n & j \\ h & -d & r & -m \\ -g & c & -q & l \end{pmatrix}, \tag{6.98}$$

and

$$\mathbf{K} = \begin{pmatrix} \mathrm{tr}(\mathbf{M})\,\mathbf{I} & \tilde{\mathbf{m}} + \mathbf{n} \\ \mathbf{m} + \tilde{\mathbf{n}} & \mathrm{tr}(\mathbf{N})\,\mathbf{I} \end{pmatrix} = \begin{pmatrix} a+f & 0 & c+p & d-k \\ 0 & a+f & g-n & h+j \\ h+j & k-d & l+r & 0 \\ n-g & c+p & 0 & l+r \end{pmatrix}. \tag{6.99}$$

The characteristic equation for $\mathbf{K}$ is then

$$
\begin{aligned}
0 &= |\mathbf{K} - \kappa\mathbf{I}| \\
&= (a+f-\kappa) \begin{vmatrix} a+f-\kappa & g-n & h+j \\ k-d & l+r-\kappa & 0 \\ c+p & 0 & l+r-\kappa \end{vmatrix} \\
&\quad + (h+j) \begin{vmatrix} 0 & c+p & d-k \\ a+f-\kappa & g-n & h+j \\ c+p & 0 & l+r-\kappa \end{vmatrix} \\
&\quad + (g-n) \begin{vmatrix} 0 & c+p & d-k \\ a+f-\kappa & g-n & h+j \\ k-d & l+r-\kappa & 0 \end{vmatrix} \\
&= [(a+f-\kappa)(l+r-\kappa) - |\mathbf{m} + \tilde{\mathbf{n}}|]^2 , \tag{6.100}
\end{aligned}
$$

which is just the square of the simple quadratic equation: [†]

$$0 = \kappa^2 - \mathrm{tr}(\mathbf{M} + \mathbf{N})\kappa + \mathrm{tr}(\mathbf{M})\mathrm{tr}(\mathbf{N}) - |\mathbf{m} + \tilde{\mathbf{n}}|. \tag{6.101}$$

[†] It is perhaps of passing interest to note that this is the characteristic equation of the matrix:

$$\begin{pmatrix} \mathrm{tr}(\mathbf{M}) & |\mathbf{m} + \tilde{\mathbf{m}}| \\ 1 & \mathrm{tr}(\mathbf{N}) \end{pmatrix}.$$

Solving for κ gives

$$\kappa = \lambda + \lambda^{-1} = \frac{\text{tr}(M+N)}{2} \pm \sqrt{\left(\frac{\text{tr}(M+N)}{2}\right)^2 - \text{tr}(M)\text{tr}(N) + |m+\tilde{n}|}$$

$$= \frac{\text{tr}(M+N)}{2} \pm \sqrt{\left(\frac{\text{tr}(M-N)}{2}\right)^2 + |m+\tilde{n}|}. \qquad (6.102)$$

The only stable solutions require that the corresponding $\lambda_j = e^{i\mu_j}$ and $\lambda_j^{-1} = e^{-i\mu_j}$ for some real phase μ_j. In this case, then $\kappa = 2\cos\mu_j$. So for stability, both κ_1 and κ_2 must lie on the real interval $[-2, 2]$. If either $\kappa = 2$, then at least two of the eigenvalues of T are 1, and the motion may or may not be stable. Similarly, the motion may or may not be stable if either $\kappa = -2$, since at least two of the eigenvalues of T are -1. Both κ_1 and κ_2 will be real if all four eigenvalues λ_j lie on the union of the unit circle and the real line. If κ is complex but not real, then the motion is unstable with four distinct eigenvalues λ_j in the complex plane lying neither on the unit circle nor on the real line. For the unstable case

$$|m+\tilde{n}| < -\left(\frac{\text{tr}(M-N)}{2}\right)^2, \qquad (6.103)$$

κ_1 and κ_2 will be complex with $\lambda^* \neq \lambda^{-1}$.

Having found solutions for κ, we can now solve for the λ's

$$0 = \lambda^2 - \kappa_j\lambda + 1, \qquad (6.104)$$

$$\lambda = \frac{\kappa_j}{2} \pm \sqrt{\left(\frac{\kappa_j}{2}\right)^2 - 1}, \qquad (6.105)$$

for $j = 1, 2$. Stable motion will yield the four eigenvalues of T

$$\lambda = e^{\pm i\mu_j}, \quad \text{with} \quad \kappa_j = 2\cos\mu_j \qquad (6.106)$$

for real phases μ_1 and μ_2.

It is interesting to examine what happens to the off-diagonal blocks of the 1-turn matrix for a ring with two pairs of degenerate eigenvalues on the unit circle. So long as the four eigenvalues are on the unit circle, the matrix may be decoupled via a canonical transformation to normal mode coordinates, i. e., via some symplectic similarity transformation

$$R = \begin{pmatrix} P & q \\ p & Q \end{pmatrix} \qquad (6.107),$$

where $\mathbf{P}$, $\mathbf{q}$, $\mathbf{p}$, and $\mathbf{Q}$ are 2×2-block matrices. The transformation will have the form

$$
\begin{aligned}
\mathbf{T} = \begin{pmatrix} \mathbf{M} & \mathbf{n} \\ \mathbf{m} & \mathbf{N} \end{pmatrix} &= \begin{pmatrix} \mathbf{P} & \mathbf{q} \\ \mathbf{p} & \mathbf{Q} \end{pmatrix} \begin{pmatrix} \mathbf{A} & 0 \\ 0 & \mathbf{B} \end{pmatrix} \begin{pmatrix} \tilde{\mathbf{P}} & \tilde{\mathbf{p}} \\ \tilde{\mathbf{q}} & \tilde{\mathbf{Q}} \end{pmatrix} \\
&= \begin{pmatrix} \mathbf{P}\mathbf{A}\tilde{\mathbf{P}} + \mathbf{q}\mathbf{B}\tilde{\mathbf{q}} & \mathbf{P}\mathbf{A}\tilde{\mathbf{p}} + \mathbf{q}\mathbf{B}\tilde{\mathbf{Q}} \\ \mathbf{p}\mathbf{A}\tilde{\mathbf{P}} + \mathbf{Q}\mathbf{B}\tilde{\mathbf{q}} & \mathbf{p}\mathbf{A}\tilde{\mathbf{p}} + \mathbf{Q}\mathbf{B}\tilde{\mathbf{Q}} \end{pmatrix},
\end{aligned}
\tag{6.108}
$$

so

$$
\begin{aligned}
\mathbf{m} + \tilde{\mathbf{n}} &= \mathbf{p}(\mathbf{A} + \tilde{\mathbf{A}})\mathbf{P} + \mathbf{Q}(\mathbf{B} + \tilde{\mathbf{B}})\tilde{\mathbf{q}} \\
&= \mathbf{p}\mathbf{P}\,\mathrm{tr}(\mathbf{A}) + \mathbf{Q}\tilde{\mathbf{q}}\,\mathrm{tr}(\mathbf{B}).
\end{aligned}
\tag{6.109}
$$

The symplectic condition for $\mathbf{R}$ requires that

$$
\mathbf{p} = -\frac{\mathbf{Q}\tilde{\mathbf{q}}\tilde{\mathbf{P}}}{|\mathbf{Q}|}
\tag{6.110}
$$

so we may now write

$$
\begin{aligned}
\mathbf{m} + \tilde{\mathbf{n}} &= -\mathbf{Q}\tilde{\mathbf{q}}\,\mathrm{tr}(\mathbf{A}) + \mathbf{Q}\tilde{\mathbf{q}}\,\mathrm{tr}(\mathbf{B}) \\
&= 2(\cos\mu_B - \cos\mu_A)\mathbf{Q}\tilde{\mathbf{q}},
\end{aligned}
\tag{6.111}
$$

since $\tilde{\mathbf{P}}\mathbf{P} = |\mathbf{P}| = |\mathbf{Q}|$. If $\cos\mu_A = \cos\mu_B$ then we find $\mathbf{m} = -\tilde{\mathbf{n}}$ which implies the determinant $|\mathbf{m} + \tilde{\mathbf{n}}| = 0$. Substituting this back into Eq. 6.102, it should also be obvious that since equal tunes require $\kappa_1 = \kappa_2$ we must also have $\mathrm{tr}(\mathbf{M}) = \mathrm{tr}(\mathbf{N})$ as well when $\cos\mu_A = \cos\mu_B$.

In Chapter 10 we will revisit the case of degenerate eigenvalues of the 1-turn matrix with regard to coupling resonances.

6.10 Chromaticity

The focal properties of a FODO lattice depend upon the quadrupoles' strengths which, in turn, depend on the particle's momentum, since

$$
\begin{aligned}
k &= \frac{e}{p}\frac{\partial B_y(s)}{\partial x} \quad \text{(in general)} \\
&= \frac{eG}{p} \quad \text{(in the quad)},
\end{aligned}
\tag{6.112}
$$

where $p = p_0 \pm \Delta p$. Then

$$k \simeq k_0 \left(1 - \frac{\Delta p}{p}\right), \tag{6.113}$$

with

$$k_0 = \frac{e}{p_0} \frac{\partial B_y(s)}{\partial x}, \tag{6.114}$$

since dispersion does exist, and beams are not monoenergetic or monochromatic. The aim of this section is to study the effects of a nonzero $\Delta p/p$ over the betatron tunes.

In an infinitesimal stretch, ds, of the orbit, the chromatic effect may be expressed as

$$d\mathbf{M}(\Delta p) = \begin{pmatrix} 1 & 0 \\ -k\,ds & 1 \end{pmatrix}, \tag{6.115}$$

which looks like the transformation through a thin lens. In order to insert this perturbation in the lattice, described by the usual matrix, we have to take away the corresponding unperturbed piece of orbit, namely:

$$d\mathbf{M}(0) = \begin{pmatrix} 1 & 0 \\ -k_0\,ds & 1 \end{pmatrix}, \tag{6.116}$$

with k_0 given by Eq. (6.114). Thus, at the generic location s, we have

$$\mathbf{M} = [d\mathbf{M}(\Delta p)]\,[d\mathbf{M}(0)]^{-1}[\mathbf{I}\cos\mu_0 + \mathbf{J}\sin\mu_0], \tag{6.117}$$

where μ_0 is the unperturbed phase advance of the whole machine with $\Delta p = 0$. Then

$$\mathbf{M}(s) = \begin{pmatrix} 1 & 0 \\ -k\,ds & 1 \end{pmatrix} \begin{pmatrix} 1 & 0 \\ k_0\,ds & 1 \end{pmatrix}$$
$$\begin{pmatrix} \cos\mu_0 + \alpha\sin\mu_0 & \beta\sin\mu_0 \\ -\gamma\sin\mu_0 & \cos\mu_0 - \alpha\sin\mu_0 \end{pmatrix}. \tag{6.118}$$

The product of the first two matrices on the right side of Eq. (6.118) is

$$\begin{pmatrix} 1 & 0 \\ -(k - k_0)ds & 1 \end{pmatrix} = \begin{pmatrix} 1 & 0 \\ k_0 \frac{\Delta p}{p}ds & 1 \end{pmatrix} = \begin{pmatrix} 1 & 0 \\ k_0\delta\,ds & 1 \end{pmatrix}, \tag{6.119}$$

and $\mathbf{M}$ becomes

$$\begin{pmatrix} \cos\mu_0 + \alpha\sin\mu_0 & \beta\sin\mu_0 \\ -\gamma\sin\mu_0 + k_0\delta(\cos\mu_0 + \alpha\sin\mu_0)ds & \cos\mu_0 - \alpha\sin\mu_0 + k_0\delta\beta\sin\mu_0\,ds \end{pmatrix}.$$
$$\tag{6.120}$$

As usual,

$$\cos\mu = \frac{1}{2}\mathrm{tr}(\mathbf{M}) = \cos\mu_0 + \frac{k_0\delta}{2}\beta\sin\mu_0\,ds. \qquad (6.121)$$

On the other hand,

$$\cos\mu = \cos(\mu_0 + d\mu) \simeq \cos\mu_0 - \sin\mu_0\,d\mu. \qquad (6.122)$$

Comparison of the last two equations yields

$$d\mu = 2\pi\,dQ = -\frac{1}{2}\beta(s)k_0(s)\delta\,ds, \qquad (6.123)$$

which when integrated around the full machine gives

$$\Delta Q = -\frac{1}{4\pi}\oint\beta(s)k_0(s)\frac{\Delta p}{p}\,ds. \qquad (6.124)$$

Then

$$\frac{\Delta Q}{Q} = \frac{\Delta p}{p}\left[-\frac{1}{4\pi Q}\oint\beta(s)k_0(s)\,ds\right]. \qquad (6.125)$$

The *natural chromaticity* is defined as

$$\xi_{x\mathrm{N}} = \left(\frac{\Delta Q_{\mathrm{H}}}{Q_{\mathrm{H}}}\right)\Big/\left(\frac{\Delta p}{p}\right) = -\frac{1}{4\pi Q_{\mathrm{H}}}\oint\beta_x(s)k_0(s)\,ds, \qquad (6.126)$$

or using Eq. (6.112),

$$\xi_{x\mathrm{N}} = -\frac{1}{4\pi Q_{\mathrm{H}}}\frac{e}{p}\oint\beta_x(s)G(s)\,ds. \qquad (6.127a)$$

Actually there are two different natural chromaticities for the horizontal and vertical planes with the vertical given by

$$\xi_{y\mathrm{N}} = \frac{1}{4\pi Q_{\mathrm{V}}}\oint\beta_y(s)k_0(s)\,ds, \qquad (6.127b)$$

since a horizontally focusing quadrupole defocuses in the vertical plane.

If some magnetic field imperfections give rise to a perturbation of the kind

$$B_y = \sum_{n=2}^{\infty}b_n x^n, \qquad (6.128)$$

then in Eq. (6.119) $k - k_0$ must be replaced by

$$\left(-\frac{qG}{p_0} + 2\frac{q}{p_0}b_2\eta_x\right)\frac{\Delta p}{p} + \dots, \qquad (6.129)$$

which yields the additional *residual chromaticity* in the horizontal plane:

$$\xi_{xR} = \frac{1}{4\pi Q_H} \frac{q}{p_0} \oint \beta_x(s)\, 2b_2(s)\eta_x(s)\, ds + \ldots . \tag{6.130}$$

The total chromaticity will then be

$$\xi_{x\text{total}} = -\frac{1}{4\pi Q} \frac{q}{p} \oint \beta(s)\, [G(s) - 2b_2(s)\eta(s)]\, ds, \tag{6.131}$$

which may vanish, at least in principle, if

$$b_2(s) = \frac{G(s)}{2\eta(s)}. \tag{6.132}$$

Appropriately placed sextupole lenses may then compensate the natural chromaticity. For the vertical direction in a planar accelerator, the vertical residual chromaticity will be

$$\xi_{yR} = -\frac{1}{2\pi Q} \frac{q}{p_0} \oint \beta_y(s) b_2(s)\eta_x(s)\, ds, \tag{6.133}$$

since the field components of a normal sextupole are

$$B_y = b_2(x^2 - y^2), \quad \text{and} \tag{6.134}$$
$$B_x = 2b_2 xy, \tag{6.135}$$

as was shown in Chapter 4. Skew quadrupoles can also add to the residual chromaticity. If there are vertical bends, the vertical dispersion function η_y may not be zero, in which case, the residual chromaticities will also have terms with integrals of η_y.

As a final remark, we stress that the low beta insertion dictates having very strong quadrupoles which give rise to a high natural chromaticity.

Problems

6–1 a) What is the maximum possible phase advance in a drift?
b) What is the maximum possible phase advance in a FODO cell?

6–2 Design a lattice with four identical FODO cells. Let the basic cell consist of a thin lens, a sector magnet, another thin lens, and another sector magnet. Pick suitable values for the two quadrupole strengths, drift length, and bending radius, so that there is stability in both dimensions.

a) What are the horizontal and vertical betatron tunes? b) Plot $\beta_x(s)$, $\beta_y(s)$, and $\eta(s)$ for the periodic cell.

c) Calculate the momentum compaction.

6–3   Given a full turn transfer matrix for (x, x', z, δ), i. e.,

$$
M = \begin{pmatrix} C & S & 0 & D \\ C' & S' & 0 & D' \\ E & F & 1 & G \\ 0 & 0 & 0 & 1 \end{pmatrix},
$$

find an expression for the momentum compaction. Assume the circumference is L, and the design momentum is p_0.

6–4   a) Calculate the luminosity in CESR, assuming $\beta_x^* = 1.1$ m, $\beta_y^* = 2$ cm, $\eta_x^* = 1.1$ m, $\sigma_p/p = 6 \times 10^{-4}$, and rms emittances of $\pi\epsilon_x = \pi \times 10^{-7}$ m and $\pi\epsilon_y = \pi \times 6 \times 10^{-8}$ m. (See Problem 1-1.) b) What is the value if we change β_y^* to 1 cm? (You may want to do a numerical integration.)

6–5   Derive the conditions in Eqs. (6.84–6.88).

6–6   Show that any real symplectic $2n \times 2n$ matrix has a determinant of $+1$.
   Note: This result means that there are no *improper* symplectic groups as there are with the rotation groups, $SO(n)$.

6–7   Calculate the natural chromaticities of the lattice designed in Problem 6.2, for both horizontal and vertical motion. Use two families of thin sextupole magnets to cancel the natural chromaticity of the lattice with the residual chromaticity of the sextupoles. Where and how large are the sextupoles?

6–8   Consider a ring made of N identical FODO cells with equally spaced quadrupoles. Assume that the two quadrupoles are both of length l_q, but their strengths may differ. Calculate the natural chromaticities for this machine, and show that for short quadrupoles,

$$
\xi_N \simeq -\frac{2\tan\frac{\mu}{2}}{\mu},
$$

where μ is the betatron phase advance per cell.

References for Chapter 6

[1] E. J. N. Wilson, "Proton Synchrotron Accelerator Theory", CERN 77-07 (1977).

[2] E. Keil, "Linear Machines Lattices", *Proc. of the 1st Course of the International School of Particle Accelerators of the "Ettore Majorana" Center for Scientific Culture*, Erice, p. 22, CERN 77-13 (1977).

[3] K. Steffen, "Basic Course on Accelerator Optics", *CAS General Accelerator Physics*, p. 25., Gif-sur-Yvette, Paris, CERN 85-19 (1985).

[4] P. Schmüser, "Basic Course on Accelerator Optics", *CAS Second General Accelerator Physics Course*, p.1, CERN 87-10, (1987).

[5] R. Brinkmann, "Insertions", *CAS Second General Accelerator Physics Course*, p. 45, CERN 87-10 (1987).

[6] L. C. Teng, "Concerning n-Dimensional Coupled Motions", FN-229 0100, National Accelerator Laboratory, Batavia, Ill. (1971).

[7] D. A. Edwards and L. C. Teng, "Parametrization of Linear Coupled Motion in Periodic Systems", *IEEE Trans. on Nucl. Sci.*, NS-20, no. 3, (1973).

7

Synchrotron Oscillations

With the exception of Chapter 1 and § 5.4, we have only treated particle motion without the influence of electric fields which are at the heart of the acceleration process. In this chapter, we examine the effects of rf electric fields on the motion of particles. The process is much like the popular sport of surfing. The particles are pushed along by the electric force $q\vec{E}$, just as the surfboard is pushed by the wave. There are stable regions for the particles to move, and there are unstable regions. In the unstable regions, the particles are lost, just as the surfer who finds himself in the wrong place will "wipe out". The particles will tend to bunch up in the stable regions which are called *buckets*. We examine the oscillations, take a closer look at adiabatic damping and look at some of the problems associated with the transfer of particles from one accelerator to another.

7.1 Transition energy

The angular frequency of a particle revolving in a synchrotron is

$$\omega = \frac{2\pi}{\tau} = \frac{2\pi\beta c}{L}, \tag{7.1}$$

where τ is the revolution period, and L is the circumference of the orbit. Differentiating $\ln(\omega)$ yields

$$\frac{d\omega}{\omega} = -\frac{d\tau}{\tau} = \frac{d\beta}{\beta} - \frac{dL}{L} = \left(\frac{1}{\gamma^2} - \alpha_p\right)\frac{dp}{p}. \tag{7.2}$$

The expression within parentheses is usually written as

$$\eta_{\mathrm{tr}} = \frac{1}{\gamma^2} - \alpha_p = \frac{1}{\gamma^2} - \frac{1}{\gamma_{\mathrm{tr}}^2}, \tag{7.3}$$

and is frequently called the *phase slip factor*. Note that there is a *transition* at $\eta_{\mathrm{tr}} = 0$ zero when

$$\gamma = \frac{1}{\sqrt{\alpha_p}} = \gamma_{\mathrm{tr}}. \tag{7.4}$$

It can immediately be seen that:

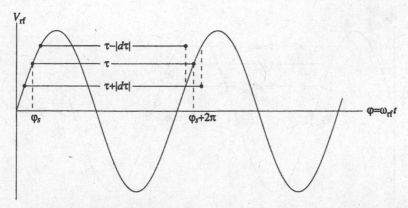

Figure. 7.1 A graphical demonstration of the phase-stability principle for $\eta_{tr} > 0$ (below transition).

$\eta_{tr} < 0$ for weak focusing machines and for the strong focusing case, when the
beam energy is larger than $U_{tr} = \gamma_{tr} mc^2$;

$\eta_{tr} > 0$ for strong focusing synchrotrons when $U < U_{tr}$ (injection) and for linear accelerators, where the average radius is infinite, and consequently
$\alpha_p = 0$.

7.2 The phase stability principle

The aim of this section is to discuss the synchronism between the circulating particles and the rf field in the accelerating structure. In fact, apart from the
synchronous particle which is in perfect synchronism with the sinusoidally varying rf field, all other particles will cross the rf cavity (or cavities) with different
phases relative to the rf oscillation, arriving either early or late with respect to the
synchronous particle.

Thus below transition $d\tau/\tau = -|\eta_{tr}| dp/p$, meaning that more energetic par-
ticles orbit more quickly than the less energetic ones, exhibiting quite a natural
behavior. Hence the ramp of the accelerating voltage, $V \sin(\omega_{rf} t + \phi_s)$, must have
a positive slope. The phase of the synchronous particle, ϕ_s, must lie between $0°$
and $90°$, so that an early (late) particle receives a smaller (larger) energy kick at
the next crossing, as shown in Fig. 7.1. This, in effect, provides a restoring force to
produce energy oscillations about the synchronous particle.

The opposite occurs above transition where $d\tau/\tau = |\eta_{tr}| dp/p$, implying that
particles with higher momentum go slower, contrary to any physical intuition, acting
as though their masses were negative. This effect is called the *negative mass effect*.

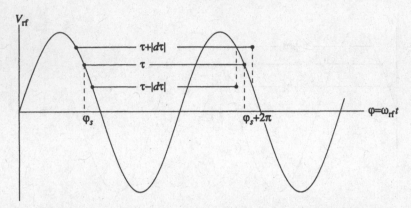

Figure. 7.2 A graphical demonstration of the phase-stability principle for $\eta_{tr} < 0$ (above transition).

Fig. 7.2 illustrates the phase-stability principle in this situation, with $90° < \phi_s < 180°$.

During the accelerating cycle, if the injection energy is less than the transition energy as occurs in most proton synchrotrons, there will be a stage where *transition crossing* has to take place. Setting aside all the considerations involved in such a process, one must say, at least, that a phase jump has to be performed from ϕ_s to $\pi - \phi_s$, in order to maintain the stability about the synchronous particle.

7.3 Resonant acceleration

The acceleration process may be studied by considering the standing wave set up in the rf cavity as a series of waves traveling in both directions around the ring. Only the wave which moves synchronously (or resonantly) with the beam will produce any net acceleration in the long term. The other waves will sometimes accelerate and sometimes decelerate the beam, but the average effect will produce no net acceleration. To quantify this, we will make a few simplifying assumptions:

1) There is only one accelerating gap of length g, located at $s = 0$.

2) The accelerating gap is much shorter than the distance traveled by the beam during one rf period, i. e., $g \ll \beta\lambda_{rf}$.

3) The rf angular frequency is an integer multiple of the angular revolution frequency, ω_s, i. e., $\omega_{rf} = h\omega_s$ for some integer, h.

4) The synchronous particle crosses the gap at time $t = 0$, when the rf phase is ϕ_s, and the voltage across the gap is $V \sin \phi_s$.

The energy gained by the synchronous particle per revolution is

$$\Delta U_s = qV \sin\phi_s, \tag{7.5}$$

and the effective electric field may be written as

$$\vec{E}(s,t) = \hat{s}\, E(s,t) = \hat{s}\, V \sin(\omega_{rf} t + \phi_s) \sum_{n=-\infty}^{\infty} \delta(s - nL), \tag{7.6}$$

where L is the circumference of the synchronous particle's orbit. The delta function may be expanded in a Fourier series, so that

$$E(s,t) = \frac{V}{L} \sin(\omega_{rf} t + \phi_s) \sum_{n=-\infty}^{\infty} \cos\left(\frac{2\pi n s}{L}\right)$$

$$= \frac{V}{L} \sum_{n=-\infty}^{\infty} \sin\left[\omega_s\left(ht - \frac{n}{v}s\right) + \phi_s\right], \tag{7.7}$$

with the synchronous particle's velocity being $v = L\omega_s/2\pi$. It is easy to see that each term in the last equation represents a traveling wave with a different velocity. The only wave with the same velocity as the synchronous particle is the wave with $n = h$. This is the only wave which gives a net acceleration. The time that the synchronous particle passes a point s may be written

$$t_s = \frac{s}{v}, \tag{7.8}$$

and the time for a generic particle will be

$$t = t_s + \delta t, \tag{7.9}$$

where the generic particle lags behind the synchronous particle by an amount δt. If we assume that the longitudinal oscillations have a much lower frequency* than ω_s, then the electric field may be averaged over one revolution, giving

$$\langle E(\delta t)\rangle = \frac{V}{L} \sin(\omega_{rf}\,\delta t + \phi_s) \tag{7.10}$$

for the effective field seen by a generic particle whose time lag is δt. A generic particle will then gain

$$\Delta U = qV \sin(\omega_{rf}\delta t + \phi_s) \tag{7.11}$$

per turn which agrees with Eq. (7.5) for $\delta t = 0$.

---

* Typical frequencies for longitudinal oscillations are more than a hundred times smaller than the orbital frequencies.

7.4 The phase oscillation equation

The physical quantities related to a generic particle and to the synchronous one (labeled with the subscript s) are linked by the following relations:

$$\text{total energy} \quad U = U_s + \delta U \tag{7.12}$$

$$\text{momentum} \quad p = p_s + \delta p \tag{7.13}$$

$$\text{angular frequency} \quad \omega = \omega_s + \delta\omega \tag{7.14}$$

$$\text{revolution period} \quad \tau = \tau_s + \delta\tau \tag{7.15}$$

with $\text{sign}(\delta\omega) = -\text{sign}(\delta\tau)$. Since the synchronous particle must arrive at the rf section with the same phase, we may write

$$\omega_{\text{rf}} = h\omega_s, \tag{7.16}$$

for some integer h, which is called the *harmonic number* because it is the number of rf cycles occurring during one orbit of the synchronous particle. The phase of the rf voltage when the synchronous particle arrives at the cavity is ϕ_s, and the phase that the generic particle sees is ϕ. The relative phase difference between the generic and synchronous particles is then

$$\varphi = \delta\phi = \phi - \phi_s. \tag{7.17}$$

Assuming that the particles pass through the accelerating gap much faster than the rf period, we may define as energy gain per turn

$$\Delta U = qV \sin\phi, \quad \text{and} \tag{7.18}$$

$$\Delta U_s = qV \sin\phi_s, \tag{7.19}$$

with V being the peak voltage. The energy deviation of the generic particle from that of the synchronous particle at the beginning of the n-th turn may be written as

$$(\delta U)_n = U - U_s. \tag{7.20}$$

After the n-th turn the deviation is

$$(\delta U)_{n+1} = (U + \Delta U) - (U_s + \Delta U_s). \tag{7.21}$$

The change in δU after one turn is then

$$\Delta(\delta U) = \Delta U - \Delta U_s = qV(\sin\phi - \sin\phi_s). \tag{7.22}$$

For a slowly varying oscillation about the synchronous energy, we may write

$$\frac{d(\delta U)}{dt} \simeq \frac{\Delta(\delta U)}{\tau_s} = \frac{qV}{2\pi}\omega_s(\sin\phi - \sin\phi_s). \tag{7.23}$$

Defining the variable

$$W = -\delta U/\omega_{rf} = -\frac{U - U_s}{\omega_{rf}}, \tag{7.24}$$

this becomes

$$\frac{dW}{dt} = \frac{qV}{2\pi h}(\sin\phi_s - \sin\phi). \tag{7.25}$$

In § 7.6, we will show that the variables, $\varphi = \phi - \phi_s$ and W, are canonical variables; although, it should be noted that a simple transformation from the coordinate, t, with canonical momentum, $-U$, to $\delta\phi = \omega_{rf}\delta t$ would require dividing the energy by ω_{rf}. (Some authors define W differently as will be discussed in Appendix A.)

Again remembering that the oscillations are slow, after one revolution we have

$$\Delta\varphi \simeq \frac{d\varphi}{dt}\tau_s = \omega_{rf}\delta t, \tag{7.26}$$

where δt is the difference in the arrival times at the cavity of the generic and synchronous particles. After one revolution the change in δt is

$$\Delta(\delta t) = \tau - \tau_s = \delta\tau = -\eta_{tr}\tau\frac{dp}{p}, \tag{7.27}$$

where we have used Eq. (7.3). Solving for $\dot\varphi$ yields

$$\frac{d\varphi}{dt} \simeq \frac{\omega_{rf}^2\eta_{tr}}{\beta^2 U_s}W, \tag{7.28}$$

where the momentum deviation is related to the energy deviation by

$$\frac{\delta p}{p} = \frac{1}{\beta^2 U}\delta U. \tag{7.29}$$

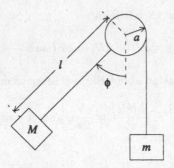

Figure. 7.3 A biased pendulum swinging about the axis of a cylinder.

Substitution of the equation for $\dot{W}$ into that for $\dot{\phi}$ produces an equation for the oscillations of the rf phase for a generic particle:

$$\ddot{\varphi} + \frac{h\omega_s^2 \eta_{tr} qV}{2\pi\beta^2 U_s}(\sin\phi - \sin\phi_s) = 0. \tag{7.30}$$

For small oscillations of the phase, $\sin\phi = \sin(\phi_s + \varphi) \simeq \varphi\cos\phi_s + \sin\phi_s$, and Eq. (7.30) becomes the equation for a linear harmonic oscillator;

$$\ddot{\varphi} + \Omega_s^2\varphi \simeq 0, \quad \text{with} \tag{7.31}$$

$$\Omega_s = \omega_s\sqrt{\frac{h\eta_{tr}\cos\phi_s}{2\pi\beta^2\gamma}\frac{qV}{mc^2}} \tag{7.32}$$

being the *synchrotron oscillation* frequency. The *synchrotron tune* is then given by $Q_s = \Omega_s/\omega_s$. Notice that the product $\eta_{tr}\cos\phi_s$ must be positive to have real frequencies and ensure phase-stability. Bearing in mind that it is hard to exceed a few MV per turn in the rf cavities, one can easily infer that $Q_s \ll 1$.

7.5 Large oscillations

In an rf system the amplitude of the phase oscillations can be rather large, so that the small angle approximation of Eq. (7.31) is no longer valid. For the case of large oscillations Eq. (7.30) can be written as

$$\ddot{\varphi} + \frac{\Omega_s^2}{\cos\phi_s}[\sin(\varphi + \phi_s) - \sin\phi_s] = 0, \tag{7.33}$$

where Ω_s is the oscillation frequency for small oscillations as given by Eq. (7.32).

A simple mechanical model with an identical differential equation is the biased pendulum made with a rigid rod attached to a massless cylinder of radius a as shown in Fig. 7.3. The equilibrium angle is biased by hanging a weight of mass m from the cylinder using a massless cord wrapped around the cylinder several times. If we assume a massless rod of length $l - a$ with a bob of mass M to form the pendulum, the differential equation for the angle ϕ of the pendulum is found to be

$$\ddot{\phi} + \frac{g}{l}\left(\sin\phi - \frac{ma}{Ml}\right) = 0. \tag{7.34}$$

The equilibrium angle ϕ_s is given by

$$\sin\phi_s = \frac{ma}{Ml}, \tag{7.35}$$

and the frequency Ω_s for small oscillations may be found from

$$\frac{\Omega_s^2}{\cos\phi_s} = \frac{g}{l}. \tag{7.36}$$

Even though Eq. (7.33) cannot be solved exactly, it can be converted into a first order equation by realizing that

$$\frac{d(\dot{\phi}^2)}{dt} = 2\ddot{\phi}\frac{d\phi}{dt}. \tag{7.37}$$

Using this, we get

$$d\left(\dot{\phi}^2\right) = \frac{2\Omega_s^2}{\cos\phi_s}(-\sin\phi\,d\phi) + 2\Omega_s^2\tan\phi_s\,d\phi, \tag{7.38}$$

which after integration becomes

$$\frac{1}{\Omega_s}\dot{\phi} = \pm\sqrt{\frac{2(\cos\phi - \cos\phi_0)}{\cos\phi_s} + 2(\phi - \phi_0)\tan\phi_s + \frac{1}{\Omega_s^2}\dot{\phi}_0^2}\,, \tag{7.39}$$

where ϕ_0 is the initial condition for ϕ at $t = 0$. This may easily be integrated with a simple computer program. Fig. 7.4 shows phase flow plots of $\dot{\phi}$ versus ϕ for different values of ϕ_s. The plots of W versus ϕ for the rf case will look identical since W is proportional to $\dot{\phi}$. The closed curves show the stable solutions where the pendulum (charged particle) oscillates about the equilibrium angle (synchronous phase), and the open curves are unstable trajectories.

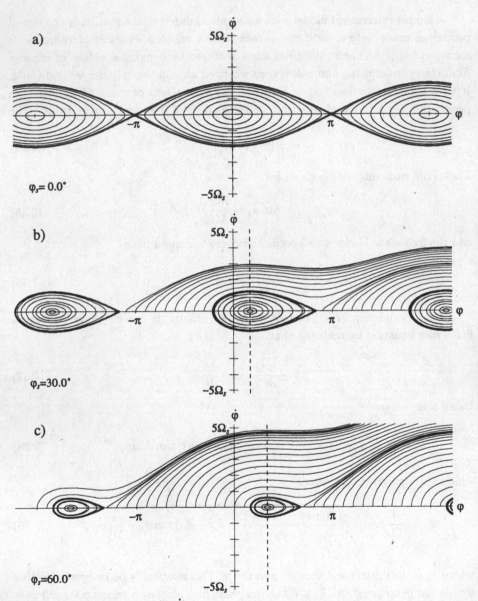

Figure. 7.4 Phase flow plots in $(\phi, \dot{\phi})$ space for the biased pendulum. The initial angular velocity of the displayed trajectories are zero. Plots are shown for synchronous phases of a) $\phi_s = 0°$, b) $\phi_s = 30°$, and c) $\phi_s = 60°$. The trajectories move left for $\dot{\phi} < 0$ and to the right for $\dot{\phi} > 0$. The separatrices are indicated by bold curves.

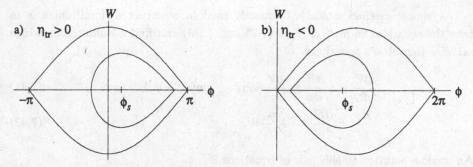

Figure. 7.5 Comparison of separatrices for the stationary bucket ($\sin \phi_s = 0$) and an accelerating bucket for a) below transition, $\eta_{tr} > 0$, and b) above transition $\eta_{tr} < 0$.

Considering the pendulum, it is easy to see that for an initial angle ϕ_0 between $\pi - \phi_s$ and π with $\dot{\phi}_0 = 0$, the initial force does not restore the pendulum but will cause an increase in ϕ, which will increase without bound (or at least until the cord unwinds and rewinds back about the cylinder in the other direction.) For an angle ϕ_0 just less than $\pi - \phi_s$ force is restorative and the pendulum will oscillate stably. The point $(\phi_1, \dot{\phi}_1) = (\pi - \phi_s, 0)$ in phase space is clearly a hyperbolic fixed point which lies on the separatrix. Another point $(\phi_2, 0)$ on the separatrix which intersects the $W = 0$ axis is a solution to the transcendental equation

$$\cos \phi_2 + \cos \phi_s + (\phi_2 + \phi_s - \pi) \sin \phi_s = 0, \qquad (7.40)$$

which is easily obtainable from Eq. (7.39). The point at $(\phi, 0) = (\phi_s, 0)$ is an elliptical fixed point. The pattern of separatrices and fixed points repeats every 360°. In the rf case, the stable phase regions within the separatrices are called *buckets*.

Fig. 7.5 shows how the relative phases of stable buckets must be shifted when the transition energy is crossed.

7.6 Hamiltonian formalism

In this section, we give two different approaches for obtaining a Hamiltonian function for the longitudinal motion. The first is a relatively simple but somewhat arbitrary integration of the equations of motion, using time as the independent variable. In the second approach, we start with a Hamiltonian derived in Chapter 3 and make several canonical transformations to obtain a Hamiltonian which may used to study the coupled *synchro-betatron* motion of the beam. This second approach may be skipped without much loss of continuity in the rest of the book.

A simple method which is frequently used to construct a Hamiltonian is to take the equations of motion (Eqs. (7.25 and 7.28)) and find a Hamiltonian which satisfies Hamilton's equations:

$$\frac{dW}{dt} = -\frac{\partial H}{\partial \varphi} = \frac{qV}{2\pi h}[\sin \phi_s - \sin(\phi_s + \varphi)], \quad \text{and} \tag{7.41}$$

$$\frac{d\varphi}{dt} = \frac{\partial H}{\partial W} = \frac{\omega_{\rm rf}^2 \eta_{\rm tr}}{\beta^2 U_s} W. \tag{7.42}$$

An obvious solution to this pair of equations is

$$H = \frac{1}{2}\frac{\omega_{\rm rf}^2 \eta_{\rm tr}}{\beta^2 U_s} W^2 - \frac{qV}{2\pi h}[\varphi \sin \phi_s + \cos(\varphi + \phi_s)]. \tag{7.43}$$

For small amplitudes this becomes

$$H \simeq \frac{1}{2}\frac{\omega_{\rm rf}^2 \eta_{\rm tr}}{\beta^2 U_s} W^2 + \frac{qV \cos \phi_s}{4\pi h}\varphi^2 + \text{constant}, \tag{7.44}$$

which is just the Hamiltonian for a harmonic oscillator.

For the second method, we will start with the Hamiltonian derived in Chapter 3, Eq. (3.45):

$$H = -P_s(x, P_x, y, P_y, t, -U; s)$$
$$= -qA_s - \left(1 + \frac{x}{\rho}\right)\sqrt{\left(\frac{U - q\Phi}{c}\right)^2 - (mc)^2 - (P_x - qA_x)^2 - (P_y - qA_y)^2}. \tag{7.45}$$

For simplicity let us assume that there are no longitudinal magnetic fields and no static electric fields. In this case the external fields may all be described by $A_s(t, s)$, with the electric field given by

$$E_s = -\frac{\partial A_s}{\partial t}, \tag{7.46}$$

and the transverse magnetic fields by

$$B_x = -\frac{1}{1 + x/\rho}\frac{\partial A_s}{\partial y}, \quad \text{and} \quad B_y = \frac{1}{1 + x/\rho}\frac{\partial A_s}{\partial x}. \tag{7.47}$$

The Hamiltonian is then

$$H = -qA_s(t, s) - \left(1 + \frac{x}{\rho}\right)\sqrt{\left(\frac{U}{c}\right)^2 - (mc)^2 - p_x^2 - p_y^2}. \tag{7.48}$$

In a flat horizontal ring, the vertical motion is not directly coupled to the longitudinal motion, and the results for y are the same as was studied in Chapter 5. We will therefore ignore the y and p_y variables and just treat the coupled horizontal betatron and synchrotron motions.

The time that a generic particle passes a position s, may be written as

$$t(s) = t_s(s) + \delta t(s), \tag{7.49}$$

where the synchronous particle passes this position at

$$t_s = \frac{2\pi h}{\omega_{\rm rf} L} s, \tag{7.50}$$

and the time lag of the generic particle is given by δt. As before the electric field is

$$E_s = V \sum_{n=-\infty}^{\infty} \delta(s - nL) \sin(\omega_{\rm rf} t + \phi_s). \tag{7.51}$$

The vertical component of the magnetic field may be expanded in a power series giving

$$B_y = B_0(s) + \left(\frac{\partial B}{\partial x}\right)_0 x + \dots. \tag{7.52}$$

Using Eqs. (7.46) and (7.47), we find that

$$A_s \approx -\frac{p_s x}{q\rho} - \frac{p_s K}{2q} x^2 + \frac{V}{\omega_{\rm rf}} \sum_{n=-\infty}^{\infty} \delta(s - nL) \cos(\omega_{\rm rf} t + \phi_s), \tag{7.53}$$

where

$$K = \frac{1}{\rho^2} + \frac{1}{B\rho}\left(\frac{\partial B}{\partial x}\right)_0. \tag{7.54}$$

Assuming that the synchronous particle's energy gain per turn is much smaller than the typical deviation in energy of a generic particle from the synchronous particle, we may assume a smooth function for the synchronous particle, called the adiabatic approximation:

$$U_s = U_0 + \frac{qV \sin \phi_s}{L} s, \tag{7.55}$$

where U_0 is the initial energy of the synchronous particle at $s = 0$. The energy deviation of the generic particle may be defined by

$$\delta U = U - U_s. \tag{7.56}$$

Converting the time deviation, δt, into a deviation of phase,

$$\delta\phi = \omega_{\rm rf}\delta t, \tag{7.57}$$

we might expect the corresponding canonical momentum to be

$$W = -\frac{\delta U}{\omega_{\rm rf}}. \tag{7.58}$$

This expectation may be realized by use of the generating function

$$F_2(x,p_x,t,W;s) = xp_x + \omega_{\rm rf}Wt - \left(U_0 + \frac{qV\sin\phi_s}{L}s\right)t - \frac{2\pi hs}{L}W + \frac{qV\pi h\sin\phi_s}{L^2\omega_{\rm rf}}s^2. \tag{7.59}$$

The non-trivial transformations are given by

$$-U = \frac{\partial F_2}{\partial t}, \quad \text{and} \quad \delta\phi = \frac{\partial F_2}{\partial W}, \tag{7.60}$$

and the Hamiltonian is transformed to

$$H_1(x,p_x,\delta\phi,W;s) = H + \frac{\partial F_2}{\partial s}$$
$$\approx -p_s + \frac{p_sK}{2}x^2 - \frac{qV}{\omega_{\rm rf}}\sum_{n=-\infty}^{\infty}\delta(s-nL)\cos\left(\phi_s + \delta\phi + \frac{2\pi hs}{L}\right) - \frac{2\pi h}{L}W$$
$$- \frac{qV\sin\phi_s}{L\omega_{\rm rf}}\delta\phi + \left(1 + \frac{x}{\rho}\right)\left[\frac{U_s\omega_{\rm rf}}{p_sc^2}W + \frac{m^2\omega_{\rm rf}^2}{2p_s^3}W^2 + \frac{p_x^2}{2p_s}\right]. \tag{7.61}$$

Next we would like to subtract the energy dependent closed orbit from the radial coordinate. This closed orbit is described by $x_p = \eta_x\delta$, with the dispersion function satisfying the differential equation

$$\eta_x'' + K(s)\eta_x = \frac{1}{\rho(s)}. \tag{7.62}$$

The fractional momentum deviation is

$$\delta = \frac{\delta p}{p} = -\frac{U_s\omega_{\rm rf}}{(p_sc)^2}W. \tag{7.63}$$

The betatron part of the radial motion can be written as

$$x_\beta = x + \frac{U_s\omega_{\rm rf}}{(p_sc)^2}\eta_xW. \tag{7.64}$$

Similarly we may write the betatron part of the radial momentum as

$$p_\beta = p_x + \frac{U_s \omega_{\text{rf}}}{p_s c^2} W. \tag{7.65}$$

In the paraxial approximation this last equation is equivalent to

$$x'_\beta = x' - \eta'_x \delta. \tag{7.66}$$

In order to decouple the energy component of the x-motion from the betatron motion, we must also modify the $(\delta\phi, W)$-coordinates to remove the small component of the betatron motion. This may be done, keeping W unchanged by allowing $\delta\phi$ to transform to a new phase deviation φ. It is easy to see that this transformation of the phase coordinate is necessary, since the path length subtending a distance Δs along the design orbit will be shorter for $x_\beta < 0$ than for $x_\beta > 0$. A new generating function $\mathcal{F}_2(x, p_\beta, \delta\phi, W; s)$ can be constructed from the relations

$$x_\beta = \frac{\partial \mathcal{F}_2}{\partial p_\beta}, \tag{7.67}$$

$$p_x = \frac{\partial \mathcal{F}_2}{\partial x}, \quad \text{and} \tag{7.68}$$

$$W = \frac{\partial \mathcal{F}_2}{\partial (\delta\phi)}. \tag{7.69}$$

The new phase is then obtained from the remaining equation

$$\varphi = \frac{\partial \mathcal{F}_2}{\partial W}. \tag{7.70}$$

This generating function is not unique and may have an arbitrary function of s and W added to it. We will use the function

$$\mathcal{F}_2 = x p_\beta + \frac{\omega_{\text{rf}} U_s}{(p_s c)^2} (\eta_x p_\beta - p_s \eta'_x x) W - \left(\frac{\omega_{\text{rf}} U_s}{(p_s c)^2} \right)^2 \frac{p_s}{2} \eta_x \eta'_x W^2. \tag{7.71}$$

The transformations to the new system may be summarized as

$$x = x_\beta - \frac{\omega_{\text{rf}} U_s}{(p_s c)^2} \eta_x W, \tag{7.72}$$

$$p_x = p_\beta - \frac{\omega_{\text{rf}} U_s}{(p_s c)^2} p_s \eta'_x W, \tag{7.73}$$

$$\delta\phi = \varphi - \frac{\omega_{\text{rf}} U_s}{(p_s c)^2} (\eta_x p_\beta - p_s \eta'_x x_\beta), \tag{7.74}$$

$$W = W, \quad \text{and} \tag{7.75}$$

$$H_2 = H_1 + \frac{\partial \mathcal{F}_2}{\partial s}. \tag{7.76}$$

The partial derivative $\partial \mathcal{F}_2 / \partial s$ is actually quite messy, since p_s and U_s are functions of s; however, in the adiabatic approximation we may assume that these derivatives are small enough to be negligible, giving

$$\frac{\partial \mathcal{F}_2}{\partial s} \approx \frac{\omega_{\rm rf} U_s}{(p_s c)^2} (\eta_x' p_\beta - p_s \eta_x'' x_\beta) W + \left(\frac{\omega_{\rm rf} U_s}{(p_s c)^2} \right)^2 \frac{p_s}{2} (\eta_x \eta_x'' - \eta_x'^2) W^2. \qquad (7.77)$$

With the help of Eq. (7.62), the new Hamiltonian becomes

$$\begin{aligned}
H_2 = &-p_s + \frac{p_s K}{2} x_\beta^2 + \frac{p_\beta^2}{2 p_s} + \frac{\omega_{\rm rf}^2}{2 \beta^3 U_s c} \left(\frac{1}{\gamma^2} - \frac{\eta_x}{\rho} \right) W^2 \\
&- \frac{qV}{\omega_{\rm rf}} \sum_{n=-\infty}^{\infty} \delta(s - nL) \cos \left[\phi_s + \varphi + \frac{2\pi hs}{L} - \frac{\omega_{\rm rf} U_s}{(p_s c)^2} (\eta_x p_\beta - p_s \eta_x' x_\beta) \right] \\
&- \frac{qV}{L \omega_{\rm rf}} \sin \phi_s \left[\varphi - \frac{\omega_{\rm rf} U_s}{(p_s c)^2} (\eta_x p_\beta - p_s \eta_x' x_\beta) \right],
\end{aligned} \qquad (7.78)$$

where γ and β are the usual relativistic quantities for the synchronous particle.

The equations for synchrotron motion may be obtained by ignoring the betatron coordinates* and averaging the Hamiltonian over one revolution, since synchrotron oscillations typically have periods much longer than the revolution period of the beam. (Synchrotron tunes are usually of the order of 0.01 or less.) The averaged Hamiltonian becomes

$$H_2 \simeq -p_s + \frac{\omega_{\rm rf}^2 \eta_{\rm tr}}{2 \beta^3 U_s c} W^2 - \frac{qV}{L \omega_{\rm rf}} [\cos(\phi_s + \varphi) + \varphi \sin \phi_s], \qquad (7.79)$$

where $\eta_{\rm tr}$ is the phase slip factor. The equations of motion for φ and W are

$$\varphi' = \frac{\partial H_2}{\partial W} = \frac{\omega_{\rm rf}^2 \eta_{\rm tr}}{\beta^3 U_s c} W, \quad \text{and} \qquad (7.80)$$

$$W' = -\frac{\partial H_2}{\partial \varphi} = \frac{qV}{L \omega_{\rm rf}} [\sin \phi_s - \sin(\phi_s + \varphi)], \qquad (7.81)$$

which may be combined to give

$$\varphi'' + \frac{qV \omega_{\rm rf} \eta_{\rm tr}}{\beta^3 U_s c L} [\cos \phi_s \sin \varphi + \sin \phi_s (\cos \varphi - 1)] = 0. \qquad (7.82)$$

---

* It is not sufficient to replace x and p_x in H_1 by their purely energy dependent parts $\eta_x \delta p / p$ and $\eta_x' \delta$ respectively, since the second term in Eq. (7.77) is independent of x_β and p_β, and will still make H_2 different from H_1.

Since $\ddot{\varphi} = (\beta c)^2 \varphi''$, this may be written as

$$\ddot{\varphi} + \Omega_s^2 [\sin \varphi + (\cos \varphi - 1) \tan \phi_s] = 0, \tag{7.83}$$

with

$$\Omega_s^2 = \frac{qVh\eta_{tr}\omega_s^2 \cos \phi_s}{2\pi\beta^2 U_s}, \tag{7.84}$$

and we again have the result of § 7.4. Comparing Eqs. (7.43 and 7.79) we see that they differ by an irrelevant constant, $-p_s$, and an overall factor of βc, since $L\omega_{rf} = 2\pi h\beta c$. This factor of velocity just reflects the difference in independent variables, since

$$\frac{d}{dt} = \frac{ds}{dt}\frac{d}{ds} = \beta c \frac{d}{ds}. \tag{7.85}$$

7.7 Adiabatic invariant

In the adiabatic approximation the Poincaré–Cartan invariant becomes

$$I_L = \oint p \, dq = \oint W \, d\phi = \oint W \frac{d\phi}{dt} dt \tag{7.86}$$

with the integral $\oint$ to be carried out over at least one cycle of the synchrotron oscillations. Using Eq. (7.28) for $d\varphi/dt$ gives

$$I_L = \frac{h^2 \eta_{tr}\omega_s^2}{\beta^2 \gamma mc^2} \oint W^2 dt \tag{7.87}$$

which must be solved numerically in the case of large oscillations; nevertheless, for small oscillations

$$\varphi(t) = \varphi_m \sin(\Omega_s t + \psi_0), \tag{7.88}$$

where ψ_0 is the phase of the oscillation at $t = 0$, and φ_m is the amplitude of the oscillation. Using Eq. (7.25), we get

$$W = W_m \cos(\Omega_s t + \psi_0) = \frac{\Omega_s \beta^2 U_s}{\omega_{rf}^2 \eta_{tr}} \varphi_m \cos(\Omega_s t + \psi_0), \tag{7.89}$$

and comparison shows that

$$W_m = \frac{\Omega_s \beta^2 U_s}{\omega_{rf}^2 \eta_{tr}} \varphi_m. \tag{7.90}$$

The invariant may now be written as

$$I_L = \frac{h^2\omega_s^2\eta_{\rm tr}}{\beta^2 U_s} \oint \frac{\Omega_s\beta^2 U_s}{h^2\omega_s^2\eta_{\rm tr}} \varphi_m W_m \cos^2(\Omega_s t + \psi_0)\, dt = \pi\varphi_m W_m, \qquad (7.91)$$

which is the third motion-invariant to be added to I_H and I_V of § 5.4. On the other hand, we may write this as

$$I_L = \frac{\pi\omega_{\rm rf}^2\eta_{\rm tr}}{\Omega_s\beta^2 U_s} W_m^2. \qquad (7.92)$$

Squaring I_L gives

$$W_m^4 = \frac{qV\cos\phi_s\beta^2 U_s}{2\pi^3 h\omega_{\rm rf}^2\eta_{\rm tr}} I_L^2. \qquad (7.93)$$

If V and ϕ_s are kept constant, and the transition energy is far away, W_m is seen to grow as the beam is adiabatically accelerated. With the help of Eq. (7.91), we also have

$$\varphi_m^2 = I_L\sqrt{\frac{2h\omega_{\rm rf}^2\eta_{\rm tr}}{\pi qV\cos\phi_s\beta^2 U_s}}, \qquad (7.94)$$

which shows that the bunch length shortens as the energy is adiabatically increased, if the rf voltage and synchronous phase are kept constant. More obviously, the bunch length will lengthen if the voltage is decreased to zero.

In the small oscillation case, the phase taken up by the bunch is

$$\Delta\varphi_b = 2\varphi_m = \frac{2\pi h l_b}{L}, \qquad (7.95)$$

where l_b is the bunch length, and L is the circumference of the ring. Solving for φ_m and inserting this in the relation $I_L = \pi\varphi_m W_m$, yields

$$I_L = -\frac{\pi l_b}{2\beta c}\delta U = -\frac{\pi}{2}l_b\delta p, \qquad (7.96)$$

where δp is the momentum spread of the beam.

7.8 Bunch manipulations

Stationary rf buckets are required in order to capture injected particles and to manipulate bunches of particles. Such manipulations may require bunching or debunching a beam or rotating the bunches in the bucket to exchange phase-space area between the bunch length and momentum spread.

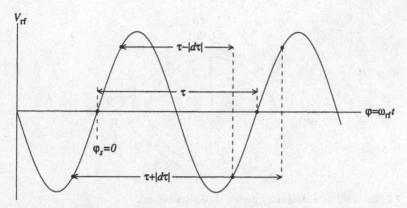

Figure. 7.6 Phase-stability below transition in stationary buckets.

The main feature of such buckets is that the synchronous particle **is not accelerated**, since

$$\sin \phi_s = 0, \qquad (7.97)$$

with ϕ_s either being 0 or π, according to whether the machine is below or above transition, respectively. All other particles oscillate around the synchronous one, but on average, keep the same energy, if the magnetic guide fields remain constant.

The phase-stability principle, in such a case, is illustrated in Fig. 7.6 for the case of $\eta_{tr} > 0$, below transition. The condition $\sin \phi_s = 0$ simplifies the phase-oscillation equation; in fact Eq. (7.39) becomes

$$\frac{1}{\Omega_s} \frac{d\phi}{dt} = \pm \sqrt{2(\cos \phi - \cos \phi_m)} \qquad (7.98)$$

having chosen $\pm \phi_m$ as the phase where the trajectory crosses the axis on the phase-space plot, i. e., with $\dot{\phi} = 0$.

The equation of the *separatrix*, with $\phi_m = \pi$, is

$$\frac{1}{\Omega_s} \frac{d\phi}{dt} = \pm \sqrt{2(1 + \cos \phi)} = \pm 2 \cos \frac{\phi}{2} \qquad (7.99)$$

as illustrated in Fig. 7.7.

Notice that synchrotron frequency for the stationary bucket is

$$\Omega_s = \omega_s \sqrt{\frac{h|\eta_{tr}|}{2\pi \beta^2 \gamma} \frac{V}{V_m}}, \qquad (7.100)$$

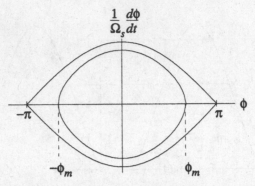

Figure. 7.7 Bunch within a stationary bucket (below transition).

where $V_m = mc^2/q$. For a stationary bucket it is possible to evaluate the bucket area analytically, which instead must be found numerically in the previous example of accelerated buckets.

Before proceeding, it is worth while to reintroduce the canonical variable W so that

$$\frac{1}{\Omega_s} \frac{d\phi}{dt} = \frac{2\pi c}{L} \sqrt{\frac{2\pi h^3 \eta_{tr}}{U_s q V \cos \phi_s}} W. \tag{7.101}$$

Combining this with the equation of the separatrix, yields

$$W = \pm \frac{L}{\pi c} \sqrt{\frac{qVU_s}{2\pi h^3 |\eta_{tr}|}} \cos \frac{\phi}{2}. \tag{7.102}$$

The area of a bucket is then

$$A_{bk} = 2 \int_{-\pi}^{\pi} W \, d\phi$$

$$= \frac{8L}{\pi c} \sqrt{\frac{qVU_s}{2\pi h^3 |\eta_{tr}|}}. \tag{7.103}$$

The phase oscillation equation, Eq. (7.98), may now be written as

$$W = \pm \frac{A_{bk}}{8} \sqrt{\cos^2 \frac{\phi}{2} - \cos^2 \frac{\phi_m}{2}}. \tag{7.104}$$

Setting $\phi = 0$, we may evaluate the heights of the bucket and the bunch as

$$W_{bk} = \frac{A_{bk}}{8}, \tag{7.105}$$

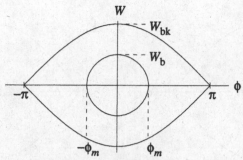

Figure. 7.8 Bunch and stationary bucket in the (ϕ, W) plane.

and

$$W_{\rm b} = \frac{A_{\rm bk}}{8} \sin \frac{\phi_m}{2}, \qquad (7.106)$$

respectively. (See Fig. 7.8.) These two equations allow us to evaluate the required voltage of the crest to obtain a bunch of a certain length. For a bunch with a momentum spread of $\pm\delta p$,

$$|W_{\rm b}| = \frac{\beta^2 U_s}{\omega_{\rm rf}} \frac{\delta p}{p}. \qquad (7.107)$$

Recalling Eq. (7.95),

$$\frac{L}{\pi c} \sqrt{\frac{qVU_s}{2\pi h^3 |\eta_{\rm tr}|}} = \frac{\beta^2 U_s}{\omega_{\rm rf} \sin(\pi h l_b/2L)} \frac{\delta p}{p}, \qquad (7.108)$$

which can be manipulated to give

$$V = \frac{\pi}{2} \frac{\beta^2 \gamma h |\eta_{\rm tr}|}{\sin^2(\pi h l_b/2L)} \left(\frac{\delta p}{p}\right)^2 V_m. \qquad (7.109)$$

For a given $\delta p/p$, a shorter bunch length requires a higher voltage, and the simple scaling relation,

$$\frac{V}{h} \sin^2 \frac{\pi h l_b}{2L} = \text{constant}, \qquad (7.110)$$

can be used.

Notice also that the maximum momentum spread accepted by a bucket is

$$\left(\frac{\delta p}{p}\right)_{\rm Max} = \sqrt{\frac{2V}{\pi \beta^2 \gamma h |\eta_{\rm tr}| V_m}}. \qquad (7.111)$$

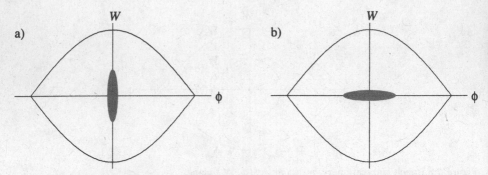

Figure. 7.9 a) Injected bunch with a large momentum spread and small phase spread. b) Rotated bunch with a small momentum spread and large phase spread.

For small phases, Eq. (7.104) becomes

$$W \simeq \pm \frac{1}{16} A_{\text{bk}} \sqrt{2(1 - \frac{1}{2}\phi^2 - 1 + \frac{1}{2}\phi_m^2)} = \frac{A_{\text{bk}}}{16} \sqrt{\phi_m^2 - \phi^2}, \qquad (7.112)$$

or

$$\left(\frac{16W}{A_{\text{bk}}\phi_m}\right)^2 + \left(\frac{\phi}{\phi_m}\right)^2 = 1. \qquad (7.113)$$

This is the equation of an ellipse whose area is

$$A_b = \frac{\pi}{16} A_{\text{bk}} \phi_m^2. \qquad (7.114)$$

If the rf voltage is switched off keeping the momentum spread $(\delta p/p)_m$ constant, the bunch diffuses, filling the whole bucket as ϕ_m slowly approaches π with the rate

$$\frac{d\phi}{dt} = -\omega_{\text{rf}}\eta_{\text{tr}}\left(\frac{\delta p}{p}\right)_m. \qquad (7.115)$$

In order to allow the beam to debunch into a continuous beam, we must wait somewhat longer than it takes for the bunch length to grow to the length of the rf bucket, i. e., for ϕ_m to grow at least as large as π. A reasonable minimum time for debunching, τ_d, can be estimated from

$$\tau_d > \frac{\pi}{-\frac{d\phi}{dt}} = \frac{\pi}{h\omega_s \eta_{\text{tr}}\left(\frac{\delta p}{p}\right)_m} = \frac{\tau_s}{2h\eta_{\text{tr}}\left(\frac{\delta p}{p}\right)_m}. \qquad (7.116)$$

If a bunch, short in time but high in momentum spread, is injected into a bucket, fulfilling the condition $W_b < W_{\text{bk}}$ as shown in Fig. 7.9, the bunch experiences suddenly a rather strong rf force and begins to rotate within the bucket. In fact, the crest voltage must be very high in order to match the bucket-height with the bunch momentum-spread.

After a quarter of a period of synchrotron oscillations,

$$\tau_{1/4} = \frac{1}{4}\left(\frac{2\pi}{\Omega_s}\right) = \frac{1}{4}\frac{L}{2\pi c}\sqrt{\frac{2\pi\beta^2\gamma}{h\eta_{tr}}\frac{V_m}{V}}, \qquad (7.117)$$

the bunch lies down in the bucket with a longer phase spread but exhibiting a reduced momentum spread[7], as shown in Fig. 7.9.

If at this instant the rf voltage is abruptly switched off, the bunch maintains the latter characteristics, and an actual reduction of the momentum spread is permanently achieved.

Problems

7–1 Show that for a ring with η_{tr} = constant, the following scaling relations hopefully apply:

$$\varphi_m(U) = \left(\frac{U_i}{U}\right)^{\frac{1}{4}}\varphi_m(U_i),$$

$$W_m(U) = \left(\frac{U}{U_i}\right)^{\frac{1}{4}}W_m(U_i), \quad \text{and}$$

$$\left(\frac{\Delta p}{p}\right) = \left[\frac{(\gamma_i^2 - 1)^2\gamma}{(\gamma^2 - 1)^2\gamma_i}\right]^{\frac{1}{4}}\left(\frac{\Delta p}{p}\right)_i,$$

where the subscript i indicates the initial conditions.

7–2 Consider a ring with a thin rf cavity whose linear transfer matrix just after the cavity is given by

$$\mathbf{M} = \begin{pmatrix} C & S & 0 & D \\ C' & S' & 0 & D' \\ E & F & 1 & G \\ 0 & 0 & 0 & 1 \end{pmatrix}\begin{pmatrix} 1 & 0 & 0 & 0 \\ 0 & 1 & 0 & 0 \\ 0 & 0 & 1 & 0 \\ 0 & 0 & Q & 1 \end{pmatrix}$$

Show that the dispersion functions are still given by

$$\eta = \frac{(1 - S')D + SD'}{2(1 - \cos\mu)}, \quad \text{and} \quad \eta' = \frac{(1 - C)D' + C'D}{2(1 - \cos\mu)}.$$

Hint: The eigenvector equation of Eq. (5.86) must be modified to allow for momentum compaction:

$$\mathbf{M}\begin{pmatrix} \eta \\ \eta' \\ 0 \\ 1 \end{pmatrix}\delta = \begin{pmatrix} \eta \\ \eta' \\ 0 \\ 1 \end{pmatrix}\delta + \begin{pmatrix} 0 \\ 0 \\ \Delta L \\ 0 \end{pmatrix}.$$

7–3

a) Calculate the synchrotron tune for RHIC for fully stripped $^{197}\text{Au}^{79+}$ (gold ions)

$$\gamma_{\text{inj}} = 10.4$$
$$\gamma\text{tr} = 22.8$$
$$L = 3834 \text{ m}$$
$$h = 360$$
$$\phi_s = 0°$$
$$mc^2 = 197 \times 0.93113 \text{ GeV}$$
$$Z = 79 \quad (\text{protons})$$
$$A = 179 \quad (\text{neutrons + protons})$$
$$V_{\text{rf}} = 300 \text{ kV}$$

b) What is the synchrotron frequency?

c) For a synchronous phase of $\phi_s = 5.5°$, how much energy does the synchronous particle gain per turn?

d) How long would it take to accelerate to $\gamma = 107.4$ (100 GeV/nucleon)? Assume that the phase jump at transition has been performed correctly (i.e., ignore it).

e) Plot the synchrotron frequency as a function of energy.

References for Chapter 7

[1] E. Persico, E. Ferrari, and S. E. Segre, *Principles of Particle Accelerators*, W. A. Benjamin, New York (1968).

[2] H. Bruck, *Accélérateurs circulaires de particules*, Presses Universitaires de France, Paris (1966).

[3] J. Le Duff, "Longitudinal beam dynamics in circular accelerators", *CERN Accelerator School General Accelerator Physics*, at Gif-sur-Yvette, CERN 85-19, Eds: P. Bryant and S. Turner, CERN, Geneva (1985).

[4] M. S. Livingston and J. P. Blewett, *Particle Accelerators*, McGraw-Hill, New York (1962).

[5] A. A. Kolomensky and A. N. Lebedev, *Theory of Cyclic Accelerators*, North-Holland, Amsterdam (1966).

[6] T. Suzuki, "Synchrobetatron Resonance Driven by Dispersion in rf Cavities", *Particle Accelerators*, <u>18</u>, 115 (1985).

[7] J. Griffin, J. MacLachlan, A. G. Ruggiero, and K. Takayama, "Time and Momentum Exchange for Production and Collection of Intense Antiproton Beams at Fermilab", IEEE Trans. on Nucl. Sci., Vol. NS-30, No. 4, 2630(1983).

8

Synchrotron Radiation

Synchrotron radiation from accelerated charges has a large impact on particle motion in accelerators at ultrarelativistic energies. This is particularly true for electrons and positrons in many circular machines. For protons and heavier ions, the extremely small amount of synchrotron radiation can usually be neglected, at least for energies less than the SSC.

The effect of radiation clearly is a nonconservative interaction, and therefore the conditions for Liouville's theorem are not satisfied. It will be shown that the radiation can lead to a shrinking of the beam's volume in phase space. The radiation is not continuous, but consists of individual photons with random directions and momenta. These quantum fluctuations tend to increase the phase space volume and compete with the damping. This results in equilibrium values for the transverse and longitudinal sizes of the beam.

The radiation also tends to drive synchrotron oscillations. Since the longitudinal and horizontal motion are coupled, as demonstrated in the previous chapter, the radiation can be seen to drive the horizontal betatron oscillations.

8.1 Radiated power

A nonrelativistic charged particle which is being accelerated radiates power as electromagnetic radiation according to the Larmor formula,[1-7]

$$P_\gamma = \frac{q^2}{6\pi\epsilon_0 m^2 c^3} \left(\frac{d\vec{p}}{dt} \right)^2,$$ (8.1)

where $\vec{p}$ is the particle's momentum. In the nonrelativistic limit, where $v/c \to 0$, the square of the momentum derivative may be written in terms of a four-vector using the proper time, τ:

$$\left(\frac{d\vec{p}}{dt} \right)^2 = -\frac{dp^\mu}{d\tau} \frac{dp_\mu}{d\tau},$$ (8.2)

where we have used the implied summation over μ with the usual conventions of sign giving

$$p^\mu p_\mu = (U/c)^2 - (\vec{p})^2,$$ (8.3)

and τ is the proper time in the rest system of the particle.

Since Eq. (8.1) is written as a constant times the scalar product of two four-vectors, the total radiated power must be Lorentz invariant. Furthermore, in the lab frame,

$$\frac{dU}{dt} = \gamma^3 m\vec{v} \cdot \dot{\vec{v}}, \quad \text{and} \tag{8.4}$$

$$\frac{d\vec{p}}{dt} = \gamma m\dot{\vec{v}} + \gamma^3 m \left(\frac{\vec{v} \cdot \dot{\vec{v}}}{c}\right)\frac{\vec{v}}{c}, \tag{8.5}$$

so that

$$P_\gamma = \frac{q^2\gamma^4}{6\pi\epsilon_0 c^3}\left[\dot{\vec{v}}^2 + \gamma^2\left(\frac{\vec{v} \cdot \dot{\vec{v}}}{c}\right)^2\right]. \tag{8.6}$$

If the acceleration is in the same direction as the velocity,

$$P_\gamma = \frac{q^2\gamma^6}{6\pi\epsilon_0 c^3}\dot{\vec{v}}^2, \tag{8.7}$$

or by realizing that the accelerating force $F = dU/dz = \gamma^3 m\dot{v}$,

$$P_\gamma = \frac{q^2}{6\pi\epsilon_0 m^2 c^3}F^2. \tag{8.8}$$

This radiation from linear acceleration is extremely small, e. g., an accelerating gradient of 100 MV/m would cause an electron to radiate power of about 1.1keV/s, which for a 1 TeV accelerator of 10 km length, the total radiated power would be less than 0.04 eV.

For the case of circular motion with very small changes in energy, such as in electron rings, we may assume that $\dot{\vec{v}}$ is perpendicular to $\vec{v}$, and

$$P_\gamma = \frac{q^2\gamma^4}{6\pi\epsilon_0 c^3}\dot{\vec{v}}^2, \quad \text{with} \quad \dot{\vec{v}} = \frac{qvB}{\gamma m} \tag{8.9}$$

for motion in a transverse magnetic field. The instantaneous radiated power for an electron can now be written in the form,

$$P_\gamma = \frac{2}{3}r_e mc^3\frac{\gamma^4\beta^4}{\rho^2}, \tag{8.10}$$

where the classical radius of the electron is given by

$$r_e = \frac{q^2}{4\pi\epsilon_0 mc^2} = 2.82 \times 10^{-15} \text{ m}, \tag{8.11}$$

and $\rho = p/qB$ is the usual cyclotron radius. The radiation is negligible unless the particles are ultrarelativistic, so we may make the approximation that $\beta = 1$ for the rest of this chapter. It is important to realize that the radiation from heavier particles is greatly reduced; a useful proportionality is

$$P_\gamma \propto \left(\frac{q}{m}\right)^4 U^2 B^2. \tag{8.12}$$

Obviously the radiation from a proton of the same energy as an electron in identical magnetic fields will be suppressed by the ratio of masses to the fourth power, i. e., about 9×10^{-14} smaller!

In an electron synchrotron, the particles loose energy as they traverse the bending magnets, and the lost energy must be added back by means of rf cavities. The amount of energy lost in one turn can be obtained by integrating the radiated power through one turn,

$$U_\gamma = \oint \frac{P_\gamma}{c}\, ds. \tag{8.13}$$

Generally $U_\gamma \ll U$, and

$$U_\gamma \simeq \frac{C_\gamma U^4}{2\pi} \oint \frac{ds}{\rho^2}, \tag{8.14}$$

where

$$C_\gamma = \frac{4\pi}{3} \frac{r_e}{(mc^2)^3} = 8.85 \times 10^{-5} \frac{m}{(GeV)^3}. \tag{8.15}$$

The integration, in effect, can be performed just over the bending magnets, and if all dipole magnets have identical curvatures, $U_\gamma = C_\gamma U^4/\rho$. For example, a 5 GeV electron in an accelerator, whose dipoles have 90m bending radii, will loose about 0.6 MeV per revolution. Other useful formulations are

$$U_\gamma = C_\rho \frac{\gamma^4}{\rho} = C_B B \gamma^3, \quad \text{with} \tag{8.16}$$

$$C_\rho = \frac{q^2}{3\epsilon_0} = 6.03 \text{ neVm}, \quad \text{and} \tag{8.17}$$

$$C_B = \frac{q^3}{3\epsilon_0 mc} = 3.54 \ \mu\text{eVT}^{-1}. \tag{8.18}$$

8.2 Radiation damping of energy oscillations

Simplifying the notation of the last chapter, the deviation of energy of the particle in question from that of the synchronous particle will be written as

$$u = U - U_s. \tag{8.19}$$

With $\beta = 1$ and for small oscillations, the longitudinal equations of motion now become

$$\frac{du}{dt} \simeq \frac{\omega_s}{2\pi} qV \cos\phi_s\, \varphi - \frac{1}{\tau_s}[U_\gamma(U_s + u) - U_\gamma(U_s)], \quad \text{and} \tag{8.20}$$

$$\frac{d\varphi}{dt} \simeq -\frac{\omega_{\mathrm{rf}}\eta_{\mathrm{tr}}}{U_s} u, \tag{8.21}$$

where the energy lost in a turn has been included in the difference equation. The radiated energy is a function of the particles energy and may be expanded in a Taylor series about U_s:

$$U_\gamma(U) = U_\gamma(U_s) + \left(\frac{dU_\gamma}{dU}\right)_s u + \dots. \tag{8.22}$$

Combining the last three equations yields

$$\frac{d^2u}{dt^2} + \frac{\omega_s}{2\pi}\left(\frac{dU_\gamma}{dU}\right)_s \frac{du}{dt} + \Omega_s^2 u = 0, \quad \text{with} \tag{8.23}$$

$$\Omega_s = \omega_s\sqrt{\frac{qVh\eta_{\mathrm{tr}}\cos\phi_s}{2\pi U_s}} \tag{8.24}$$

again, as in §7.4. The general solution of the damped oscillator equation can be written as

$$u(t) = u_0 e^{-t/\tau_u}\sin(\Omega_s' t + \psi_0), \quad \text{with} \tag{8.25}$$

$$\Omega_s' = \sqrt{\Omega_s^2 - \frac{1}{\tau_u^2}}, \quad \text{and} \tag{8.26}$$

$$\frac{1}{\tau_u} = \frac{1}{2\tau_s}\left(\frac{dU_\gamma}{dU}\right)_s, \tag{8.27}$$

and where u_0 and ψ_0 depend on the initial conditions.

The derivative, $(dU_\gamma/dU)_s$, can be calculated from the formula,

$$U_\gamma = \oint P_\gamma \frac{dt}{ds} ds = \frac{1}{c} \oint P_\gamma \left(1 + \frac{\eta}{\rho} \frac{u}{U_s}\right) ds, \qquad (8.28)$$

on account of the particle having a velocity,

$$c = \frac{d\sigma}{dt} = \left(1 + \frac{\eta}{\rho} \frac{u}{U_s}\right) \frac{ds}{dt}, \qquad (8.29)$$

as in §5.6, with η being the horizontal dispersion function. The damping is very slow compared to the betatron oscillations, so an average of the radial position, $x = \eta u/U_s$, may be used in the last equation. (This, of course, assumes that the synchrotron and horizontal betatron motions are not driving a coupling resonance, in which case, the beam will probably blow up, and this treatment becomes invalid.) Differentiating with respect to energy gives

$$\left(\frac{dU_\gamma}{dU}\right)_s = \frac{1}{c} \oint \left[\left(\frac{dP_\gamma}{dU}\right)_s + P_\gamma \frac{\eta}{\rho U_s} + \left(\frac{dP_\gamma}{dU}\right)_s \frac{\eta}{\rho} \frac{u}{U_s}\right] ds. \qquad (8.30)$$

The third term inside the integral is negligible since the synchrotron oscillations are also considerably faster than the damping; u may be effectively averaged over all possible phases, ψ_0.

Recalling Eq. (8.12), and realizing that on average, $dx/dU = \eta/U_s$, the derivative of P_γ becomes

$$\left(\frac{dP_\gamma}{dU}\right)_s = 2\frac{P_\gamma}{U_s} + 2\frac{P_\gamma}{\rho} \frac{\rho}{B_0} \left(\frac{\partial B_y}{\partial x}\right)_s \frac{\eta}{U_s}. \qquad (8.31)$$

Merging the last two equations produces

$$\left(\frac{dU_\gamma}{dU}\right)_s = \frac{U_\gamma}{U_s}(2 + \mathcal{D}), \quad \text{where} \qquad (8.32)$$

$$\mathcal{D} = \frac{1}{cU_\gamma} \oint P_\gamma \eta \frac{1 - 2n}{\rho} ds, \quad \text{and} \qquad (8.33)$$

$$n = -\frac{\rho}{B_0} \left(\frac{\partial B_y}{\partial x}\right)_0 \qquad (8.34)$$

is the field index.

Usually in modern electron rings, the bending magnets have minimal gradients, and the radiation in the quadrupoles is negligible, so that

$$\mathcal{D} \simeq \frac{1}{cU_\gamma} \oint P_\gamma \frac{\eta}{\rho} ds. \qquad (8.35)$$

The damping time for synchrotron oscillations is now

$$\tau_u = \frac{2}{2 + \mathcal{D}} \frac{U_s}{U_\gamma} \tau_s. \qquad (8.36)$$

8.3 Damping of the vertical oscillations

The synchrotron radiation is emitted in the general direction of the electron's motion. The distribution of directions can be reasonably approximated by a two dimensional random distribution of angles, relative to the particle's momentum, with a standard deviation of about $1/\gamma$. The quantum nature of the radiation will be examined in later sections of this chapter, but for the damping of betatron oscillations, it is only necessary to use the behavior of the radiation. This means that we consider the radiation to be continuous, rather than quantized, inside the bending magnets, and that the radiation is emitted along the direction of the particle's momentum. Since the particle is undergoing betatron oscillations, with $y' = p_y/p_z$, the radiation decreases the vertical component of momentum, p_y, along with the longitudinal component, p_z. The rf cavities, however, only add back momentum in the longitudinal direction, so that y' is smaller after passing through the rf section than it was just before the cavities. This is, in effect, like the adiabatic damping discussed in §5.4.

For simplicity, assume that the rf section is lumped in a single cavity as was done in Chapter 7. A particle leaving the cavity with momentum, p_0, will loose energy as it orbits, until it comes back around to the cavity with a momentum, $p_0 - U_\gamma/c$, at the entrance of the cavity. If the electron has a vertical divergence of

$$y_0' = \frac{p_{y0}}{p_0} \tag{8.37}$$

before a photon of momentum, δp, is emitted, then, after the emission, the divergence is still y_0'. The slope remains unchanged, since it has been emitted along the direction of motion, but the vertical component of momentum must now be

$$p_{y1} = y_0' p_0 \left(1 - \frac{\delta p}{p_0}\right). \tag{8.38}$$

With this approximation, the position and slope do not change every time a photon is emitted, but the momentum components are reduced. Integrating around the ring and stopping just in front of the cavity, the vertical component of momentum is

$$p_y \simeq p_0 y' \left(1 - \frac{U_\gamma}{U_s}\right), \tag{8.39}$$

and the longitudinal component is

$$p_z \simeq p_0 \left(1 - \frac{U_\gamma}{U_s}\right). \tag{8.40}$$

Now the rf kick changes only p_z to

$$p_z \simeq p_0 \left(1 - \frac{U_\gamma}{U_s}\right) \left(1 + \frac{U_\gamma}{U_s}\right), \tag{8.41}$$

so that to first order in U_γ/U_s after one complete revolution,

$$y'_{\mathrm{rf}} \simeq y' \left(1 - \frac{U_\gamma}{U_s}\right). \tag{8.42}$$

The trajectory of a particle may be written as

$$y(s) = A\sqrt{\beta_{\mathrm{V}}(s)} \cos\psi(s), \tag{8.43}$$

and the slope of the trajectory as

$$y'(s) = -\frac{A}{\sqrt{\beta_{\mathrm{V}}(s)}} [\sin\psi(s) + \alpha_{\mathrm{V}}(s) \cos\psi(s)], \tag{8.44}$$

where A and $\psi(0)$ may be obtained from the initial conditions. Recall from Chapter 5 the Courant-Snyder invariant,

$$\mathcal{W} = A^2 = \beta_{\mathrm{V}} y'^2 + 2\alpha_{\mathrm{V}} y y' + \gamma_{\mathrm{V}} y^2. \tag{8.45}$$

Converting this to a difference equation for A with changes in y and y', gives

$$\begin{aligned}
2A\Delta A + (\Delta A)^2 &= (A + \Delta A)^2 - A^2 \\
&= 2[\beta_{\mathrm{V}} y' \Delta y' + \alpha_{\mathrm{V}}(y\Delta y' + y'\Delta y) + \gamma_{\mathrm{V}} y\Delta y] \\
&\quad + [\beta_{\mathrm{V}}(\Delta y')^2 + 2\alpha_{\mathrm{V}}\Delta y\Delta y' + \gamma_{\mathrm{V}}(\Delta y)^2].
\end{aligned} \tag{8.46}$$

For calculating the damping rate, the quadratic deviations may be ignored if the initial value of A is much larger than the equilibrium value. We will consider the quadratic terms along with the quantum nature of the radiation in a later section when the equilibrium emittance is calculated.

In the approximation of continuous radiation exactly in the direction of the particle's motion, $\Delta y = 0$ always, and $\Delta y' = 0$ except at the rf cavity, where

$$\Delta y' = -\frac{U_\gamma}{U_s} y'. \tag{8.47}$$

Assuming that the damping time of the vertical oscillations is much slower than the revolution period, τ_s can conveniently be used as the time step for the difference equation, and

$$A\Delta A = -\frac{U_\gamma}{U_s}(\beta_V y'^2 + \alpha y y'). \tag{8.48}$$

In typical machines the synchrotron oscillations are much slower than the betatron oscillations $(Q_s/Q_H \ll 1)$, so we can average over all possible betatron phases. From Eqs. (8.43) and (8.44), we may calculate the averages:

$$\langle yy' \rangle = -A^2 \langle \cos\psi(\sin\psi + \alpha_V \cos\psi) \rangle = -A^2\frac{\alpha_V}{2}, \quad \text{and} \tag{8.49}$$

$$\langle y'^2 \rangle = A^2\frac{1}{\beta_V}\langle \sin^2\psi + 2\alpha_V \sin\psi\cos\psi + \alpha_V{}^2 \cos^2\psi \rangle = A^2\frac{\gamma_V}{2}. \tag{8.50}$$

The difference equation simplifies to

$$A\Delta A = -\frac{1}{2}A^2\frac{U_\gamma}{U_s}, \tag{8.51}$$

which can be transformed to a differential equation by dividing by the revolution period:

$$\frac{dA}{dt} \simeq \frac{\Delta A}{\tau_s} = -\left(\frac{1}{2\tau_s}\frac{U_\gamma}{U_s}\right) A. \tag{8.52}$$

The solution of this equation is just an exponential,

$$A = A_0 e^{-t/\tau_v}, \quad \text{with} \tag{8.53}$$

$$\tau_y = \frac{2\tau_s U_s}{U_\gamma} = 2\frac{U_s}{\langle P_\gamma \rangle}, \tag{8.54}$$

where $\langle P_\gamma \rangle$ is the radiated power averaged over one full revolution.

8.4 Damping of the horizontal oscillations

The procedure for calculating the damping of the horizontal betatron oscillations is similar to that of the previous section, but is complicated by the dispersion, η, since the momentum of the electron changes as energy is radiated.

In the horizontal plane, the trajectory for a particle may be written as

$$x(s) = x_\beta(s) + \eta(s)\frac{u}{U_s}, \tag{8.55}$$

and the corresponding slope as

$$x'(s) = x'_\beta(s) + \eta'(s)\frac{u}{U_s}, \quad \text{where} \tag{8.56}$$

$$x_\beta(s) = A\sqrt{\beta_{\mathrm{H}}(s)}\cos\psi(s), \quad \text{and} \tag{8.57}$$

$$x'_\beta(s) = -\frac{A}{\sqrt{\beta_{\mathrm{H}}(s)}}[\sin\psi(s) + \alpha_{\mathrm{H}}(s)\cos\psi(s)]. \tag{8.58}$$

Now when a photon of energy, dU_γ, is emitted, $\Delta x = 0$, and $\Delta x' = 0$, similar to Δy and $\Delta y'$ in the previous section; however, there will be discontinuous steps in both x_β and x'_β, so that

$$d(\Delta x_\beta) = -\eta\frac{-dU_\gamma}{U_s}, \quad \text{and} \tag{8.59}$$

$$d(\Delta x'_\beta) = -\eta'\frac{-dU_\gamma}{U_s}, \tag{8.60}$$

where the extra minus signs are due to the recoil of the electron. The amount of energy radiated in a length ds depends on the horizontal coordinate since both dt/ds and B can depend on x. Remembering that $P_\gamma \propto U^2 B^2$,

$$\begin{aligned}
\frac{dU_\gamma}{ds} &= P_\gamma\left(1 + \frac{2}{B}\frac{\partial B}{\partial x}x\right)\frac{1}{c}\left(1 + \frac{x}{\rho}\right), \\
&= \frac{P_\gamma}{c}\left(1 - \frac{2n}{\rho}x\right)\left(1 + \frac{x}{\rho}\right).
\end{aligned} \tag{8.61}$$

Much the same as Eq. (8.46), a difference equation for the horizontal betatron amplitude coefficient, A, can be written as

$$\begin{aligned}
2A\,d(\Delta A) + [d(\Delta A)]^2 &= [A + d(\Delta A)]^2 - A^2 \\
&= 2\{\beta_{\mathrm{H}}x'_\beta d(\Delta x'_\beta) + \alpha_{\mathrm{H}}[x_\beta d(\Delta x'_\beta) + x'_\beta d(\Delta x_\beta)] + \gamma_{\mathrm{H}}x_\beta d(\Delta x_\beta)\} \\
&\quad + \{\beta_{\mathrm{H}}[d(\Delta x'_\beta)]^2 + 2\alpha_{\mathrm{H}}d(\Delta x_\beta)d(\Delta x'_\beta) + \gamma_{\mathrm{H}}[d(\Delta x_\beta)]^2\},
\end{aligned} \tag{8.62}$$

for a small increase, $d(\Delta A)$, from an emission in length, ds. Again, the terms with quadratic deviations may be ignored. Assuming that the damping is much slower than both the longitudinal and horizontal oscillations, and that the two oscillations are uncorrelated, * then for each emission, dU_γ, the value of $A\,d(\Delta A)$

---

* That is, the accelerator is not running at a synchro-betatron resonance.

can be averaged over both the betatron and synchrotron phases. These infinitesimal contributions to ΔA can be integrated around the ring to obtain the total difference, $(\Delta A)_{\text{rad}}$, due to radiation in one turn.

In averaging $A\,d(\Delta A)$ over the betatron phase, only terms with quadratic forms of $d(\Delta x_\beta)$ and $d(\Delta x'_\beta)$ are nonzero. Thus Eq. (8.62) becomes

$$A\,d(\Delta A) = \frac{P_\gamma}{U_s c} \left\langle \left[\frac{1-2n}{\rho} x_\beta - \frac{4n\eta}{\rho^2} \frac{u}{U_s} x_\beta \right] \right.$$
$$\left. \times \left[\beta_H \eta' x'_\beta + \alpha_H(\eta' x_\beta + \eta x'_\beta) + \gamma_H \eta x_\beta \right] \right\rangle ds. \qquad (8.63)$$

The term with u to the first power drops out after averaging over the synchrotron phase, so that

$$A\,d(\Delta A) = \frac{P_\gamma}{U_s c} \frac{1-2n}{\rho} [(\beta_H \eta' + \alpha_H \eta)\langle x_\beta x'_\beta \rangle + (\gamma_H \eta + \alpha_H \eta')\langle x_\beta^2 \rangle] ds. \qquad (8.64)$$

Bearing in mind Eqs. (8.57) and (8.58), this simplifies to

$$d(\Delta A) = \frac{A}{2} \frac{P_\gamma \eta}{U_s c} \frac{1-2n}{\rho} ds, \quad \text{since} \qquad (8.65)$$

$$\langle x_\beta x'_\beta \rangle = -\frac{A^2 \alpha_H}{2}, \quad \text{and} \quad \langle x_\beta^2 \rangle = \frac{A^2 \beta_H}{2}. \qquad (8.66)$$

The total contribution due to radiation is then arrived at by integrating around the ring and using Eq. (8.33);

$$(\Delta A)_{\text{rad}} = \frac{A}{2} \frac{U_\gamma}{U_s} \mathcal{D}. \qquad (8.67)$$

Now the contribution from acceleration through the rf cavity must be calculated. If we have

$$x_1 = x_{\beta 1} + \eta \frac{u_1}{U_s}, \quad \text{and} \qquad (8.68)$$

$$x'_1 = x'_{\beta 1} + \eta' \frac{u_1}{U_s}, \qquad (8.69)$$

just in front of the cavity, then the values just after the cavity will be

$$x_2 = x_{\beta 2} + \eta \frac{u_2}{U_s} = x_1, \quad \text{and} \qquad (8.70)$$

$$x'_2 = x'_{\beta 2} + \eta' \frac{u_2}{U_s} = \left(1 - \frac{U_\gamma}{U_s}\right) x'_1, \qquad (8.71)$$

where $u_2 = u_1 + U_\gamma$. The contribution to Δx_β from the rf cavity is just

$$(\Delta x_\beta)_{\text{rf}} = x_{\beta 2} - x_{\beta 1} = -\eta \frac{U_\gamma}{U_s}, \tag{8.72}$$

and the contribution to $\Delta x'_\beta$ is

$$(\Delta x'_\beta)_{\text{rf}} = x'_{\beta 2} - x'_{\beta 1} \simeq -\frac{U_\gamma}{U_s} x'_{\beta 1} - \eta' \frac{U_\gamma}{U_s} - \eta' \frac{U_\gamma u_1}{U_s^2}, \tag{8.73}$$

(If the acceleration section of the ring is long, then an integral should be performed, as in the case of the radiation component of ΔA.) Substituting these into the difference equation and averaging over phases leaves only

$$A(\Delta A)_{\text{rf}} = -\frac{U_\gamma}{U_s} [\beta_{\text{H}} \langle x'^2_\beta \rangle + \alpha_{\text{H}} \langle x_\beta x'_\beta \rangle]$$

$$= -\frac{U_\gamma}{U_s} \frac{A^2}{2}. \tag{8.74}$$

Combining $(\Delta A)_{\text{rad}}$ with $(\Delta A)_{\text{rf}}$ and dividing by τ_s gives

$$\frac{dA}{dt} = \frac{\Delta A}{\tau_s} = -\frac{U_\gamma}{2U_s \tau_s}(1 - \mathcal{D})A, \tag{8.75}$$

having a solution of the form:

$$A = A_0 e^{-t/\tau_x}, \quad \text{with} \tag{8.76}$$

$$\tau_x = \frac{2}{1 - \mathcal{D}} \frac{U_s}{U_\gamma} \tau_s,$$

for the damping time of horizontal betatron oscillations.

8.5 Damping partition numbers

The damping times, Eqs. (8.36, 8.54, and 8.76), all contain the factor,

$$\tau_0 = \frac{2U_s}{U_\gamma} \tau_s, \tag{8.77}$$

and may be expressed collectively as

$$\tau_i = \frac{\tau_0}{J_i}, \tag{8.78}$$

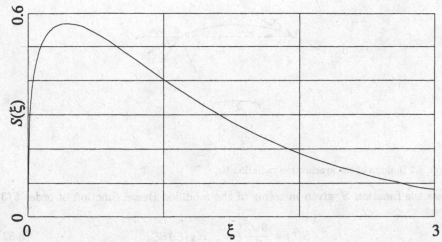

Figure. 8.1 A plot of the function, $S(\xi)$.

where i is any of the three dimensions, x, y, or u, and

$$J_x = 1 - \mathcal{D}, \quad J_y = 1, \quad \text{and} \quad J_u = 2 + \mathcal{D}. \tag{8.79}$$

Notice that $J_x + J_y + J_u = 4$, and $J_x + J_u = 3$, so the sum of the damping rates of all three dimensions is

$$\frac{1}{\tau_x} + \frac{1}{\tau_y} + \frac{1}{\tau_u} = \frac{4}{\tau_0}. \tag{8.80}$$

The time, τ_0, only depends on the beam energy and energy loss per turn, whereas the J_i's depend also on the focusing and coupling of the three motions. In effect, the J_i's divide the quantity $4/\tau_0$ into the individual damping rates; hence the J_i's are called partition numbers.

8.6 Frequency spectrum of the radiation

Before continuing with the radiation excitation calculations, it is necessary to recall some properties about synchrotron radiation. Much of what is mentioned in this section will be given without proof, since excellent treatments are given in standard texts on electrodynamics.[1-6] For the student wishing a profound knowledge of synchrotron radiation, including a quantum mechanical treatment, the book[7] by Sokolov and Ternov is recommended.

The amount of power radiated in the frequency interval $d\omega$ is

$$dP_\gamma = \frac{P_\gamma}{\omega_c} S\left(\frac{\omega}{\omega_c}\right) d\omega, \tag{8.81}$$

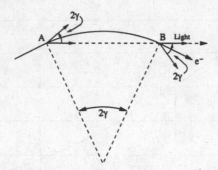

Figure. 8.2 Buildup of the synchrotron radiation fan.

where the function S, given in terms of the modified Bessel function of order 5/3, is

$$S(\xi) = \frac{9\sqrt{3}}{8\pi}\xi \int_{\xi}^{\infty} K_{\frac{5}{3}}(\xi')\,d\xi', \tag{8.82}$$

with normalization

$$\int_{0}^{\infty} S(\xi)d\xi = 1. \tag{8.83}$$

A plot of the function $S(\xi)$ is shown in Fig. 8.1. The critical frequency* is defined by

$$\omega_c = \frac{3c}{2\rho}\gamma^3, \tag{8.84}$$

so that the critical photon energy is $u_c = \hbar\omega_c$. It can be shown[16] that half the power is radiated at energies less than the critical energy, and the other half, above.

Synchrotron radiation is emitted over quite a broad spectrum of frequencies whose band-width can be estimated in the following way. It can be shown that most of the radiation is emitted within quite a narrow cone of aperture $2/\gamma$; therefore an electron which travels along an arc of length $2\rho/\gamma$ (see Fig. 8.2) emits the first photon from its right side, i.e. from point A, and the last photon from its left side, i.e. from point B. The resulting flash of light will last a time

$$\delta t = \Delta t_e - \Delta t_\gamma, \tag{8.85}$$

with electron's time of flight given by

$$\Delta_e = \frac{2\rho}{\beta\gamma c} = \frac{2\rho}{c}(\gamma^2 - 1)^{-1/2} \simeq \frac{2\rho}{c}\left(\frac{1}{\gamma} + \frac{1}{2\gamma^3} + \cdots\right), \tag{8.86}$$

---

* Note that in his first[2] and second[3] editions, Jackson defines the critical frequency as being a factor of two larger. In his third edition[4] he uses the more common normalization of Schwinger[10] as do we.

and the time of flight of the light by

$$\Delta_\gamma = \frac{2\rho}{c} \sin \frac{1}{\gamma} \simeq \frac{2\rho}{c} \left(\frac{1}{\gamma} - \frac{1}{6\gamma^3} + \cdots \right). \tag{8.87}$$

The difference in time is then

$$\delta t \simeq \frac{4}{3} \frac{\rho}{c\gamma^3}, \tag{8.88}$$

and we see that typical radiated frequencies will scale as

$$\omega \propto \frac{1}{\delta t} \propto \frac{\rho}{\gamma^3}. \tag{8.89}$$

By using Nyquist's theorem we would expect the central frequency to be near

$$f \simeq \frac{1}{2\delta t} = \frac{3c\gamma^3}{8\rho} = \frac{\pi}{4} \frac{\omega_c}{2\pi}. \tag{8.90}$$

Photons are emitted in the energy range from u_γ to $u_\gamma + du_\gamma$ at the rate

$$n_\gamma(u_\gamma) du_\gamma = \frac{P_\gamma}{c} F \left(\frac{u_\gamma}{u_c} \right), \tag{8.91}$$

with $F(\xi) = S(\xi)/\xi$, whose normalization is given by

$$\int_0^\infty F(\xi) d\xi = \frac{15\sqrt{3}}{8}. \tag{8.92}$$

The total number of photons emitted per second can be calculated by integrating Eq. (8.91) over all energies:

$$N_\gamma = \int_0^\infty n_\gamma(u_\gamma) du_\gamma = \frac{5}{2\sqrt{3}} \frac{\alpha_f c}{\rho} \gamma, \tag{8.93}$$

having written α_f for the fine structure constant. Remarkably, this gives the mean number of photons emitted per radian of orbit to be

$$N_r = \frac{5\alpha_f}{2\sqrt{3}} \gamma, \tag{8.94}$$

which only depends on the energy of the electron.

In addition to the total number of photons per second, the first and second moments of the distribution $n_\gamma(u_\gamma)$ will also be needed:

$$\langle u_\gamma \rangle = \frac{1}{N_\gamma} \int_0^\infty u n_\gamma(u) du = \frac{8}{15\sqrt{3}} u_c \simeq 0.32 u_c; \tag{8.95}$$

$$\langle u_\gamma^2 \rangle = \frac{1}{N_\gamma} \int_0^\infty u^2 n_\gamma(u) du = \frac{11}{27} u_c^2 \simeq 0.41 u_c^2; \tag{8.96}$$

Table I lists values of U_s, $\langle u_\gamma \rangle$, N_γ, and N_r for several electron storage rings.

Table I. Estimate of radiation for several rings.

Machine	U_s[GeV]	$\langle u_\gamma \rangle$[eV]	N_γ[s^{-1}]	N_r
Adone	1.5	4.6×10^2	1.8×10^9	31
SPEAR	4.5	4.9×10^3	2.2×10^9	93
CESR	5.0	5.0×10^2	4.0×10^8	160
PETRA	19	2.5×10^4	6.1×10^8	390
LEP I	50	1.4×10^4	5.0×10^7	1030
LEP II	100	1.1×10^5	1.0×10^8	2060

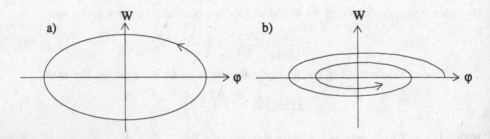

Figure. 8.3 Longitudinal motion. a) A particle travels on a closed path in side the bucket, if there is no radiation. b) With continuous damping and no quantum fluctuations, the particle spirals in towards the synchronous particle. (Here the depiction of damping is greatly exaggerated.)

8.7 Energy spread and bunch length

The core of the beam is usually much smaller than the size of the rf bucket, so the linear approximation of the rf kicks is reasonable. Recall, from §7.4, the canonical longitudinal phase space variables, φ, and

$$W = -\frac{u}{\omega_{\rm rf}} = -\frac{U_s}{\omega_{\rm rf}^2 \alpha_p} \frac{d\varphi}{dt}, \tag{8.97}$$

where the phase slip factor has been replaced by $-\alpha_p$ in the ultrarelativistic limit. In most electron rings, α_p is positive, and u will have the same sign as $d\varphi/dt$.

For a case with no radiation, a particle in the linear region of the bucket will have an elliptical trajectory about the synchronous particle, like that shown in Fig. 8.3a. With smooth radiation and no quantum fluctuations, the particle spirals towards the fixed point of the synchronous particle (see Fig. 8.3b).

In reality the radiation is not continuous, but consists of individual photons emitted at random times with varying energies. The rf phase, φ, does not appre-

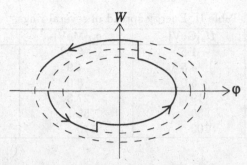

Figure. 8.4 Effect of quantum fluctuations on the longitudinal oscillations. A fluctuation makes the particle jump from one ellipse to another. The phase does not change with the emanation of a quantum, but the energy is decreased by the energy of the photon. (Note that $W = -u/\omega_{\mathrm{rf}}$, so that lower energy corresponds to more positive W.)

ciably change during the emission of a photon, since the longitudinal position of the electron must not have any discontinuities. Consider an electron moving on a trajectory with energy given by

$$u = -\omega_{\mathrm{rf}} W = A \cos[\Omega_s(t - t_0)], \qquad (8.98)$$

and rf phase by,

$$\varphi = \frac{\omega_{\mathrm{rf}} \alpha_p}{U_s \Omega_s} A \sin[\Omega_s(t - t_0)], \qquad (8.99)$$

before the emission of a photon at time, t_1. For simplicity, define the rescaled phase

$$\xi = \frac{U_s \Omega_s}{\omega_{\mathrm{rf}} \alpha_p} \varphi. \qquad (8.100)$$

Then, without radiation, the equivalent of the Courant-Snyder invariant for longitudinal motion may be written as

$$A^2 = \xi^2 + u^2. \qquad (8.101)$$

When a photon of energy u_γ is emitted, the change in A^2 will be

$$\Delta(A^2) = [(u - u_\gamma)^2 + \xi^2] - [u^2 + \xi^2]$$
$$= -2uu_\gamma + u_\gamma^2. \qquad (8.102)$$

Summing over all emissions around the ring will yield, on average,

$$\Delta(A^2) \simeq -2u \oint N_\gamma \langle u_\gamma \rangle \frac{ds}{c} + \oint N_\gamma \langle u_\gamma^2 \rangle \frac{ds}{c}. \qquad (8.103)$$

Table II. Energy spread of several rings.

Machine	U_s[GeV]	σ_u[MeV]	σ_u/U_s
Adone	1.5	0.86	5.7×10^{-4}
SPEAR	4.5	4.9	1.1×10^{-3}
CESR	5.0	3.0	5.8×10^{-4}
PETRA	19	27.3	1.2×10^{-3}
LEP I	50	55	1.1×10^{-3}
LEP II	100	110	1.1×10^{-3}

The first term vanishes after averaging over the synchrotron oscillation phase, so that

$$\left(\frac{d\langle A^2\rangle}{dt}\right)_{QF} \simeq \frac{\Delta\langle A^2\rangle}{\tau_s} = \frac{1}{c\tau_s} \oint N_\gamma \langle U_\gamma^2\rangle ds \tag{8.104}$$

for the effect of the quantum fluctuations.

The effect of damping comes from the rf kicks to replace the radiated energy and may be calculated from

$$A = A_0 e^{-t/\tau_u}, \tag{8.105}$$

by squaring this equation and differentiating with respect to time, to establish

$$\left(\frac{d\langle A^2\rangle}{dt}\right)_{damping} = -\frac{2}{\tau_u} A^2. \tag{8.106}$$

Clearly the effects of damping and quantum excitation work in opposite directions; therefore, an equilibrium may be reached when the sum of the two rates is zero:

$$0 = -\frac{2}{\tau_u}\langle A^2\rangle + \frac{1}{c\tau_s} \oint N_\gamma \langle u_\gamma^2\rangle ds, \tag{8.107}$$

where $\langle A^2\rangle$ denotes the equilibrium value of A^2. The variance of the energy oscillation, $\sigma_u^2 = \frac{1}{2}\langle A^2\rangle$ since $\langle \cos^2\rangle = \frac{1}{2}$, so that

$$\sigma_u^2 = \frac{\tau_u}{4c\tau_s} \oint N_\gamma \langle u_\gamma^2\rangle ds. \tag{8.108}$$

It is interesting to note that σ_u only depends on the electron's energy and the magnet fields of the magnets — not on the rf voltage or frequency. Table II gives the energy spread for several accelerators.

The rms amplitude of the phase variable ξ is the same as for u, so that

$$\sigma_\varphi^2 = \left(\frac{\omega_{rf}\alpha_p}{U_s\Omega_s}\right)^2 \sigma_u^2.$$

This can be converted into an rms bunch size, σ_z, by realizing that

$$\sigma_z = \frac{\sigma_\varphi}{2\pi}\lambda_{\rm rf}, \tag{8.109}$$

where the rf wavelength is given by

$$\lambda_{\rm rf} = \frac{2\pi}{\omega_{\rm rf}}c. \tag{8.110}$$

The rms bunch length, l_b, can now be written as

$$l_b = 2\sigma_z = \frac{\alpha_p c}{U_s \Omega_s}\sigma_u, \tag{8.111}$$

remembering from Chapter 7 that

$$\frac{1}{\Omega_s^2} = \frac{2\pi U_s}{\omega_s \omega_{\rm rf} \alpha_p q V \cos\phi_s}. \tag{8.112}$$

Even though the energy spread does not depend on the rf voltage and frequency, the bunch length clearly does, since

$$\sigma_z \propto (\omega_{\rm rf} V \cos\phi_s)^{-\frac{1}{2}}. \tag{8.113}$$

8.8 Transverse excitations

In §8.4 it was shown that emission of a photon of energy, u_γ, results in a discontinuous shift of both x_β and x'_β. Fig. 8.5 shows a photon of momentum $\vec{k}_\gamma = u_\gamma/c$ being emitted by an electron of initial momentum $\vec{p}_i$. The final momentum of the electron after the emission of the photon is

$$\vec{p}_f = \vec{p}_i - \vec{k}_\gamma, \tag{8.114}$$

and the horizontal projection of the electron's recoil angle must be

$$\chi = -\frac{u_\gamma}{U_s}\theta_x. \tag{8.115}$$

The shifts in betatron coordinates from this emission then become

$$\delta x_\beta = \eta \frac{u_\gamma}{U_s}, \quad \text{and} \quad \delta x'_\beta = (\eta' - \theta_x)\frac{u_\gamma}{U_s}. \tag{8.116}$$

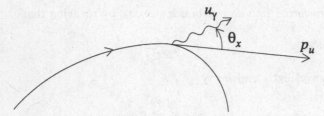

Figure. 8.5 A photon of energy, u_γ is emitted with a horizontal angular component, θ_x, relative to the direction of motion of the electron. The electron must recoil in order to preserve the total momentum.

As before, the incremental change in the Courant-Snyder invariant is

$$\Delta(A^2) = 2[\beta_{\rm H} x'_\beta \delta x'_\beta + \alpha_{\rm H}(x_\beta \delta x'_\beta + x'_\beta \delta x_\beta) + \gamma_{\rm H} x_\beta \delta x_\beta]$$
$$+ [\beta_{\rm H}(\delta x'_\beta)^2 + 2\alpha_{\rm H}(\delta x'_\beta)(\delta x_\beta) + \gamma_{\rm H}(\delta x_\beta)^2]. \qquad (8.117)$$

The first line of this formula is linear in the variables, u_γ and θ_x, and has already been included in the calculations of §8.4. The second line was previously ignored in the damping calculation and has quadratic terms in the radiation variables. After substituting Eq. (8.116), the extra increment in A^2 from the quantum emission is

$$\Delta(A^2) = \left(\frac{u_\gamma}{U_s}\right)^2 [\beta_{\rm H}\eta'^2 + 2\alpha_{\rm H}\eta\eta' + \gamma_{\rm H}\eta^2 + \beta_{\rm H}\theta_x^2 - 2(\beta_{\rm H}\eta' + \alpha_{\rm H}\eta)\theta_x]. \quad (8.118)$$

Averaging over all possible photons,

$$\langle\Delta(A^2)\rangle \simeq \frac{\langle u_\gamma^2 \rangle}{U_s^2}[\beta_{\rm H}\eta'^2 + 2\alpha_{\rm H}\eta\eta' + \gamma_{\rm H}\eta^2] + \frac{1}{\gamma^2}\frac{\langle u_\gamma^2 \rangle}{U_s^2}\beta_{\rm H}, \qquad (8.119)$$

since $\langle\theta_x\rangle = 0$, and $\langle\theta_x^2\rangle \simeq 1/\gamma^2$. The term proportional to γ^{-2} is many orders of magnitude smaller then the rest, so it may be neglected.

It is convenient to define the function,

$$\mathcal{H}(s) = \beta_{\rm H}(s)\eta'^2(s) + 2\alpha_{\rm H}(s)\eta(s)\eta'(s) + \gamma_{\rm H}(s)\eta^2(s), \qquad (8.120)$$

which only depends on the radius of curvature and focusing strengths around the ring. As before the $\Delta(A^2)$ must be added up around the ring and divided by the revolution time to obtain the excitation rate from the quantum fluctuations;

$$\left(\frac{d(A^2)}{dt}\right)_{\rm QF} \simeq \frac{1}{c\tau_s} \oint \frac{\langle u_\gamma^2 \rangle}{U_s^2}\mathcal{H}(s)N_\gamma ds. \qquad (8.121)$$

Since the damping of A can be written as

$$A = A_0 e^{-t/\tau_x}, \tag{8.122}$$

the rate of damping is

$$\left(\frac{d(A^2)}{dt}\right)_{\text{damping}} = -\frac{2}{\tau_x} A^2. \tag{8.123}$$

Once more, at equilibrium the excitation and damping rates must cancel, so

$$\langle A^2 \rangle = \frac{\tau_x}{2\tau_s c} \oint \frac{\langle u_\gamma^2 \rangle}{U_s^2} \mathcal{H}(s) N_\gamma ds. \tag{8.124}$$

The rms spread of the betatron oscillation may be found by squaring and averaging Eq. (8.57);

$$\sigma_{x_\beta}^2 = \frac{1}{2}\beta_{\text{H}}\langle A^2 \rangle = \beta_{\text{H}}\epsilon_{\text{rms}}, \tag{8.125}$$

where

$$\epsilon_{\text{rms}} = \frac{\tau_x}{4LU_s^2} \oint N_\gamma \langle u_\gamma^2 \rangle \mathcal{H}(s) ds. \tag{8.126}$$

All the particles in the beam have the same equilibrium value for A^2, but they populate a distribution which is roughly Gaussian; a given particle explores all amplitudes of oscillations due to the excitations and damping. The quantity $\pi\epsilon_{\text{rms}}$ is the area of the rms beam envelope in the $x_\beta x_\beta'$-plane. When dealing with electron storage rings, the emittance is defined as ϵ_{rms}, i. e., the area of the rms ellipse divided by π. This convention is different than that of Chapter 5.

For vertical motion, the treatment is similar to that for horizontal motion, except that in flat rings, there is no vertical dispersion, and only the term proportional to γ^{-2} in Eq. (8.119) is nonzero. However, there is always some amount of coupling between the vertical and horizontal betatron motion, so that the vertical emittance is much larger than would be calculated from this treatment. It is possible, but tedious, to use the treatment for coupled motion, as discussed in § 6.8, to find damping times and emittances for the normal mode coordinates, u and v. In typical operation the ratio of vertical to horizontal emittance, $\epsilon_{\text{V}}/\epsilon_{\text{H}}$, may range from a few percent in synchrotron light sources to 10–20% in e^+e^- colliders.

8.9 Beam lifetime considerations

Due to the coupling of radial and longitudinal motion, the rms beam size in the radial direction is really a convolution of two independent distributions; therefore,

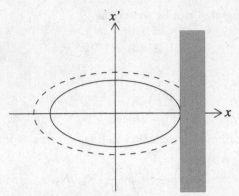

Figure. 8.6 A plot of the xx'-phase plane at a location where an obstruction scrapes the beam.

the overall rms amplitude is

$$\sigma_x = \sqrt{\beta_H \epsilon_H + \left(\eta \frac{\sigma_u}{U_s}\right)^2}. \tag{8.127}$$

First consider an accelerator with a very large beam pipe, but having a protrusion at an azimuthal position, s_1, where the dispersion is equal to zero, so that $x = x_\beta$. The protrusion scrapes off particles with

$$x > X = \sqrt{\beta_H E}, \tag{8.128}$$

as shown in Fig. 8.6. If there were no radiation, any particle outside the ellipse whose Courant-Snyder invariant is E would be scraped away, and any particle inside would survive. Without the obstruction the rate at which particles would cross from the inside to the outside of this ellipse must be equal to the rate of particles crossing from outside to inside.

As a simplifying measure, assume that $\alpha_H(s_1) = 0$, so that ellipse is oriented with its major and minor axes aligned along the x and x'-axes.* Then the density of particles in the thin ring bounded by E and $E + dE$ is given by

$$\frac{1}{\pi}\frac{dN}{dE} = \frac{N}{2\pi\sigma_x\sigma_{x'}} \exp\left[-\left(\frac{x^2}{2\sigma_x^2} + \frac{x'^2}{2\sigma_{x'}^2}\right)\right] = \frac{N}{2\pi\epsilon} \exp\left[-\frac{E}{2\epsilon}\right], \tag{8.129}$$

---

* This is not really necessary since the same density distribution in terms of E can be obtained with $\alpha_H \neq 0$ by using the results of Problem 5–4.

where N is the number of particles in the beam. Rearranging this and dividing by dt, the rate of particles crossing to the outside of the ellipse is

$$\frac{dN}{dt} = \frac{N}{2\epsilon} \exp\left(-\frac{E}{2\epsilon}\right) \frac{dE}{dt}, \qquad (8.130)$$

but the excitation rate of the Courant-Snyder invariant is just

$$\frac{dE}{dt} = -\frac{2}{\tau_x} E, \qquad (8.131)$$

so that the loss rate becomes

$$\frac{dN}{dt} = -\frac{NE}{\tau_x \epsilon} e^{-E/2\epsilon} = -\frac{N}{\tau_N}, \qquad (8.132)$$

and the lifetime of the beam is

$$\tau_N = \tau_x \frac{\epsilon}{E} e^{E/2\epsilon}. \qquad (8.133)$$

Since $E = X^2/\beta_H$, the worst places for beam loss would obviously be where β_H is large, e. g., in the center of a horizontally focusing quadrupole. Lifetimes calculated by Eq. (8.133) for single beam can be of the order of a few days; however, the actual measured lifetime is usually on the order of a few tens of hours. With colliding beams in e^+e^- colliders, the lifetime is frequently no more than a few hours. A more thorough treatment of lifetime due to quantum fluctuations should consider the effects of nonzero dispersion, and particles being lost in the vertical direction as well as migrating out of the stable rf bucket. The current can also be limited by the amount of available rf power, since the beam requires power of NU_γ/τ_s to remain at a constant energy.

Of course the whole calculation of beam lifetime is complicated by the nonlinearities in a real machine. Additionally, the lifetime can be reduced by collisions with residual gas in the vacuum chamber, and by collisions between different electrons within the beam.

Problems

8–1   For a separated function ring with identical dipole magnets of bending radius, ρ, show that

$$\mathcal{D} = \frac{L\alpha_p}{2\pi\rho},$$

where L is the circumference, and α_p is the momentum compaction.

8–2   Show that the damping time for the six dimensional phase space volume of a beam is just $\tau_0/8 = U_s/(4U_\gamma)$.

8–3   An achromatic bend (the double bend achromat) may be made from two dipoles with a horizontally focusing quadrupole between them. The transfer matrix through the achromat is of the form:

$$\mathbf{M} = \mathbf{B}(\theta)\mathbf{L}\left(\tfrac{1}{2}\mathbf{Q}\right)\left(\tfrac{1}{2}\mathbf{Q}\right)\mathbf{L}\mathbf{B}(\theta).$$

a) Use thin the lens approximation for quads and small angle approximation for bends to find the dispersion in the middle of the quad. Write the focal length in terms of the drift length and bend parameters.

$$\text{Hint:}\qquad \begin{pmatrix} \eta_c \\ 0 \\ 1 \end{pmatrix} = \left(\tfrac{1}{2}\mathbf{Q}\right)\mathbf{L}\mathbf{B}\begin{pmatrix} 0 \\ 0 \\ 1 \end{pmatrix}.$$

b) Show that the dispersion is again zero ($\eta = \eta' = 0$) after the bend.

8–4   A light source ring has eight equal double achromat bends (16 dipoles). Each dipole is 2.7 m long, and the circumference is 176 m. The energy of the beam is 2.5 GeV.

a) Calculate the critical energy of photons radiated in the dipoles.
b) Calculate the total energy lost per turn.
c) Calculate the momentum compaction of the ring.
d) Calculate the damping times τ_x, τ_y, and τ_u.

8–5   According to its design, the Large Hadron Collider (LHC) will be capable of accelerating protons to 7 TeV in each of two rings. The circumference will be 26.7 km, and the arc dipole field at 7 TeV will be 8.33 T.

a) Calculate the critical energy of photons.
b) Calculate the energy loss per turn per proton.
b) Calculate the total power radiated by synchrotron radiation for a beam with an average current of 0.56 A.

References for Chapter 8

[1] L. D. Landau and E. M. Lifshitz, *The Classical Theory of Fields*, Pergamon Press, Oxford (1975).

[2] J. D. Jackson, *Classical Electrodynamics*, John Wiley, New York, (1962).

[3] J. D. Jackson, *Classical Electrodynamics*, 2nd Ed., John Wiley, New York, (1975).

[4] J. D. Jackson, *Classical Electrodynamics*, 3rd Ed., John Wiley, New York, (1999).

[5] W. K. H. Panofsky and M. Phillips, *Classical Electricity and Magnetism*, Addison-Wesley, Reading, Mass. (1962).

[6] P. C. Clemmow and J. P. Dougherty, *Electrodynamics of Particles and Plasmas*, Addison-Wesley, Redwood City, CA (1990).

[7] A. A. Sokolov and I. M. Ternov, *Synchrotron Radiation*, Pergamon Press, Oxford (1968).

[8] M. Sands, "The Physics of Electron Storage Rings", SLAC Report 121 (1970).

[9] M. Sands, *Phys. Rev.*, $\underline{97}$, 470 (1955).

[10] J. Schwinger, *Phys. Rev.* $\underline{75}$, 1912 (1949).

[11] D. Neuffer, AIP Conf. Proc. $\underline{87}$, p. 45, New York (1982).

[12] A. Renieri, CERN 77-13, p. 82 (1977).

[13] K. Hübner, CERN 85-19, p. 226 (1985).

[14] K. Robinson, *Phys. Rev.*, $\underline{111}$, 373 (1958).

[15] R. Helm et al., *IEEE Trans. on Nucl. Sci.*, Vol. NS-20, #3, 900 (1973).

[16] G. K. Green, "Spectra and Optics of Synchrotron Radiation," BNL-50522, Brookhaven National Lab report (1976).

9

RF Linear Accelerators

One of the most important purposes of an accelerator is to increase the energy of particles in the beam. This requires having an electric field. Even though a plane wave of electromagnetic radiation carries momentum which can be imparted to the charged particles, this classical version of Compton scattering is much less efficient than the use of a longitudinal electric field. Classically the interaction can be roughly explained in two steps: first, the transverse electric field accelerates the particles in a direction perpendicular to the flow of momentum; second, the magnetic field, which is perpendicular to both the electric field and the momentum flow, bends the particles around in the general direction of propagation of the wave. This requires a much more intense field than would be needed for the same amount of acceleration from a longitudinal electric field. (See Problem 9–1.)

To have a longitudinal component of the electric field, the direction of propagation of the wave must not coincide with the direction of the beam. By placing some sort of reflectors for the wave, a net longitudinal component of electric field may be realized with the average direction of propagation coinciding with the beam's velocity. A simple waveguide would bring about such reflections; however, the phase velocity, v_{ph}, in a straight waveguide will be shown (in the next section) to be greater than the speed of light in vacuo. In order to accelerate over many rf cycles, the phase velocity of the rf wave must match the velocity of the particles in the beam.

9.1 Maxwell's equations and waves

It is assumed that the reader is familiar with the usual treatment (see, for example, any of Refs. 1–7) of electromagnetic waves and waveguides, so that only a brief review will be necessary. Maxwell's equations, written in their microscopic form, are

$$\nabla \times \vec{E} = -\frac{\partial \vec{B}}{\partial t}, \tag{9.1}$$

$$\nabla \times \vec{H} = \vec{J} + \frac{\partial D}{\partial t}, \tag{9.2}$$

$$\nabla \cdot \vec{D} = \rho_v, \quad \text{and} \tag{9.3}$$

$$\nabla \cdot \vec{B} = 0, \tag{9.4}$$

with the variables being

$\vec{E}$ — *electric field intensity,*

$\vec{H}$ — *magnetic field intensity,*

$\vec{D}$ — *electric flux density,*

$\vec{B}$ — *magnetic flux density,*

$\vec{J}$ — *electric current density,* and

ρ_v — *electric charge density.*

The microscopic version of Ohm's law,

$$\vec{J} = \sigma \vec{E}, \tag{9.5}$$

expresses the current density in terms of the conductivity, σ, and electric field. The electric and magnetic field variables are related by the permittivity, ϵ, and the permeability, μ, i. e.,

$$\vec{D} = \epsilon \vec{E}, \quad \text{and} \tag{9.6}$$

$$\vec{B} = \mu \vec{H}. \tag{9.7}$$

The boundary conditions on the fields at an interface between two media are:
1. $E_{\parallel}$ is continuous across the boundary.
2. $D_{\perp}$ is discontinuous by an amount equal to the surface charge density at the boundary.
3. $H_{\parallel}$ is discontinuous by the amount of surface current density at the boundary.
4. $B_{\perp}$ is continuous across the boundary.

For the rest of this chapter, no dielectric or magnetic materials will be considered, so that

$$\mu = \mu_0, \quad \text{and} \quad \epsilon = \epsilon_0. \tag{9.8}$$

Collin[4] has written a good treatment of microwave interactions with dielectrics and ferrites.

Maxwell's equations (9.1 and 9.2) may be combined to produce the electromagnetic wave equation

$$\nabla^2 \Psi = \mu \sigma \frac{\partial \Psi}{\partial t} + \mu \epsilon \frac{\partial^2 \Psi}{\partial t^2}, \tag{9.9}$$

with $\vec{B} = \mu \vec{H}$, $\vec{J} = \sigma \vec{E}$, and Ψ equal to either of $\vec{E}$ or $\vec{H}$.

For a wave of single frequency, the fields can be written in complex form to isolate the time dependence;

$$\Psi(\vec{r}, t) = \hat{\Psi}(\vec{r})e^{-i\omega t}. \tag{9.10}$$

The new time independent function, $\hat{\Psi}(\vec{r})$, is really just the temporal Fourier transform of $\Psi(\vec{r}, t)$. A superposition of such functions for different frequencies may be used to construct more complicated solutions, since Eq. (9.9) is linear in Ψ. The wave equation simplifies to Helmholtz's equation,

$$\nabla^2 \hat{\Psi} = \Gamma^2 \hat{\Psi}, \tag{9.11}$$

where

$$\Gamma^2 = (-\mu\epsilon\omega^2 + i\mu\sigma\omega). \tag{9.12}$$

With the conductivity sufficiently high, i. e., $\sigma \gg \omega\epsilon$, the charges inside the medium may readjust to keep the external electric field lines perpendicular to the boundary much more quickly than the rf oscillations. For example, in copper $\sigma \simeq 6 \times 10^7 \Omega^{-1} \mathrm{m}^{-1}$, and this condition holds for frequencies well above the range of visible light ($\sim 10^{15}\mathrm{Hz}$). Typical frequencies for accelerators vary anywhere from dc up to a few gigahertz. In this regime of good conductors,

$$\Gamma \simeq \sqrt{\frac{\mu\omega\sigma}{2}}(1 + i). \tag{9.13}$$

When an electromagnetic wave penetrates a plane of metal with a boundary at $z = 0$, the fields inside the conductor drop off exponentially;

$$\hat{\Psi}(z) = \hat{\Psi}(0)e^{-z/\delta}, \tag{9.14}$$

where the *skin depth* is defined by

$$\delta = \sqrt{\frac{2}{\mu\omega\sigma}}. \tag{9.15}$$

Currents flowing through the metal dissipate energy through ohmic heating. It can be shown (see Problem 9–2) that the time-averaged power loss in a conductor from rf fields is

$$\langle P_{\mathrm{loss}} \rangle = \frac{R_s}{2} \iint_S |H_\parallel|^2 \, dS, \tag{9.16}$$

integrated over the boundary, with the surface resistance of the conductor defined by

$$R_s = \sqrt{\frac{\mu\omega}{2\sigma}} = \frac{1}{\sigma\delta}, \tag{9.17}$$

and where $\vec{H}_\parallel$ is the component of magnetic field intensity parallel to the interface.

Any physical quantity propagating as a wave along the z-axis without attenuation may be represented by a complex function,

$$F(z, t) = F_0 e^{i(\omega t - kz)}, \tag{9.18}$$

with the wave number

$$k = \frac{2\pi}{\lambda_{\rm rf}}. \tag{9.19}$$

The actual value will of course be given by $\text{Re}(F)$. The phase velocity is

$$v_{\rm ph} = \frac{\omega}{k}, \tag{9.20}$$

and the group velocity,

$$v_{\rm g} = \frac{d\omega}{dk}. \tag{9.21}$$

For a superposition of waves Eq. (9.18) becomes

$$F(z, t) = \int_{-\infty}^{\infty} A(\omega) e^{i(\omega t - kz)} d\omega, \tag{9.22}$$

where $A(\omega)$ is the spectral intensity of the waves, and $k = k(\omega)$.

9.2 Cylindrical waveguides

Consider a cylindrical waveguide along the z-axis with radius, a. Since a propagation of energy along the axis is needed to accelerate particles down the waveguide, a traveling wave solution for the electric field will be of the form

$$\vec{E} = \vec{E}(r, \theta) e^{i(\omega t - k_g z)}, \tag{9.23}$$

where the k_g is the wave number or propagation constant of the wave in the guide. Attenuation along the guide, due to power loss in the walls, can produce an imaginary part to k_g. Substitution of E_z into the Helmholtz equation, gives

$$\left[\frac{1}{r} \frac{\partial}{\partial r} \left(r \frac{\partial}{\partial r} \right) + \frac{1}{r^2} \frac{\partial^2}{\partial \theta^2} + \frac{\partial^2}{\partial z^2} + k^2 \right] E_z = 0, \tag{9.24}$$

where $k^2 = \omega^2/c^2$ is the free-space wave number. Eq. (9.24) may be separated into the transverse and longitudinal equations:

$$\frac{1}{r}\frac{\partial}{\partial r}\left(r\frac{\partial E_z}{\partial r}\right) + \frac{1}{r^2}\frac{\partial^2 E_z}{\partial \theta^2} = -k_c^2 E_z, \quad \text{and} \tag{9.25}$$

$$\frac{\partial^2 E_z}{\partial z^2} = -k_g^2 E_z, \tag{9.26}$$

where the new constant

$$k_c^2 = k^2 - k_g^2. \tag{9.27}$$

An identical pair of equations for H_z may be written:

$$\frac{1}{r}\frac{\partial}{\partial r}\left(r\frac{\partial H_z}{\partial r}\right) + \frac{1}{r^2}\frac{\partial^2 H_z}{\partial \theta^2} = -k_c^2 H_z, \quad \text{and} \tag{9.28}$$

$$\frac{\partial^2 H_z}{\partial z^2} = -k_g^2 H_z. \tag{9.29}$$

Once E_z and H_z have been found, the other components of $\vec{E}$ and $\vec{H}$ may then be found with help from Eqs. (9.1 and 9.2);

$$E_r = -\frac{1}{k_c^2}\left[ik_g\frac{\partial E_z}{\partial r} + \frac{i\omega\mu}{r}\frac{\partial H_z}{\partial \theta}\right], \tag{9.30}$$

$$E_\theta = \frac{1}{k_c^2}\left[-\frac{ik_g}{r}\frac{\partial E_z}{\partial \theta} + i\omega\mu\frac{\partial H_z}{\partial r}\right], \tag{9.31}$$

$$H_r = \frac{1}{k_c^2}\left[\frac{i\omega\epsilon}{r}\frac{\partial E_z}{\partial \theta} - ik_g\frac{\partial H_z}{\partial r}\right], \quad \text{and} \tag{9.32}$$

$$H_\theta = -\frac{1}{k_c^2}\left[i\omega\epsilon\frac{\partial E_z}{\partial r} + \frac{ik_g}{r}\frac{\partial H_z}{\partial \theta}\right]. \tag{9.33}$$

If both E_z and H_z are zero, then either the other transverse components of E and H must also be zero, or

$$k_c^2 = k^2 - k_g^2 = 0, \tag{9.34}$$

so that the wave is similar to a transverse wave in free space. This type of solution is called a *transverse electromagnetic mode* (TEM), and can only exist if the conductor is made of more than one disconnected piece, such as in a coaxial transmission line. For TEM modes Eqs. (9.25 and 9.28) reduce to Laplace equations in two dimensions and may be solved as simple potential problems. The conductor is an equipotential surface, so that the potential inside such a conductor will be constant unless there is another separate conductor of different potential inside the first. Therefore, a hollow waveguide with a single conductor cannot support a TEM mode.

Since the wave equations are linear, the case with E_z and H_z both nonzero may be treated as a superposition of *transverse magnetic* (TM with $H_z = 0$ and $E_z \neq 0$) and *transverse electric* (TE with $E_z = 0$ and $H_z \neq 0$) modes. For acceleration the TM mode is obviously what is wanted, so we may look for solutions with

$$H_z = 0. \tag{9.35}$$

At the wall of the waveguide the tangential components of the electric field must vanish;

$$E_z = E_\theta = 0, \quad \text{for} \quad r = a. \tag{9.36}$$

The usual method of separation of variables can be used to solve Eq. (9.25) for E_z, if we put

$$E_z(r, \theta) = R(r)\Theta(\theta). \tag{9.37}$$

With this substitution Eq. (9.25) may be rearranged to

$$\frac{R''}{R}r^2 + \frac{R'}{R}r + r^2 k_c^2 = -\frac{\Theta''}{\Theta}, \tag{9.38}$$

where the primes indicate differentiation with respect to the appropriate variable. The last equation has the two variables r and θ separated so that functions $R(r)$ and $\Theta(\theta)$ may be solved independently by introducing a new parameter, n; the separated equations then become

$$R'' + \frac{R'}{r} + \left(k_c^2 - \frac{n^2}{r^2}\right)R = 0, \quad \text{and} \tag{9.39}$$

$$\Theta'' + n^2\Theta = 0. \tag{9.40}$$

Eq. (9.40) is equivalent to the equation of a simple harmonic oscillator and has the general solution

$$\Theta(\theta) = C_1 \cos n\theta + C_2 \sin n\theta, \tag{9.41}$$

with C_1 and C_2 being arbitrary constants. Since the field should be single-valued at any given point, n must be an integer, and $n^2 \geq 0$. The other differential equation (9.39) with integral n has Bessel functions of the first and second kind as independent solutions, thus the general form of the radial function should be

$$R(r) \sim A_1 J_n(k_c r) + A_2 N_n(k_c r), \tag{9.42}$$

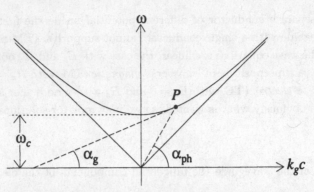

Figure. 9.1 The dispersion or Brillouin diagram for a uniform cylindrical waveguide of radius a. The angles α_{ph} and α_g for the point P are defined by the relations $\tan \alpha_{\text{ph}} = v_{\text{ph}}/c$, and $\tan \alpha_g = v_g/c$. The asymptotic diagonal lines are the dispersion curve for a wave traveling in free space.

where A_1 and A_2 are again arbitrary. Since the electric field should be finite on the axis, and $N_n(k_c r)$ is singular at $r = 0$, A_2 must be zero. Bearing in mind the boundary condition of Eq. (9.36),

$$J_n(k_c a) = 0, \tag{9.43}$$

which implies

$$k_c = \frac{X_{nj}}{a}, \tag{9.44}$$

where X_{nj} is the jth nonzero root of J_n. The general solution E_z for a given n may then be written as

$$E_z = (C_1 \cos n\theta + C_2 \sin n\theta) \, J_n\left(\frac{X_{nj}}{a}r\right) e^{i(\omega t - k_g t)}. \tag{9.45}$$

This mode is frequently referred to as the TM_{nj} mode.

Using the new value for k_c from Eq. (9.44) in Eq. (9.27) produces the dispersion relation

$$k^2 - k_g^2 = \frac{\omega^2}{c^2} - k_g^2 = \left(\frac{X_{nj}}{a}\right)^2, \tag{9.46}$$

which is illustrated in Fig. 9.1. In order to have a propagating wave, $k_g^2 > 0$, and

$$\frac{\omega}{c} \geq \frac{X_{nj}}{a}; \tag{9.47}$$

that is, the angular frequency ω must be greater than the *cutoff frequency*

$$\omega_c = \frac{c X_{nj}}{a}. \tag{9.48}$$

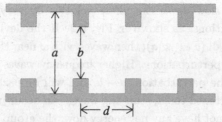

Figure. 9.2 An iris loaded waveguide.

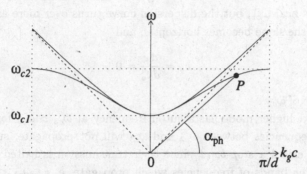

Figure. 9.3 Brillouin diagram for a loaded cylindrical waveguide. The point P has phase velocity less than c, i. e., $\alpha_{\text{ph}} < 45°$. As $k_g c$ approaches π/d, the group velocity goes to zero.

The lowest root X_{nj} occurs for J_0 in the case of azimuthal symmetry;

$$X_{01} \simeq 2.405. \tag{9.49}$$

The wavelength of the lowest cutoff frequency is then

$$\lambda_c = \frac{2\pi}{k_c} \simeq 2.61a, \tag{9.50}$$

which is a little larger than the diameter of the waveguide. Naively one should expect wavelengths shorter than $2a$ to pass through the waveguide, while wavelengths much larger than the diameter would meet with considerable resistance.

It should be noted that for a simple cylindrical waveguide, the phase velocity,

$$v_{\text{ph}} = \frac{\omega}{k_g} = \frac{c}{\sqrt{1 - \left(\frac{\omega_c}{\omega}\right)^2}}, \tag{9.51}$$

is always greater than c for ω above the cutoff frequency; i. e., the angle $\alpha_{\text{ph}} > 45°$ in Fig. 9.1. This type of solution will clearly not accelerate a charged particle along a linac for more than a few wavelengths since the crest of the wave will travel faster than the particle which is limited to velocities less than c.

It is possible to slow down the phase velocity by varying the radius of the cylinder with constrictions, as shown in Fig. 9.2. If the deviations from a smooth cylinder are rather small $(a-b \ll a)$, then waves with ω near the cutoff frequency ω_{c1} will barely notice the perturbations. Higher frequency waves will have wavelengths closer to the size of the perturbation, $a - b$, and will be partially reflected. When the wavelength is shortened to $2d$, standing waves can occur in the waveguide; since a standing wave does not have any net energy flow, the group velocity must be zero. This is shown in Fig. 9.3; the loaded waveguide behaves similar to the uniform guide for frequencies near ω_{c1}, but the dispersion curve turns over more as $k_g c$ increases to π/d where the slope becomes horizontal, and

$$v_g = \frac{d\omega}{dk_g} = 0, \tag{9.52}$$

at a frequency of ω_{c2}.

If the next higher mode has a cutoff frequency at $\omega_{c3} > \omega_{c2}$ for $k_g = 0$, then waves with frequencies between ω_{c2} and ω_{c3} will not propagate; such a range of frequencies is called a *stop-band*, since their transmission is halted by the loaded waveguide. The bands of frequencies which propagate, e. g., $\omega_{c1} < \omega < \omega_{c2}$, are called *pass-bands*

9.3 Electron capture

At the front end of a linac, the particles typically travel with velocities considerably less than the speed of light. It may still be possible to capture light particles, such as electrons, with a traveling wave whose phase velocity is c, if the electric field is large enough. We may write the electric field seen by an electron as

$$E_z = E_{z0} \sin \varphi, \tag{9.53}$$

where φ is the phase angle between the peak of the traveling wave and the electron. The force on the particle is then

$$\frac{d}{dt}\left(\frac{\beta}{\sqrt{1-\beta^2}}\right) mc = qE_{z0} \sin \theta. \tag{9.54}$$

For a particle with velocity equal to βc, the difference in path lengths traveled by the wave and the electron in a time dt is

$$dl = (c - v)dt = \frac{d\varphi}{k_g}, \tag{9.55}$$

where $d\varphi$ is the corresponding change in φ. This equation can be solved for the rate of change of phase;

$$\frac{d\varphi}{dt} = k_g c (1 - \beta).$$ (9.56)

Temporarily introducing the variable α, by the equation

$$\beta = \cos \alpha,$$ (9.57)

will allow Eq. (9.54) to be integrated. With this substitution for β, the equation of motion becomes

$$\frac{d\alpha}{dt} = -\frac{qE_{z0}}{mc} \sin \varphi \sin^2 \alpha.$$ (9.58)

Employing the chain rule

$$\frac{d\varphi}{dt} = \frac{d\varphi}{d\alpha}\frac{d\alpha}{dt},$$ (9.59)

Eqs. (9.56 and 9.58) may be combined to give

$$-\sin \varphi \, d\varphi = \frac{k_g mc^2}{qE_{z0}}\frac{1 - \cos \alpha}{\sin^2 \alpha} d\alpha,$$ (9.60)

which can be easily integrated to yield

$$\cos \varphi - \cos \varphi_0 = \frac{k_g mc^2}{qE_{z0}} \left(\tan \frac{\alpha}{2} - \tan \frac{\alpha_0}{2}\right).$$ (9.61)

Here the initial value of α_0 is determined by the velocity $\beta_0 c$ at $t = 0$ when the particle enters the linac. The half-angle tangents may be replaced with the appropriate identity,

$$\tan \frac{\alpha}{2} = \left(\frac{1 - \cos \alpha}{1 + \cos \alpha}\right)^{\frac{1}{2}} = \left(\frac{1 - \beta}{1 + \beta}\right)^{\frac{1}{2}},$$ (9.62)

so that

$$\cos \varphi - \cos \varphi_0 = \frac{k_g mc^2}{qE_{z0}} \left(\frac{1 - \beta_0}{1 + \beta_0}\right)^{\frac{1}{2}},$$ (9.63)

if we assume that $\beta \simeq 1$ at time t. The maximum value of the left-hand expression is two. If the electron is to be successfully captured, then the amplitude of the electric field

$$|E_{z0}| \geq \frac{k_g mc^2}{2|q|} \left(\frac{1 - \beta_0}{1 + \beta_0}\right)^{\frac{1}{2}}.$$ (9.64)

Higher velocity electrons require less field for capture. For example, with a field $E_{z0} = 10$ MV/m and wavelength $\lambda_g = 0.1$ m, electrons with kinetic energy of 60 keV can be captured even though their velocity is only about $0.44c$. On the other hand, protons of the same velocity would need to have $E_{z0} > 18$ GV/m or 1.8 V/Å, which is not currently achievable!

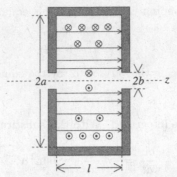

Figure. 9.4 A simplified rf cavity shaped like a tuna fish can with two small holes in either end for the beam. The electromagnetic fields are indicated for the TM_{010} mode.

9.4 Cylindrical cavities

The previous sections consider acceleration by traveling waves. In Chapter 7 it was noted that a standing wave is composed of similar waves traveling in opposite directions, and thus can also sustain a coherent acceleration of particles over the full length of the linac. For standing waves the waveguide is usually replaced by a quasi-periodic structure, e. g., a series of drift tubes as in an Alvarez linac, or a series of resonant cavities.

Fig. 9.4 shows a simplified version of such a cavity. (Actual cavities usually have rounded corners and reentrant hollow nose cones on the axis.) Assuming that the aperture $2b$ is considerably less than the diameter $2a$ and the rf wavelength λ_{rf}, the fields may be reasonably approximated by the solutions for a cylindrical waveguide of radius a, with additional constraints from two flat conductors cutting through the cylinder at $z = 0$ and $z = l$. These new boundary conditions require that

$$H_z(r, \theta, 0) = H_z(r, \theta, l) = 0, \quad \text{and} \tag{9.65}$$

$$E_r(r, \theta, 0) = E_\theta(r, \theta, 0) = E_r(r, \theta, l) = E_\theta(r, \theta, l) = 0. \tag{9.66}$$

For TE modes, $E_z = 0$, and the longitudinal component of magnetic field should be of the form

$$H_z(r, \theta, z) \sim J_n(k_c r)(C_1 \cos n\theta + C_2 \sin n\theta) \sin\left(\frac{m\pi z}{l}\right), \tag{9.67}$$

where m is some positive integer, and

$$k_c^2 = \left(\frac{\omega}{c}\right)^2 - \left(\frac{m\pi}{l}\right)^2. \tag{9.68}$$

The magnetic field must run along the surface of the conductor, so remembering Eq. (9.32),

$$H_r(a, \theta, z) = 0 = -\frac{ik_g}{k_c^2}\frac{\partial H_z}{\partial r}, \qquad (9.69)$$

which implies that

$$J_n'(k_c a) = 0.$$

Writing X_{nj}' for the jth nonzero root of J_n', and solving Eq. (9.68) for ω, we get the frequency for the TE_{nmj} mode of oscillation to be

$$f = \frac{c}{2\pi}\sqrt{\left(\frac{X_{nj}'}{a}\right)^2 + \left(\frac{m\pi}{l}\right)^2}. \qquad (9.70)$$

The TM modes have $H_z = 0$ everywhere, and

$$E_z \sim J_n(k_c r)(C_1 \cos n\theta + C_2 \sin n\theta)\cos\left(\frac{m\pi z}{l}\right). \qquad (9.71)$$

Here m can be any nonnegative integer. The parallel component of E at the surface must be zero, thus at $r = a$,

$$E_z(a, \theta, z) = 0, \qquad (9.72)$$

and we must have

$$J_n(k_c a) = 0. \qquad (9.73)$$

After a similar calculation for the TE frequencies, the resonant frequency for the TM_{nmj} mode is

$$f = \frac{c}{2\pi}\sqrt{\left(\frac{X_{nj}}{a}\right)^2 + \left(\frac{m\pi}{l}\right)^2}. \qquad (9.74)$$

For acceleration the preferred mode is the TM_{010} mode with longitudinal field components of

$$E_z = E_0 J_0\left(\frac{X_{01}}{a}r\right), \quad \text{and} \qquad (9.75)$$

$$H_z = 0, \qquad (9.76)$$

with the amplitude of the electric field on the axis being given by E_0. Applying Eqs. (9.30–9.33) to E_z and H_z produces

$$E_r = 0, \qquad (9.77)$$

$$E_\theta = 0, \qquad (9.78)$$

$$H_r = 0, \quad \text{and} \qquad (9.79)$$

$$H_\theta = i\frac{E_0}{\mu_0 c} J_1\left(\frac{X_{01}}{a}r\right). \qquad (9.80)$$

The cavity behaves somewhat like a lumped LRC circuit in the TM_{010} mode with the flat ends behaving like circular capacitor plates and the rounded cylinder like an inductor; the currents move back and forth along the resistive walls between the two ends of the cavity. The stored energy in the cavity at any instant is given by

$$U = \frac{\epsilon_0}{2} \iiint |\vec{E}(r, \theta, z, t)|^2 d^3r + \frac{\mu_0}{2} \iiint |\vec{H}(r, \theta, z, t)|^2 d^3r \qquad (9.81)$$

integrated over the volume of the cavity. The electric and magnetic fields in the cavity are 90° out of phase inside the cavity, as indicated by the factor of i in Eq. (9.80), so the total energy may be obtained by integrating the square of the electric field when the magnetic field is zero.

For the TM_{010} mode the stored energy becomes

$$U = \frac{\epsilon_0}{2} E_0^2 \int_0^l \int_0^{2\pi} \int_0^a \left[J_n\left(\frac{X_{01}}{a} r \right) \right]^2 r\, dr\, d\theta\, dz$$
$$= \frac{\epsilon_0}{2} E_0^2 \pi a^2 l \, [J_1(X_{01})]^2, \qquad (9.82)$$

where we have used the identities

$$\int_0^1 t\, J_n(X_{nj}t)\, J_n(X_{nk}t)\, dt = \frac{1}{2}[J_n'(X_{nj})]^2\, \delta_{jk}, \quad \text{and} \qquad (9.83)$$

$$J_0'(t) = -J_1(t). \qquad (9.84)$$

The average rate of energy dissipation in the cavity walls can be obtained from Eq. (9.16). For the TM_{010} mode this is

$$\langle P_{\text{loss}} \rangle = \frac{1}{2\sigma\delta} \left\{ 2 \times 2\pi \left(\frac{E_0}{\mu_0 c} \right)^2 \int_0^a \left[J_1\left(\frac{X_{01}}{a} r \right) \right]^2 r\, dr \right.$$
$$\left. + \pi a l \left(\frac{E_0}{\mu_0 c} \right)^2 [J_1(X_{01})]^2 \right\}$$
$$= \frac{\pi a (a + l)}{\sigma \delta} \left(\frac{E_0}{\mu_0 c} \right)^2 [J_1(X_{01})]^2, \qquad (9.85)$$

with the help of the identity

$$\int_0^1 [J_0(X_{01}t)]^2\, t\, dt = \int_0^1 [J_1(X_{01}t)]^2\, t\, dt. \qquad (9.86)$$

A useful gauge of the resonator's response to external excitation is the *quality factor*, or "Q", given by the ratio of the average stored energy in the cavity to the energy dissipated in the walls during one rf cycle;

$$Q = \frac{U\omega}{\langle P_{\text{loss}} \rangle}. \tag{9.87}$$

If an isolated cavity has a value of stored energy U_0 in a single mode of frequency ω_0 at $t = 0$, the stored energy will change with the rate

$$\frac{dU}{dt} = -\langle P_{\text{loss}} \rangle = -\frac{\omega_0 U}{Q}, \tag{9.88}$$

which has the solution

$$U(t) = U_0 e^{-\omega_0 t/Q}. \tag{9.89}$$

In our simple model

$$Q = \frac{al}{\delta (a + l)} \tag{9.90}$$

for the TM_{010} mode.

Electromagnetic energy is typically transported from an rf source to the cavity by means of either a waveguide or a transmission line. Three types of connections are typically used to connect the input line to the resonant structure: a hole in the cavity wall can provide a coupling aperture to a waveguide; a coupling loop may be made on the end of a transmission line and passed through the wall of the cavity to make a transformer; or the center conductor of a transmission line may be capacitively coupled to a stub inside the cavity. Individual cavities are frequently connected together by means of additional slots in the walls between the cavities or by small waveguides from one cavity to the next.

Figure 9.5 shows an equivalent circuit of a typical loop coupling on the end of a transmission line which is inserted through a small hole in the wall of a cavity. The equivalent impedance seen by the transmission line would be

$$Z_{\text{eff}} \approx i\omega L_1 - i\frac{\omega_0^2 M^2}{2L_2 \left(\omega - \omega_0 - i\frac{\omega_0}{2Q} \right)}. \tag{9.91}$$

We will assume that the reader is familiar with the theory of transmission lines (See, for example, any of Refs. 1 to 7.) and will only make a brief reference to reflected signals and the characteristic impedance. The reflection coefficient, defined

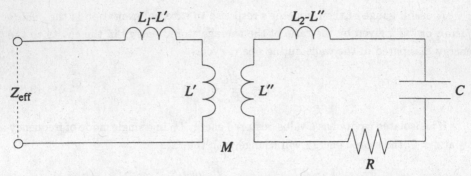

Figure. 9.5 Simple schematic of a transmission line coupled to a cavity using a loop of inductance L_1. The cavity is modeled by a lumped LRC circuit with respective values of L_2, R, and C for the inductance, resistance, and capacitance of the cavity. Part of the loop and cavity wall form a 1:1 turn transformer with mutual inductance $M = \sqrt{L'L''}$.

as the ratio of reflected to incident voltages for a signal reflecting from the end of a transmission line, can be written as

$$\Gamma = \frac{Z_e - Z_0}{Z_e + Z_0}, \tag{9.92}$$

where Z_0 is the characteristic impedance of the line (frequently 50 Ω), and Z_e is the effective impedance of the load attached to the end of the line. Clearly the power is most efficiently transferred into the cavity if $Z_e = Z_0$; for a constant input power, there is no reflection and the power dissipated in the cavity's walls must equal the power being supplied by the rf source.

Consider a cavity coupled to a correctly matched transmission line that ends in a resistor of value Z_0. Then the power converted to heat in the walls will be

$$\langle P_{\text{walls}} \rangle = \frac{\omega_0 U}{Q}, \tag{9.93}$$

and an equal amount must be dissipated in the resistor. Therefore the effective quality factor of the resonator must be reduced by a factor of two for a correctly matched line. This corrected quality factor is called the *loaded Q*,

$$Q_l = \frac{Q}{2}. \tag{9.94}$$

The Q for the cavity without the line is usually called the *unloaded Q*. With an accelerated beam passing through the cavity, the power loss increases by the amount of power being supplied to the beam.

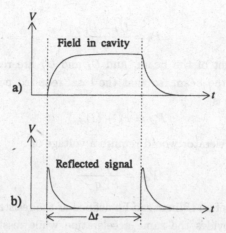

Figure. 9.6 a) The signal amplitude from a small coupling loop inserted through the wall of a cavity. b) The amplitude of the corresponding reflected wave from the cavity. The coupling of the input transmission line has been tuned to match the characteristic impedance of the line in cw operation. Note the reflected spikes at the beginning and end of an rf pulse of duration Δt.

For cw operation it is possible to adjust the coupling of a loop so that Z_{eff} is equal to the characteristic impedance of the line; however, in pulsed operation the fields in the cavity will build up from zero asymptotically to an equilibrium level at the beginning of the pulse and will decay exponentially at the end of the pulse. Figure 9.6a shows the amplitude of the accelerating voltage in a cavity for a square input pulse of duration Δt. The rise and fall times of the gap voltage are

$$\tau = \frac{Q_l}{2\omega_0}. \tag{9.95}$$

Unlike a simple resistor, the coupled cavity on the end of the transmission line has reactive components; the termination Z_e can match the characteristic impedance of the line only when the power loss in the walls is equal to the input power. Naively, at equilibrium, the energy trying to escape back down the transmission line is held in check by the incoming wave; with too little energy in the cavity, the phase of the oscillation in the cavity is slightly shifted so that there is some reflection back down the line. If there is too much energy in the cavity, the field in the cavity can start to overpower the incoming wave and produce some reflection. These effects are demonstrated by the two pulses in the amplitude of the reflected signal shown in Fig. 9.6b.

The total rf power required by the beam beyond that dissipated in Joule heating is

$$P_b = \frac{U_f - U_i}{q} I_b, \tag{9.96}$$

where I_b is the current of the beam, and U_i and U_f are respectively the initial and final energies of the beam, so that the total required power delivered to the accelerating structure is

$$P_{\text{rf}} = P_b + \langle P_{\text{loss}} \rangle. \tag{9.97}$$

The equivalent dc accelerator would require a voltage of

$$V_{\text{dc}} = \frac{U_f - U_i}{q}. \tag{9.98}$$

The *shunt impedance* of an rf linac can be defined as the resistance of the equivalent dc column which provides the same acceleration while dissipating energy at an identical rate, i. e.,

$$\mathcal{Z} = \frac{V_{\text{dc}}^2}{\langle P_{\text{loss}} \rangle}. \tag{9.99}$$

9.5 Longitudinal equations of motion in a linac

The equations for oscillation of phase and energy in a linac are similar to those derived for a circular machine in Chapter 7, except for two points. First, linear accelerators are straight without any bending magnets, so that the path length of the beam is always the length of the linac and is thus independent of momentum. Effectively, the momentum compaction is zero, and the phase slip factor becomes just

$$\eta_{\text{tr}} = \frac{1}{\gamma^2} > 0. \tag{9.100}$$

The second point is just a matter of convention; the accelerating gradient is defined with the phase of a cosine function rather than a sine function;

$$\Delta U = qV \cos(\omega_{\text{rf}} t + \phi_s), \tag{9.101}$$

with the stable phase for a synchronous particle, which gains no energy, being

$$\phi_s = -\frac{\pi}{2}. \tag{9.102}$$

Applying these differences, the equations of motion corresponding to the circular machine (7.25 and 7.28) become

$$\frac{dW}{dt} = \frac{qv}{2\pi h}(\cos \phi_s - \cos \phi), \quad \text{and} \tag{9.103}$$

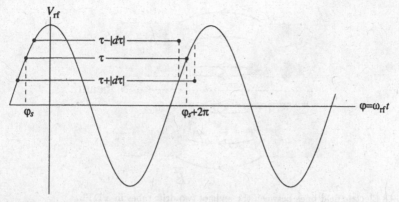

Figure. 9.7 A graphical demonstration of the phase-stability principle in a linac. The effective transition energy is always infinite, since $\eta_{tr} = 1/\gamma^2 > 0$.

$$\frac{d\varphi}{dt} = \frac{\omega_{rf}^2}{\beta^2\gamma^2 U_s} W. \qquad (9.104)$$

The transverse focusing structure of a linac is generally a FODO structure, so that the harmonic number, h, should be set equal to the number of rf wavelengths within the periodic FODO cell.

At low energies, β is rapidly changing so that the linac is only approximately periodic in the transverse and longitudinal dimensions. In this case it may be necessary to track particles through the individual components, rather than use the simplified equations of motion given above.

For small oscillations about the synchronous phase, the equations of motion describe a simple harmonic oscillator with

$$\ddot{\varphi} + \Omega_s^2\varphi \simeq 0, \qquad (9.105)$$

where the frequency of the phase oscillation is

$$\Omega_s = \frac{\omega_{rf}}{h}\sqrt{\frac{-h\sin\phi_s}{2\pi\beta^2\gamma^3}\frac{qV}{mc^2}}. \qquad (9.106)$$

The condition for stable oscillation must then be

$$\sin\phi_s < 0, \qquad (9.107)$$

as illustrated in Fig. 9.7. (This of course assumes that the phase is defined with $qV > 0$ for positive acceleration.)

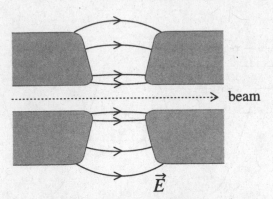

Figure. 9.8 Electric field lines between the ends of two drift tubes in a DTL.

9.6 Transverse defocusing

When particles are accelerated in a linear accelerator, there is a radial defocusing of the beam. This is illustrated in Fig. 9.8. At the entrance to the gap the lines tend to converge and actually focus the beam, but at the exit they diverge and defocus the beam. For a longitudinally stable bunch of particles, $\sin \phi_s < 0$ (see Fig. 9.7); the voltage is increasing as the bunch crosses the gap, giving rise to an overall outward kick from the electric field in the gap. Since the beam is to be accelerated, the synchronous phase should be between $-90°$ and $0°$.

Let $z = 0$ at the beginning of the gap with the phase of the rf voltage

$$\phi \sim \omega_{rf} t - \frac{\omega_{rf}}{v} z + \phi_s, \tag{9.108}$$

as seen by a particle of velocity v at position z and time t, where v has been assumed to be constant. More precisely, the electric field on axis will be

$$E_z(z, t) = E_0 \cos \left[\omega_{rf} t - \omega_{rf} \int_0^z \frac{dz'}{v(z')} + \phi_s \right], \tag{9.109}$$

allowing for a changing speed as the beam crosses the gap.

Assuming that rf structure is cylindrical and excited by a TM_{010} mode, the symmetry requires

$$E_\theta = 0, \quad \text{and} \quad B_z = 0. \tag{9.110}$$

The electric field due to the cavity must satisfy Gauss's law;

$$\nabla \cdot \vec{E} = \frac{1}{r} \frac{\partial}{\partial r}(r E_r) + \frac{\partial E_z}{\partial z} = 0, \tag{9.111}$$

which may be solved for the radial component of electric field to yield

$$E_r \simeq -\frac{r}{2}\frac{\partial E_z}{\partial z}. \tag{9.112}$$

The azimuthal component may likewise be obtained from Ampere' law:

$$\frac{1}{c^2}\frac{\partial E_z}{\partial t} = \hat{z} \cdot (\nabla \times \vec{B})$$

$$= \frac{1}{r}\frac{\partial}{\partial r}(rB_\theta) - \frac{1}{r}\frac{\partial B_r}{\partial \theta}. \tag{9.113}$$

But $B_r = 0$ since $J_\theta = 0$, so Eq. (9.113) easily integrates to

$$B_\theta = \frac{r}{2c^2}\frac{\partial E_z}{\partial t}. \tag{9.114}$$

The radial component of the Lorentz force is just

$$F_r = q(E_r - v_z B_\theta). \tag{9.115}$$

Evaluating E_r and B_θ using Eq. (9.109),

$$E_r = -\frac{\omega_{rf} r}{2v}E_0 \sin\left[\omega_{rf}t - \omega_{rf}\int_0^z \frac{dz'}{v(z')} + \phi_s\right], \tag{9.116}$$

$$B_\theta = -\frac{\omega_{rf} r}{2c^2}E_0 \sin\left[\omega_{rf}t - \omega_{rf}\int_0^z \frac{dz'}{v(z')} + \phi_s\right], \tag{9.117}$$

and the radial force simplifies to

$$F_r \simeq -qE_0\frac{\omega_{rf} r}{2v\gamma^2}\sin\phi_s, \tag{9.118}$$

when evaluated near the synchronous particle.

In the region of longitudinal stability ($qV\sin\phi_s < 0$), the radial force is positive and causes a transverse defocusing. Conversely, in the longitudinally unstable region, there is transverse focusing. In order to have stable oscillations in all three dimensions, a series of focusing and defocusing magnets may be interspersed along the rf structure.

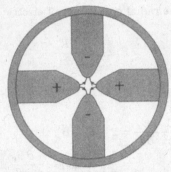

Figure. 9.9 Cross section of an RFQ with vanes located at $\theta = 0°$, $90°$, $180°$, and $270°$. The direction of the electric field lines and charge on the vanes are indicated for the peak of one half cycle of an rf oscillation.

9.7 Radio frequency quadrupoles

As described in the first chapter, a radio frequency quadrupole (RFQ) is an rf structure which can be used to accelerate low velocity particles ($\beta \ll 1$). There are several configurations which can be used to obtain the quadrupole mode. Fig. 9.9 shows a cross section of the simplest design of an RFQ with four equally spaced electrodes, called vanes, mounted inside a conducting cylinder. The quadruple mode has currents running back and forth along the inner wall of the quadrants of the cylinder and up and down the vanes so that, at any instant, the adjacent vanes have equal but opposite charges on their tips. The small region between the vanes has an alternating electric field which behaves like a FODO cell as far as an individual particle is concerned; there is net focusing in both transverse dimensions. By machining the tips of the vanes with a hyperbolic cross section, the focusing force will vary linearly with the distance from the axis, as in a quadrupole magnet.

The distance of the vanes from the axis is modulated with a slowly varying sine wave to provide an axial component to the electric field which can accelerate a beam of charged particles. The modulation of adjacent vanes must be 180° out of phase to produce this accelerating gradient. There is a trade-off between the amount of acceleration and the strength of the transverse focusing which depends on the amplitude of this modulation. The wavelength, λ_{mod}, of the modulation must also increase as the synchronous particle gains energy;

$$\lambda_{mod} = \beta \lambda_{rf}. \tag{9.119}$$

The unit cell of the RFQ structure is defined as one half of a wave of the modulation.

Magnetic field lines wrap around the ends of the vanes and go down the four large cavities formed by the cylinder and the sides of the vanes. These cavities contain very little electric field and behave mainly like inductors. The tips of the vanes are mostly capacitive with very little magnetic field between them; therefore, the electric field in the central region can be obtained from

$$\vec{E}(\vec{r}) = -\nabla\Phi, \tag{9.120}$$

where the scalar potential, Φ, satisfies the Laplace equation,

$$\nabla^2\Phi = \frac{1}{r}\frac{\partial}{\partial r}\left(r\frac{\partial\Phi}{\partial r}\right) + \frac{1}{r^2}\frac{\partial^2\Phi}{\partial\theta^2} + \frac{\partial\Phi}{\partial z^2} = 0. \tag{9.121}$$

The general solution for quadrupole symmetry, with finite values on the z-axis, may be written as

$$\Phi(r,\theta,z) = \frac{V}{2}\left[\sum_{n=0}^{\infty} A_{0n}r^{2n}\cos 2n\theta \right.$$
$$\left. + \sum_{n=0}^{\infty}\sum_{l=1}^{\infty} A_{ln}r^{2n}I_{2n}(lkr)\cos 2n\theta\cos lkz\right], \tag{9.122}$$

where V is the applied voltage between adjacent vanes, the I_{2n} are modified Bessel functions, the A_{nl} are determined by boundary conditions, and

$$k = \frac{2\pi}{\beta\lambda_{\text{rf}}}. \tag{9.123}$$

Although k is really a function of z since β increases as a particle moves along the RFQ, in most cases, it may be thought of as an adiabatic increase with k changing much more slowly than the modulation of the vanes.

The constant term is uninteresting and should be set to zero; $A_{00} = 0$. Keeping the next two low order terms,

$$\Phi = \frac{V}{2}[A_{01}r^2\cos 2\theta + A_{10}I_0(kr)\cos kz]. \tag{9.124}$$

The first term in this equation provides radial focusing, and the second term, longitudinal acceleration. (Typically the higher order terms should be kept small in designing an RFQ.)

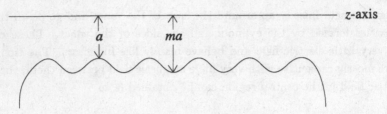

Figure. 9.10 A modulated vane of an RFQ indicating the aperture, a, and modulation parameter, m.

Application of the gradient operator gives

$$E_r(\vec{r}) = -V\left[A_{01}r\cos 2\theta + \frac{k}{2}A_{10}I_1(kr)\cos kz\right], \tag{9.125}$$

$$E_\theta(\vec{r}) = VA_{01}r^2\sin 2\theta, \quad \text{and} \tag{9.126}$$

$$E_z(\vec{r}) = \frac{kV}{2}A_{10}I_0(kr)\sin kz, \tag{9.127}$$

for the amplitude of the electric field oscillations. (Remember that the derivative of $I_0(x)$ is just $I_1(x)$.) Putting in the time dependence of the rf,

$$\vec{E}(\vec{r}, t) = \vec{E}(\vec{r})\sin(\omega_{rf}t + \phi_0), \tag{9.128}$$

where ϕ_0 is the phase of the rf oscillation, at $t = 0$, when the synchronous particle enters the RFQ.

The aperture, a, is defined as the minimum distance of closest approach of a vane to the z-axis; particles which stay within a cylinder of radius a will miss the vanes. The modulation parameter, m, is the ratio of the maximum to minimum distance from the axis of the nearest part of the vane surface, as shown in Fig. 9.10.

The electrodes are conductors and must therefore be equipotential surfaces. Arbitrarily picking $z = 0$ at one of the peaks of the vane located at $\theta = 0$, the equipotential surfaces of the vanes at $z = 0$ should satisfy the following constraints:

$$\Phi(a, 0, 0) = -\Phi\left(ma, \frac{\pi}{2}, 0\right) = \frac{V}{2}. \tag{9.129}$$

Combining these conditions with Eq. (9.124) and solving for A_{01} and A_{10} yield

$$A_{10} = \frac{m^2 - 1}{m^2 I_0(ka) + I_0(mka)}, \quad \text{and} \tag{9.130}$$

$$A_{01} = \frac{\chi}{a^2}, \tag{9.131}$$

where

$$\chi = 1 - A_{10}I_0(ka) \tag{9.132}$$

is called the *focusing efficiency* and is a measure of the strength of the transverse focusing. Similarly, A_{10} is called the *acceleration efficiency*. Note that as the modulation, m, of the vanes increases, the acceleration increases, while focusing weakens. In one cell the synchronous particle will gain an amount of energy

$$
\begin{aligned}
\Delta U &= \int_0^{\beta \lambda_{rf}/2} q E_z(0,0,z,t) dz \\
&= \frac{qkA_{10}V}{2} I_0(0) \int_0^{\beta \lambda_{rf}/2} \sin kz \sin(kz + \phi_s)\, dz \\
&= \frac{\pi q A_{10}V}{4} \cos \phi_s.
\end{aligned} \tag{9.133}
$$

Problems

9–1   How strong must the electric field intensity of a traveling plane wave be to accelerate electrons with an energy gradient of 10 MeV/m? (Hint: Use the Poynting vector.)

9–2   Show that the rf power loss in the conducting walls of a cavity is given by Eq. (9.16).

9–3   Consider a box shaped resonant cavity with a square cross section of width, w, in the transverse directions and length, l, in the longitudinal dimension. Calculate the resonant frequency and Q of the lowest order TM mode.

9–4   Derive Eq. (9.91) for the effective impedance of a coupling loop connected to a simple cavity.

9–5   Using Laplace's equation in cylindrical coordinates, show that Eq. (9.122) is the general solution for the symmetry of an RFQ.

9–6   Derive an equation of motion for the longitudinal phase oscillations about the synchronous particle in an RFQ. For what values of ϕ_0 are small oscillations stable? What is the frequency of the small oscillations? (Hint: Construct difference equations for the deviations in phase and $W = -\Delta U/\omega_{rf}$ for a test particle passing through one cell.)

References for Chapter 9

[1] J. D. Jackson, *Classical Electrodynamics*, John Wiley, New York, (1962).

[2] W. K. H. Panofsky and M. Phillips, *Classical Electricity and Magnetism*, Addison-Wesley, Reading, Mass. (1962).

[3] J. C. Slater, *Microwave Electronics*, D. Van Nostrand Co., New York (1950).

[4] R. E. Collin, *Foundations for Microwave Engineering*, McGraw-Hill, New York (1966).

[5] S. Humphries, *Principles of Charged Particle Acceleration*, John Wiley & Sons, New York (1986).

[6] S. Y. Lao, *Microwave Devices and Circuits*, Prentice Hall, Englewood Cliffs, NJ (1985).

[7] S. Y. Lao, *Microwave Electron-tube Devices*, Prentice Hall, Englewood Cliffs, NJ (1988).

[8] R. Wideröe, Arch. Electrotech., 21, 387 (1928).

[9] L. W. Alvarez, Phys. Rev., 70, 799 (1946).

[10] J. Le Duff, "Dynamics and Acceleration in Linear Structures", p. 144, CERN 85-19 (1985).

[11] E. Persico, E. Ferrari, and S. E. Segre, *Principles of Particle Accelerators*, Chapter 8 by M. Puglisi, W. A. Benjamin, New York (1968).

[12] Gregory A. Leow and Richard Talman, "Elementary Principles of Linear Accelerators", AIP Conf.Proc. No. 105, p. 1 (1983).

[13] M. Puglisi, "The Radiofrequency Quadrupole Linear Accelerator", CERN 87-03.

10

Resonances

In this chapter, we use a simple model of perturbations to examine the conditions for resonant oscillations of betatron motion which are driven by various multipole errors of the magnetic guide field.

10.1 Integer resonances

Magnetic fields in the elements constituting any lattice cannot be perfect with some small imperfections appearing here and there, giving rise to extra terms. For instance, a radial misalignment of ΔR in the bending magnets yields a term, constant with respect to x:

$$\frac{d^2x}{d\theta^2} + Q_H{}^2 x = (1-n)\Delta R. \tag{10.1}$$

Here we have chosen the horizontal equation; however, similar arguments can be made for the vertical and even longitudinal motions. The constant term may be expanded in a Fourier series with terms like $\varepsilon \cos(m\theta)$ where ε is the strength of the m-th harmonic. For a single periodic error the equation of motion becomes

$$\frac{d^2x}{d\theta^2} + Q_H{}^2 x = \varepsilon \cos(m\theta). \tag{10.2}$$

The most general solution of Eq. (10.2) may be written as

$$x = \tilde{x} + \bar{x}, \tag{10.3}$$

where the homogeneous solution is

$$\tilde{x} = A\cos(Q_H\theta) + B\sin(Q_H\theta), \tag{10.4}$$

and the particular solution is

$$\bar{x} = \frac{\varepsilon}{Q_H{}^2 - m^2}[\cos(m\theta) - \cos(Q_H\theta)], \tag{10.5}$$

213

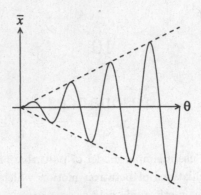

Figure. 10.1 A plot of the particular solution $\bar{x}$.

with A and B given by the initial conditions. By use of a little trigonometry, this last equation may be transformed to

$$\bar{x} = \frac{\varepsilon\theta}{Q_H + m} \sin\left(\frac{Q_H + m}{2}\theta\right) \frac{2}{(Q_H - m)\theta} \sin\left(\frac{Q_H - m}{2}\theta\right), \qquad (10.6)$$

which reduces to

$$\bar{x} \simeq \frac{\varepsilon\theta}{2Q_H} \sin(Q_H\theta), \qquad (10.7)$$

for $Q_H = m$. If the error occurs only once in the ring, as might be caused by a random error, then m can take all integral values; whereas, if the error is a systematic error occurring in each periodic cell, and there are N periodic cells in the ring, then m takes the values Nj, where j is any integer.

Fig. 10.1 illustrates the linear increase of the amplitude of the betatron oscillations referred to the particular solution, $\bar{x}$. Of course the general solution,

$$x(\theta) = A\cos(Q_H\theta) + \left(B + \frac{\varepsilon}{2Q_H}\theta\right)\sin(Q_H\theta), \qquad (10.8)$$

also exhibits the unbounded nature for $Q_H = m$.

Another simple way to look at this is to consider a small perturbation of the one-turn matrix

$$\mathbf{M} = \begin{pmatrix} \cos\mu + \alpha\sin\mu & \beta\mu \\ -\gamma\sin\mu & \cos\mu - \alpha\sin\mu \end{pmatrix}, \qquad (10.9)$$

written in terms of the Twiss parameters. On an integer resonance μ is a multiple of 2π, so the matrix becomes the identity matrix $\mathbf{M} = \mathbf{I}$. If there is a small path-length error δl in one drift section, then the one-turn matrix just after the error becomes

$$\mathbf{M} = \begin{pmatrix} 1 & \delta l \\ 0 & 1 \end{pmatrix} \begin{pmatrix} 1 & 0 \\ 0 & 1 \end{pmatrix} = \begin{pmatrix} 1 & \delta l \\ 0 & 1 \end{pmatrix}. \qquad (10.10)$$

Any particle with a nonzero slope ($x' \neq 0$) will see a linear growth in its trajectory:

$$\begin{pmatrix} x_n \\ x'_n \end{pmatrix} = \begin{pmatrix} 1 & \delta l \\ 0 & 1 \end{pmatrix}^n \begin{pmatrix} x_0 \\ x'_0 \end{pmatrix} = \begin{pmatrix} x_0 + n\, x'_0\, \delta l \\ x'_0 \end{pmatrix}. \tag{10.11}$$

A small error in a quadrupole strength can produce a similar result.

10.2 Linear coupling resonances[2]

The sharp separation between horizontal and vertical motions so far considered, is of course an approximation. Indeed some amount of coupling must exist, giving rise to a pair of modified betatron equations:

$$\frac{d^2x}{d\theta^2} + Q_H{}^2 x = \varepsilon \cos(m\theta)\, y, \quad \text{and} \tag{10.12}$$

$$\frac{d^2y}{d\theta^2} + Q_V{}^2 y = \varepsilon \cos(m\theta)\, x. \tag{10.13}$$

If ε is very small, the solutions of the homogeneous equations for x and y may be substituted into the corresponding inhomogeneous terms on the right-hand side:

$$\frac{d^2x}{d\theta^2} + Q_H{}^2 x = \frac{1}{2}\varepsilon_y \left[\cos(Q_V + m)\theta + \cos(m - Q_V)\theta\right], \quad \text{and} \tag{10.14}$$

$$\frac{d^2y}{d\theta^2} + Q_V{}^2 y = \frac{1}{2}\varepsilon_x \left[\cos(Q_H + m)\theta + \cos(m - Q_H)\theta\right], \tag{10.15}$$

where ε_x and ε_y contain the respective amplitude information of the homogeneous solutions. The same arguments as in the previous section lead to the resonance conditions

$$Q_H + Q_V = m, \quad \text{and} \tag{10.16}$$

$$|Q_H - Q_V| = m, \tag{10.17}$$

which classify linear sum and difference resonances, respectively. The sum and difference resonances behave differently as a little coupling is added to an ideal uncoupled lattice.

10.2.1 Stability of difference resonance and instability of sum resonance

Consider the uncoupled 1-turn transfer matrix:

$$\mathbf{T} = \begin{pmatrix} \mathbf{u}_1 & 0 \\ 0 & \mathbf{u}_2 \end{pmatrix} \tag{10.18}$$

$$= \begin{pmatrix} \cos\mu_1 + \alpha_1 \sin\mu_1 & \beta_1 \sin\mu_1 & 0 & 0 \\ -\gamma_1 \sin\mu_1 & \cos\mu_1 - \alpha_1 \sin\mu_1 & 0 & 0 \\ 0 & 0 & \cos\mu_2 + \alpha_2 \sin\mu_2 & \beta_2 \sin\mu_2 \\ 0 & 0 & -\gamma_2 \sin\mu_2 & \cos\mu_2 - \alpha_2 \sin\mu_2 \end{pmatrix}$$

For the difference resonance condition $\sin \mu_1 = \sin \mu_2$, and for the sum condition $\sin \mu_1 = -\sin \mu_2$.

A very common source of transverse coupling is a slight roll of a quadrupole magnet about its axis by a small angle θ. Assume that the last element in $\mathbf{T}$ is the thin quadrupole:

$$\mathbf{Q} = \begin{pmatrix} \mathbf{F} & \mathbf{0} \\ \mathbf{0} & \mathbf{D} \end{pmatrix} = \begin{pmatrix} 1 & 0 & 0 & 0 \\ -1/f & 1 & 0 & 0 \\ 0 & 0 & 1 & 0 \\ 0 & 0 & 1/f & 1 \end{pmatrix}. \tag{10.19}$$

We may estimate the effect on the eigenvalues due to a slight roll of this last quadrupole by replacing the last quadrupole in $\mathbf{T}$ by a slightly rotated quadrupole:

$$\mathbf{T}' = \begin{pmatrix} \mathbf{M} & \mathbf{n} \\ \mathbf{m} & \mathbf{N} \end{pmatrix} = \mathbf{RQR}^{-1}\mathbf{Q}^{-1}\mathbf{T} \tag{10.20}$$

$$= \begin{pmatrix} \mathbf{I}\cos\theta & \mathbf{I}\sin\theta \\ -\mathbf{I}\sin\theta & \mathbf{I}\cos\theta \end{pmatrix} \begin{pmatrix} \mathbf{F} & \mathbf{0} \\ \mathbf{0} & \mathbf{D} \end{pmatrix} \begin{pmatrix} \mathbf{I}\cos\theta & -\mathbf{I}\sin\theta \\ \mathbf{I}\sin\theta & \mathbf{I}\cos\theta \end{pmatrix} \begin{pmatrix} \mathbf{D} & \mathbf{0} \\ \mathbf{0} & \mathbf{F} \end{pmatrix} \begin{pmatrix} \mathbf{u}_1 & \mathbf{0} \\ \mathbf{0} & \mathbf{u}_2 \end{pmatrix}.$$

After a bit of multiplication, we are rewarded with

$$\mathbf{T}' = \begin{pmatrix} (\mathbf{I}\cos^2\theta + \mathbf{D}^2\sin^2\theta)\mathbf{u}_1 & (\mathbf{I}-\mathbf{F}^2)\mathbf{u}_2\cos\theta\sin\theta \\ (\mathbf{D}^2-\mathbf{I})\mathbf{u}_1\cos\theta\sin\theta & (\mathbf{I}\cos^2\theta + \mathbf{F}^2\sin^2\theta)\mathbf{u}_2 \end{pmatrix}$$

$$= \begin{pmatrix} \begin{pmatrix} 1 & 0 \\ \frac{2}{f}\sin^2\theta & 1 \end{pmatrix}\mathbf{u}_1 & \begin{pmatrix} 0 & 0 \\ \frac{2}{f}\cos\theta\sin\theta & 0 \end{pmatrix}\mathbf{u}_2 \\ \begin{pmatrix} 0 & 0 \\ \frac{2}{f}\cos\theta\sin\theta & 0 \end{pmatrix}\mathbf{u}_1 & \begin{pmatrix} 1 & 0 \\ -\frac{2}{f}\sin^2\theta & 1 \end{pmatrix}\mathbf{u}_2 \end{pmatrix}. \tag{10.21}$$

With more calculation, we find the 2×2 blocks:

$$\mathbf{M} = \mathbf{u}_1\cos^2\theta + \frac{2\sin^2\theta}{f}\begin{pmatrix} 0 & 0 \\ \cos\mu_1 + \alpha_1\sin\mu_1 & \beta_1\sin\mu_1 \end{pmatrix}, \tag{10.22}$$

$$\mathbf{N} = \mathbf{u}_2\cos^2\theta - \frac{2\sin^2\theta}{f}\begin{pmatrix} 0 & 0 \\ \cos\mu_2 + \alpha_2\sin\mu_2 & \beta_2\sin\mu_2 \end{pmatrix}, \tag{10.23}$$

$$\mathbf{m} = \begin{pmatrix} 0 & 0 \\ \cos\mu_1 + \alpha_1\sin\mu_1 & \beta_1\sin\mu_1 \end{pmatrix}\frac{\sin 2\theta}{f}, \tag{10.24}$$

$$\mathbf{n} = \begin{pmatrix} 0 & 0 \\ \cos\mu_2 + \alpha_2\sin\mu_2 & \beta_2\sin\mu_2 \end{pmatrix}\frac{\sin 2\theta}{f}. \tag{10.25}$$

Recall the treatment of eigenvalues for a 4×4-matrix in §6.9. For stability, the eigenvalues $\kappa_j = \lambda_j + \lambda_j^{-1}$ of $\mathbf{K} = \mathbf{T}' + \mathbf{T}'^{-1}$ must be real and between ± 2. For real

κ, the argument (Call it Δ.) of the radical in Eq. 6.102 must be positive. We have

$$\frac{\text{tr}(\mathbf{M}-\mathbf{N})}{2} = \frac{\beta_1 \sin \mu_1 - \beta_2 \sin \mu_2}{f} \sin^2 \theta, \tag{10.26}$$

$$|\mathbf{m}+\tilde{\mathbf{n}}| = \frac{\beta_1 \beta_2}{f^2} \sin^2(2\theta) \sin \mu_1 \sin \mu_2, \tag{10.27}$$

since $\text{tr}(\mathbf{u}_1) = \text{tr}(\mathbf{u}_2)$. Notice that $|\mathbf{m}+\tilde{\mathbf{n}}| \neq 0$ if there is a slight roll of the quadrupole, and also that the sign of this determinant is determined solely by the product $\sin \mu_1 \sin \mu_2$. As θ increases away from zero, the degenerate eigenvalues are pushed apart. For the slightly coupled $\mathbf{T}'$, the argument of the radical is

$$\begin{aligned}
\Delta_{\pm} &= \left(\frac{\text{tr}(\mathbf{M}-\mathbf{N})}{2}\right)^2 + |\mathbf{m}+\tilde{\mathbf{n}}| \\
&= \frac{\sin^4 \theta}{f^2}(\beta_1 \sin \mu_1 - \beta_2 \sin \mu_2)^2 + \frac{\beta_1 \beta_2}{f^2} \sin^2(2\theta) \sin \mu_1 \sin \mu_2 \\
&= \frac{\sin^4 \theta}{f^2}(\beta_1 \pm \beta_2)^2 \sin^2 \mu_1 \mp \frac{\beta_1 \beta_2}{f^2} \sin^2(2\theta) \sin^2 \mu_1 \\
&= \frac{\sin^2 \theta \sin^2 \mu_1}{f^2}\left[(\beta_1^2 + \beta_2^2)\sin^2 \theta \pm 2\beta_1 \beta_2(\sin^2 \theta - 2\cos^2 \theta)\right] \\
&\simeq \mp \frac{4\beta_1 \beta_2 \sin^2 \mu_1}{f^2}\theta^2
\end{aligned} \tag{10.28}$$

for the conditions $Q_1 \pm Q_2 = m$, respectively. The eigenvalues of $\mathbf{T}$ may be calculated by using Eq. 6.105. As a little coupling is added to the uncoupled $\mathbf{T}$, we find the conditions:

1. In the case of a difference resonance, $\Delta_- > 0$, and the degenerate λ_j eigenvalue pairs split apart by moving along the unit circle in the complex plane. Since the eigenvalues stay on the circle, the motion remains stable with $\lambda_j^* = \lambda_j^{-1}$.
2. For a sum resonance, $\Delta_+ < 0$, and the λ_j eigenvalues move away from the unit circle out into the complex plane resulting in unstable motion with $\lambda_j^* \neq \lambda_j^{-1}$.

10.3 Assessment of resonances

For a perturbing magnetic field $\vec{B} = (B_x, B_y, 0)$, the equations are of the form:

$$\frac{d^2 x}{d\theta^2} + Q_H{}^2 x = \varepsilon \left(\frac{\partial B_x}{\partial y}\right) x \cos(m\theta), \quad \text{and} \tag{10.29}$$

$$\frac{d^2 y}{d\theta^2} + Q_V{}^2 y = \varepsilon \left(\frac{\partial B_x}{\partial y}\right) y \cos(m\theta), \tag{10.30}$$

since $\partial B_x/\partial y = \partial B_y/\partial x$. We may use Eq. (4.8) to express the perturbing field gradient in a multipole expansion:

$$\left(\frac{\partial B_x}{\partial y}\right) = \text{Re}\left[i\sum_{n=0}^{\infty} n(a_n - ib_n)(x+iy)^{n-1}\right], \tag{10.31}$$

where we have incorporated the B_0 and a^n into the multipole constants b_n and a_n.

If the perturbing field is only due to the normal $2(n+1)$-multipole, the gradient becomes

$$\left(\frac{\partial B_x}{\partial y}\right) = \sum_{j=0}^{\left[\frac{n-1}{2}\right]} nb_n(-1)^j \binom{n-1}{2j} x^{n-1-2j} y^{2j}. \tag{10.32}$$

Substituting solutions,

$$x = A_1\cos(Q_H\theta), \quad \text{and} \quad y = A_2\cos(Q_V\theta), \tag{10.33}$$

of the linear homogeneous equations into the right-hand sides of Eqs. (10.29 and 10.30), yields

$$\frac{d^2x}{d\theta^2} + Q_H{}^2 x = \varepsilon nb_n\cos(m\theta)\sum_{j=0}^{\left[\frac{n-1}{2}\right]}\binom{n-1}{2j} A_1^{n-2j} A_2^{2j}\cos^{n-2j}(Q_H\theta)\cos^{2j}(Q_V\theta), \tag{10.34}$$

and

$$\frac{d^2y}{d\theta^2} + Q_V{}^2 y = \varepsilon nb_n\cos(m\theta)$$
$$\sum_{j=0}^{\left[\frac{n-1}{2}\right]}\binom{n-1}{2j} A_1^{n-1-2j} A_2^{2j+1}\cos^{n-1-2j}(Q_H\theta)\cos^{2j+1}(Q_V\theta). \tag{10.35}$$

A given product of cosines may be written as

$$\cos(m\theta)\cos^p(Q_H\theta)\cos^q(Q_V\theta)$$
$$= 2^{-(p+q)}\sum_{k=0}^{p}\sum_{l=0}^{q}\binom{p}{k}\binom{q}{l}\cos\{[(p-2k)Q_H + (q-2l)Q_V - m]\theta\}. \tag{10.36}$$

For the x-equation, Eq. (10.34), $p = n - 2j$, and $q = 2j$, giving resonances when

$$[n \pm 1 - 2(j+k)]Q_H + 2(j-l)Q_V = \pm m. \tag{10.37}$$

For the y-equation, Eq. (10.35), $p = n-1-2j$, and $q = 2j+1$, giving the additional conditions

$$[n - 1 - 2(j+k)]Q_H + [1 \pm 1 + 2(j-l)]Q_V = \pm m. \tag{10.38}$$

From these relations, it is easy to show that a normal quadrupole error excites the half-integer resonances,

$$2Q_{\mathrm{H}} = \pm m, \quad \text{and} \quad 2Q_{\mathrm{V}} = \pm m. \tag{10.39}$$

A normal octopole component excites the resonances

$$\pm 4Q_{\mathrm{H}} = m,$$
$$\pm 4Q_{\mathrm{V}} = m,$$
$$\pm 2Q_{\mathrm{H}} = m,$$
$$\pm 2Q_{\mathrm{V}} = m, \quad \text{and}$$
$$\pm 2Q_{\mathrm{H}} \pm 2Q_{\mathrm{V}} = m. \tag{10.40}$$

Notice that the resonances driven by the normal quadrupole are also driven by the octopole. Fig. 10.2a gives a graphical representation of these resonance relations for the quadrupole and octopole.

A normal sextupole drives the resonances given by

$$\pm 3Q_{\mathrm{H}} = m,$$
$$\pm Q_{\mathrm{H}} = m, \quad \text{and}$$
$$\pm Q_{\mathrm{H}} \pm 2Q_{\mathrm{V}} = m. \tag{10.41}$$

Normal decapoles drive the resonances given by

$$\pm 5Q_{\mathrm{H}} \pm 2Q_{\mathrm{V}} = m,$$
$$\pm 5Q_{\mathrm{H}} = m,$$
$$\pm 3Q_{\mathrm{H}} \pm 2Q_{\mathrm{V}} = m,$$
$$\pm 3Q_{\mathrm{H}} = m,$$
$$\pm Q_{\mathrm{H}} \pm 4Q_{\mathrm{V}} = m,$$
$$\pm Q_{\mathrm{H}} \pm 2Q_{\mathrm{V}} = m, \quad \text{and}$$
$$\pm Q_{\mathrm{H}} = m. \tag{10.42}$$

Similar to the case of the odd multipoles, we see that an even multipole will drive the same resonances as a lower order even multipole. The resonance relations for the normal sextupole and decapole are shown in Fig. 10.2b.

In fact, if the perturbation is strong enough, we must look to solutions which are of second order or higher in ε. For example, it has been shown[8] in second order that a sextupole can drive a quarter-integer resonance.

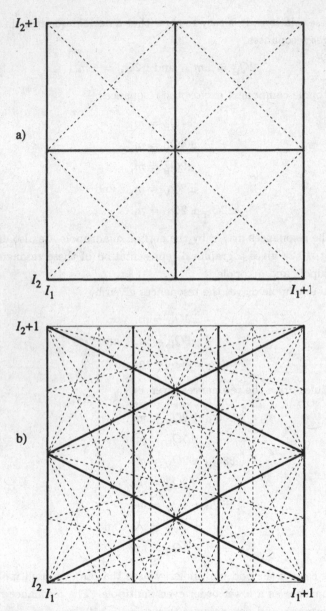

Figure. 10.2 a) The a tune plot showing the resonance lines driven by a normal quadrupole perturbation (heavy lines), and a normal octopole perturbation (all lines). I_1 and I_2 are arbitrary integers. b) A tune plot showing the resonance lines driven by a normal sextupole (heavy lines), and a normal decapole (heavy and dashed lines).

10.4 Krylov-Bogoliubov method[3]

Before proceeding, it is of primary interest to introduce a method which will enable us to solve the differential equations describing linear and nonlinear oscillators, driven by nonlinear forcing terms close to one or more resonances. A general equation for such an oscillator is

$$\frac{d^2x}{d\theta^2} + (Q_{res}^2 + \delta)x = \varepsilon F(x, \theta), \tag{10.43}$$

with

$$\delta = Q_H{}^2 - Q_{res}^2 \simeq 2Q_{res}(Q_H - Q_{res}). \tag{10.44}$$

If $\varepsilon = 0$, Eq. (10.43) is just the equation of a linear harmonic oscillator $x(\theta) = a\sin(Q_H\theta + \phi)$, with the constants a and ϕ determined by the initial conditions. For ε small, we may try a similar solution, but with slowly varying $a(\theta)$ and $\phi(\theta)$:

$$x(\theta) \simeq a(\theta)\sin\psi(\theta), \tag{10.45}$$

$$\frac{dx}{d\theta} = a(\theta)Q_{res}\cos\psi(\theta), \tag{10.46}$$

with

$$\psi(\theta) = Q_{res}\theta + \phi(\theta). \tag{10.47}$$

By writing $dx/d\theta$ as in Eq. (10.46), the two slowly varying functions, $a(\theta)$ and $\phi(\theta)$, must satisfy the equation,

$$\frac{da}{d\theta}\sin\psi + \frac{d\phi}{d\theta}a\cos\psi = 0. \tag{10.48}$$

The second derivative of x is

$$\frac{d^2x}{d\theta^2} = -Q_{res}^2 a\sin\psi + Q_{res}\frac{da}{d\theta}\cos\psi - Q_{res}\frac{d\phi}{d\theta}a\sin\psi, \tag{10.49}$$

which when substituted into Eq. (10.43) gives

$$\frac{da}{d\theta}\cos\psi - \frac{d\phi}{d\theta}a\sin\psi = -\frac{a\delta}{Q_{res}}\sin\psi + \frac{\varepsilon}{Q_{res}}F\left(a\sin\psi, \frac{\psi - \phi}{Q_{res}}\right). \tag{10.50}$$

Solving Eqs. (10.48 and 10.50) for $da/d\theta$ and $d\phi/d\theta$, via Kramer's rule, yields

$$\frac{da}{d\theta} = -\frac{\delta a}{Q_{res}} \sin\psi \cos\psi + \frac{\varepsilon}{Q_{res}} F(\psi) \cos\psi = f(\psi), \qquad (10.51)$$

$$\frac{d\phi}{d\theta} = \frac{\delta}{Q_{res}} \sin^2\psi - \frac{\varepsilon}{Q_{res}a} F(\psi) \sin\psi = g(\psi). \qquad (10.52)$$

Since a and ϕ are slowly varying relative to θ, we may treat f and g as functions which are almost periodic in ψ, and thus Fourier transforming them:

$$f(\psi) \simeq \sum_{n=-\infty}^{\infty} f_n e^{in\psi}, \quad \text{and} \qquad (10.53)$$

$$g(\psi) \simeq \sum_{n=-\infty}^{\infty} g_n e^{in\psi}, \qquad (10.54)$$

with

$$f_n = \frac{1}{2\pi} \int_0^{2\pi} f(\psi) e^{-in\psi} d\psi, \quad \text{and} \quad g_n = \frac{1}{2\pi} \int_0^{2\pi} g(\psi) e^{-in\psi} d\psi. \qquad (10.55)$$

The derivatives of a and ϕ may be approximated by their averages over one cycle of the rapid variation in ψ, giving

$$\frac{da}{d\theta} \simeq \left\langle \frac{da}{d\theta} \right\rangle = f_0 = \frac{1}{2\pi} \int_0^{2\pi} \frac{da}{d\theta} d\psi, \qquad (10.56)$$

$$\frac{d\phi}{d\theta} \simeq \left\langle \frac{d\phi}{d\theta} \right\rangle = g_0 = \frac{1}{2\pi} \int_0^{2\pi} \frac{d\phi}{d\theta} d\psi, \qquad (10.57)$$

Using Eqs. (10.51 and 10.52), we now have

$$\frac{da}{d\theta} = h_1 = \frac{\varepsilon}{2\pi Q_{res}} \int_0^{2\pi} F(\psi) \cos\psi \, d\psi, \quad \text{and} \qquad (10.58)$$

$$\frac{d\phi}{d\theta} = h_2 = \frac{\delta}{2Q_{res}} - \frac{\varepsilon}{2\pi Q_{res}a} \int_0^{2\pi} F(\psi) \sin\psi \, d\psi, \qquad (10.59)$$

which are of fundamental importance as we shall see further on.

We may anticipate that, once we have found the functions $da/d\theta = h_1(a, \phi)$ and $d\phi/d\theta = h_2(a, \phi)$, it will be possible either to solve further the system of first-order differential equations, thus finding $a(\theta)$ and $\phi(\theta)$, or to obtain useful information on the behavior of a and ϕ.

10.5 Half-integer resonance

For a half-integer resonance, we may write $Q_{res} = m/2$, for m being some odd integer. The perturbing function is

$$F(x, \theta) = x \cos(m\theta) = a \sin \psi \cos[2(\psi - \phi)], \tag{10.60}$$

where we have used Eq. (10.47). Evaluating the derivatives for a and ϕ, we have

$$\frac{da}{d\theta} = \frac{\varepsilon a}{\pi m} \int_0^{2\pi} \cos[2(\psi - \phi)] \sin \psi \cos \psi \, d\psi, \quad \text{and} \tag{10.61}$$

$$\frac{d\phi}{d\theta} = \frac{\delta}{m} - \frac{\varepsilon}{\pi m} \int_0^{2\pi} \cos[2(\psi - \phi)] \sin^2 \psi \, d\psi. \tag{10.62}$$

Carrying out the integrations gives

$$\frac{da}{d\theta} = \frac{\varepsilon a}{2m} \sin(2\phi), \quad \text{and} \tag{10.63}$$

$$\frac{d\phi}{d\theta} = \frac{\delta}{m} + \frac{\varepsilon}{2m} \cos(2\phi). \tag{10.64}$$

Combining these equations to eliminate θ yields

$$2\sin(2\phi)\frac{d\phi}{da} = -\frac{d}{da}[\cos(2\phi)] = \frac{4\delta}{\varepsilon a} + \frac{2}{a}\cos(2\phi). \tag{10.65}$$

This simple first-order linear equation may be solved for $\cos(2\phi)$, giving

$$\cos(2\phi) = \frac{A}{a^2} - \frac{2\delta}{\varepsilon}, \tag{10.66}$$

where A is a constant determined by initial conditions. Reordering this gives the invariant

$$A_0 = a^2 \left(\cos(2\phi) + \frac{2\delta}{\varepsilon} \right) = a_0^2 \left(\cos(2\phi_0) + \frac{2\delta}{\varepsilon} \right), \tag{10.67}$$

with a_0 and ϕ_0 being the initial conditions. Multiplying Eq. (10.63) by $2a$ produces the equation

$$\frac{d(a^2)}{d\theta} = \frac{\varepsilon}{m} \sqrt{a^4 - a^4 \cos^2(2\phi)}, \tag{10.68}$$

which may be written as

$$\frac{d(a^2)}{d\theta} = \frac{\varepsilon}{m} \sqrt{\left(1 - \frac{4\delta^2}{\varepsilon^2}\right)(a^2)^2 + \frac{4\delta}{\varepsilon}A_0 a^2 - A_0^2} \tag{10.69}$$

using the initial conditions for a and ϕ. This may be integrated by writing

$$\frac{\varepsilon\theta}{m} = \int_{a_0^2}^{a^2} \frac{dw}{\sqrt{\alpha w^2 + bw + c}}, \tag{10.70}$$

where we have made the substitutions

$$\alpha = 1 - \left(\frac{2\delta}{\varepsilon}\right)^2, \quad b = \frac{4\delta}{\varepsilon}A_0, \quad \text{and} \quad c = -A_0^2, \tag{10.71}$$

and have assumed that $a(\theta = 0) = a_0$.

For $\alpha > 0$, $|\delta| < \frac{1}{2}|\varepsilon|$, integration of Eq. (10.70) gives

$$\sqrt{\alpha}\frac{\varepsilon}{m}\theta = \ln\left(\frac{2\sqrt{\alpha}\sqrt{\alpha a^4 + ba^2 + c} + 2\alpha a^2 + b}{W_0}\right), \tag{10.72}$$

with

$$W_0 = 2\sqrt{\alpha}\sqrt{\alpha a_0^4 + ba_0^2 + c} + 2\alpha a_0^2 + b. \tag{10.73}$$

Solving this for a^2 gives

$$a^2 = \frac{1}{4\alpha}\left[W_0 \exp\left(\sqrt{\alpha}\frac{\varepsilon}{m}\theta\right) - 2b + \frac{4A_0^2}{W_0}\exp\left(-\sqrt{\alpha}\frac{\varepsilon}{m}\theta\right)\right]. \tag{10.74}$$

It is clear that this solution blows up exponentially for $|\delta| < \frac{1}{2}|\varepsilon|$.

For $\alpha < 0$, $|\delta| > \frac{1}{2}|\varepsilon|$, we see that Eq. (10.70) gives

$$\sqrt{-\alpha}\frac{\varepsilon}{m}\theta = \sin^{-1}\left(\frac{\alpha a_0^2 + \frac{1}{2}b}{A_0}\right) - \sin^{-1}\left(\frac{\alpha a^2 + \frac{1}{2}b}{A_0}\right). \tag{10.75}$$

Solving for a^2 gives

$$a^2 = \frac{A_0}{(2\delta/\varepsilon)^2 - 1}\left(\frac{2\delta}{\varepsilon} + \sin(\Omega\theta - S_0)\right), \tag{10.76}$$

where

$$\Omega = \frac{\sqrt{4\delta^2 - \varepsilon^2}}{m}, \tag{10.77}$$

and

$$S_0 = \sin^{-1}\left(\frac{\alpha a_0^2 + b/2}{A_0}\right). \tag{10.78}$$

In this case, $|\delta| > |\varepsilon/2|$, we see that the betatron motion of the particle is stable and is just modulated slowly with the frequency Ω.

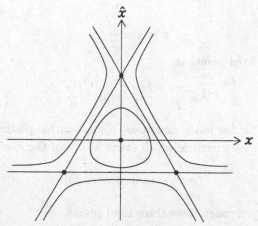

Figure. 10.3 A schematic plot of the phase plane for the $Q_{\mathrm{res}} = m/3$ resonance.

10.6 The nonlinear third-integer resonance

For the third-integers, we have $Q_{\mathrm{res}} = m/3$ with the perturbing function

$$F(x, \theta) = x^2 \cos(m\theta) = a^2 \sin^2 \psi \cos[3(\psi - \phi)]. \tag{10.79}$$

The averaged equations for a and ϕ are now

$$\frac{da}{d\theta} = -\frac{3}{8m}\varepsilon a^2 \cos(3\phi), \quad \text{and} \tag{10.80}$$

$$\frac{d\phi}{d\theta} = \frac{3\delta}{2m} + \frac{3}{8m}\varepsilon a \sin(3\phi). \tag{10.81}$$

In the representative phase-plane $(x, \hat{x})$ (see Fig. 10.3), with

$$x(\theta) = a(\theta) \sin\left[\frac{m}{3}\theta + \phi(\theta)\right], \tag{10.82}$$

and

$$\hat{x}(\theta) = \frac{3}{m}\frac{dx}{d\theta} = a(\theta) \cos\left[\frac{m}{3}\theta + \phi(\theta)\right], \tag{10.83}$$

the coordinates may be picked up at a fixed azimuth, thus realizing what is known as a *stroboscopic representation*.

A trajectory repeating itself in this representation implies that either we have the trivial case with $a = 0$ yielding an elliptical fixed point, or both $a(\theta)$ and $\phi(\theta)$ must be constant, hence

$$\cos(3\phi) = 0. \tag{10.84}$$

For $\sin(3\phi) = -1$,

$$a = \tilde{a} = \frac{4\delta}{\varepsilon}, \tag{10.85}$$

giving rise to three fixed points at

$$\tilde{\phi} = \frac{\pi}{2}, \quad \frac{7\pi}{6}, \quad \text{and} \quad \frac{11\pi}{6}. \tag{10.86}$$

Remembering that for the initial condition, the azimuthal position corresponds to $\theta = 0$, and we find that a particle which starts at one of the fixed points

$$(x, \hat{x}) = (\tilde{a}\sin\tilde{\phi}, \tilde{a}\cos\tilde{\phi}), \tag{10.87}$$

will cycle repeatedly through these three fixed points.

Expanding Eqs. (10.80 and 10.81) about a fixed point, we get

$$\frac{d(\Delta a)}{d\theta} \simeq \frac{9\varepsilon\sin(3\tilde{\phi})\tilde{a}^2}{8m}\Delta\phi, \quad \text{and} \tag{10.88}$$

$$\frac{d(\Delta\phi)}{d\theta} \simeq \frac{3\varepsilon\sin(3\tilde{\phi})}{8m}\Delta a, \tag{10.89}$$

where Δa and $\Delta\phi$ are small deviations from the fixed point, $(\tilde{a}, \tilde{\phi})$. These two equations may be combined to give

$$\frac{d^2(\Delta a)}{d\theta} - k^2\,\Delta a = 0, \tag{10.90}$$

$$\frac{d^2(\Delta\phi)}{d\theta} - k^2\,\Delta\phi = 0, \tag{10.91}$$

with

$$k^2 = \frac{27\varepsilon^2\sin^2(3\tilde{\phi})\tilde{a}^2}{64m^2}. \tag{10.92}$$

Since k^2 is always positive, we see that the nearby solutions for Δa are of the form

$$\Delta a = C_1 e^{k\theta} + C_2 e^{-k\theta}, \tag{10.93}$$

with C_1 and C_2 being constants. A similar sort of solution will result from the equation for $\Delta\phi$. These solutions show that some initial conditions will converge to the fixed point while others diverge from it. This is the condition for a hyperbolic fixed point. (It can be shown that the growth is actually faster than exponential as one moves away from the fixed point.)

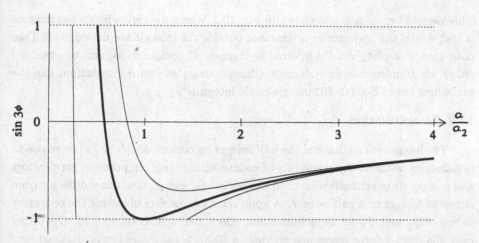

Figure. 10.4 Curves corresponding to various initial values, $(a_0, \sin 3\phi_0)$. The separatrix (heavy line) corresponds to the boundary between bounded and unbounded oscillations.

For $\sin(3\phi) = +1$,

$$a = -\tilde{a} = -\frac{4\delta}{\varepsilon}, \tag{10.94}$$

with

$$\tilde{\phi} = -\frac{\pi}{2}, \quad -\frac{7\pi}{6}, \quad \text{and} \quad -\frac{11\pi}{6}. \tag{10.95}$$

These are again hyperbolic fixed points.

Eliminating the θ dependence from Eqs. (10.80 and 10.81), we find

$$\frac{d\phi}{da} = -\frac{\tilde{a} + a\sin(3\phi)}{a^2\cos(3\phi)}, \tag{10.96}$$

which has the solution

$$\sin(3\phi) = \frac{A}{a^3} - \frac{3\tilde{a}}{2a}, \tag{10.97}$$

for some constant A. This may be rearranged to obtain the invariant condition,

$$A = a^3\left(\sin(3\phi) + \frac{3\tilde{a}}{2a}\right) = a_0^3\left(\sin(3\phi_0) + \frac{3\tilde{a}}{2a_0}\right). \tag{10.98}$$

A plot of $\sin(3\phi)$ versus a (see Fig. 10.4) shows that for some sets of initial conditions, the solutions are bounded, while for other values, the oscillation amplitudes diverge with their phases tending to assume the same value. The boundary or separatrix is given by the equation,

$$\sin(3\phi) = \frac{1}{2}\left(\frac{\tilde{a}}{a}\right)^3 - \frac{3\tilde{a}}{2a}. \tag{10.99}$$

This twofold behavior is also shown in Fig. 10.3, where the region inside the triangle is stable, and the trajectories originating outside the triangle are unbounded. Take note that this plot, usually inferred by numerical computation, can be obtained either via Hamiltonian formalism or through straightforward integration, like the one which led to Eq. (10.70) using elliptic integrals[4].

10.7 Recapitulation

The integer resonance and the half-integer resonance, for $|\delta| \leq \frac{1}{2}|\varepsilon|$, give rise to instabilities, without any possibility of counteracting them. In practice, accelerators and storage rings are built with both the horizontal and vertical tunes different from either an integer or a half-integer. A good criterion consists of setting the operating values of Q_H and Q_V far away from these values, perhaps about $n \pm 0.25$. Operating near the third-integer resonance provides a stable area, where the oscillations may remain stable for ever, if their initial amplitudes are small enough.

Since the asymptotic phases of the growing amplitudes converge toward a unique value, zero in our example, a coherent blow-up of the expanding oscillations takes place. Therefore a controlled growth of the amplitude can be realized at any wanted azimuth, thus implementing an extremely efficient extraction[5] of the circulating beam, with the aid of a thin septum at a negative abscissa in Fig. 10.3.

Apart from this useful role, nonlinear resonances are not as serious as the linear resonances, since they always give rise to a stable area which may be larger than the beam, provided that a few precautions are taken. The most general[6] stable area is a polygon with $q + 1$ vertices coinciding with the $q + 1$ fixed points, circumscribed by a circle of radius

$$a_q = 2^{\frac{q}{q-1}} \left| \frac{\delta}{\varepsilon} \right|^{\frac{1}{q-1}} . \tag{10.100}$$

If no special multipole field is implemented by purpose, the ordinary multipole-imperfections in the magnetic elements fade off with higher multipole number: i. e. the higher q is, the smaller ε is, and the broader the stable region is. It is worthwhile to note that, in the half-integer case, the search for the stable area leads to the trivial solution $a = 0$, as one can easily find by setting Eq. (10.63) to zero.

The amplitude growth does not defy Liouville's Theorem, since it occurs at the expense of the particle's total energy which, in its turn, undergoes a very small relative variation, since the energy stored in the betatron oscillations is much smaller than the longitudinal energy. In fact, the linear coupling resonance, $Q_H = Q_V$, of § 10.2 does not couple to the longitudinal energy, and the beam does not blow up in this case.

On the other hand, the horizontal emittance of a slowly extracted beam can be very small, with the slopes almost the same as the ones of the unperturbed beam, and the size is practically of the order of magnitude of the radial jump per turn. Concurrently the vertical emittance remains almost unaltered. Nevertheless the 6-dimensional density is conserved since the horizontal phase-plane decrease is compensated by a like increase of the longitudinal phase-plane. The beam length changes from $L = 2\pi R$, at most, to $\beta c \Delta t$ with Δt being the length of the extraction.

Finally we should note that the longitudinal synchrotron oscillations may couple to the betatron motion giving resonant conditions of the form

$$jQ_H + kQ_V + lQ_s = m, \tag{10.101}$$

with j, k, l, and m being integers.

Problems

10–1 Find the resonance relations for the skew multipoles and make plots of the resonance relations for the quadrupole, sextupole, octopole, and decapoles.

10–2 Show that a sextupole resonance equation of the type

$$\frac{d^2x}{d\theta^2} + Q_H{}^2 = \varepsilon x^2 \cos(m\theta),$$

will give rise to a quarter-integer resonance condition for terms of order ε^2.

10–3 Write a small computer program that tracks a particle through a periodic cell using a linear transformation. a) For various initial conditions plot the stroboscopic representation of the particle at the beginning of the periodic cell. b) Add a thin sextupole kick at the beginning of the cell and study the effect of this nonlinearity near a third-integer resonance for various initial conditions.

10–4 By treating the synchrotron oscillations as slowly varying relative to the revolutions, with $p = p_0 + A \sin(Q_s\theta)$, show that a coupling of resonances between the horizontal and vertical betatron oscillations occurs.

References for Chapter 10

[1] A. Sommerfeld, *Mechanics*, Academic Press, New York, 1952.

[2] H. Bruck, *Accélérateurs Circulaires de Particules*, Presses Universitaires de France, Paris, 1966.

[3] N. M. Krylov and N. N. Bogoliubov, *Introduction to Nonlinear Mechanics*, Princeton University Press, Princeton, N. J., 1947.

[4] M. Conte and G. Liberti, "Further Studies on Resonant Extraction", III All-Union National Conference on Particle Accelerators, Proceedings V. 2, Moscow, USSR, 1972.

[5] K. H. Reich, "Beam Extraction Techniques for Synchrotrons", *Progress in Nuclear Techniques and Instrumentations*, F. J. M. Farley ed., Vol. 2, North-Holland Pub. Co., Amsterdam, 1967.

[6] M. Conte, G. Dellavalle, and G. Rizzitelli, Il Nuovo Cimento, **57B**, 59 (1980).

[7] A. A. Kolomensky and A. N. Lebedev, *Theory of Cyclic Accelerators*, North-Holland, Amsterdam (1966).

[8] S. Ohnuma, "Quarter-Integer Resonance by Sextupoles", Proceedings of the U. S.–Japan Seminar on High Energy Accelerator Science, Tokyo and Tsukuba, (1973).

11

Space-Charge Effects

11.1 Transverse effects: tune shift

A circulating beam can be represented either as a ring of charged particles, in the case of a coasting beam, or as a set of arcs of charged particles, in the case of a bunched beam. A short bunch, or a segment of the coasting beam of the same length, is so much longer than its transverse dimensions that it may be considered as an infinite cylinder with a certain charge distribution, which travels with the speed, $v = \beta c$.

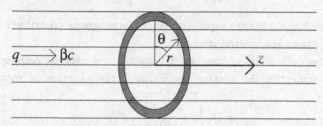

Figure. 11.1 Cylindrical beam of particles

Obviously a radial electrostatic repulsive force does exist, together with a less obvious radially attractive strength arising from the Lorentz force due to the magnetic field created by the cotraveling charged particles:

$$F(r) = q(E_r - \beta c B_\theta), \tag{11.1}$$

having been able to neglect any longitudinal component, according to the cylindrical approximation. Fig. 11.1 illustrates how these forces arise. From Maxwell's equations, we have

$$\nabla \cdot \vec{E} = \frac{1}{r} \frac{d}{dr}(rE_r) = \frac{q}{\epsilon_0} n(r), \quad \text{and} \tag{11.2}$$

$$(\nabla \times \vec{B})_z = \frac{1}{r} \frac{d}{dr}(rB_\theta) = \frac{q}{c\epsilon_0} \beta n(r), \tag{11.3}$$

where $n(r)$ is the number density of particles with respect to the radial dimension. These may be integrated to give

$$E_r = \frac{q}{\epsilon_0} \frac{1}{r} \int_0^r r\, n(r)\, dr, \quad \text{and} \tag{11.4}$$

$$B_\theta = \frac{q\beta}{c\epsilon_0} \frac{1}{r} \int_0^r r\, n(r)\, dr. \tag{11.5}$$

Assuming a Gaussian distribution for the beam, we have

$$n(r) = A \exp\left(\frac{-r^2}{2\sigma^2}\right), \tag{11.6}$$

where the normalizing constant is given by $A = N/2\pi l\sigma^2$, for N particles in a bunch of length l. Combining this with Eqs. (11.1, 11.4, and 11.5) yields

$$F(r) = \frac{Nq^2}{2\pi\epsilon_0 l}(1 - \beta^2)\frac{1 - \exp\left(-\frac{r^2}{2\sigma^2}\right)}{r}. \tag{11.7}$$

Notice that this force is always repulsive, since the magnetic contribution is smaller, by a factor β^2, than the electric one.

Due to the cylindrical symmetry, horizontal and vertical components of the force are the same; for the vertical oscillations, the equation of motion may be written as

$$\frac{d^2y}{d\theta^2} + Q_V{}^2 y = \frac{R^2}{\beta^2 \gamma mc^2} F(y), \quad \text{or} \tag{11.8}$$

$$\frac{d^2y}{d\theta^2} + Q_V{}^2 y = \frac{2Nr_0R^2}{l\beta^2\gamma^3}\frac{1 - \exp\left(-\frac{y^2}{2\sigma_V{}^2}\right)}{y}, \tag{11.9}$$

with $r_0 = q^2/4\pi\epsilon_0 mc^2$ being the classical radius of the particle. The analytic expression appearing in the right side of Eq. (11.9) can be expanded in a Maclaurin's series

$$f(y) = \frac{1 - \exp\left(-\frac{y^2}{2\sigma_V{}^2}\right)}{y} = \frac{y}{2\sigma_V{}^2} - \sum_{n=2}^{\infty}\frac{(-1)^n}{2^n n!}\frac{y^{2n-1}}{\sigma_V{}^{2n}}, \tag{11.10}$$

i. e., the space-charge force is split into a linear part and a summation of nonlinear terms. Ignoring the nonlinear terms for the moment, the vertical equation of motion may be written

$$\frac{d^2y}{d\theta^2} + \left(Q_V{}^2 - \frac{Nr_0R^2}{l\sigma_V{}^2\beta^2\gamma^3}\right) y = 0, \quad \text{or} \tag{11.11}$$

$$\frac{d^2y}{d\theta^2} + (Q_V + \delta Q_{sc})^2 y \simeq 0 \quad \text{with} \tag{11.12}$$

$$\delta Q_{sc} = -\frac{N r_0 R^2}{2 l Q_V \sigma_V^2 \beta^2 \gamma^3} = -\frac{N r_0}{4 \pi B_f \epsilon_{V,\text{rms}}^* \beta \gamma^2}. \tag{11.13}$$

Here we have used the approximation $\beta_V \simeq R/Q_V$ (see Chapter 2) and the fact that the rms normalized emittance for a Gaussian beam is

$$\pi \epsilon_{V,\text{rms}}^* = \pi \frac{(\sigma_V)^2}{\beta_V \beta \gamma} \tag{11.14}$$

(see § 5.4). The *bunching factor*, B_f, is defined by

$$B_f = \frac{l_b}{2\pi R} = \frac{I_{\text{ave}}}{I_{\text{peak}}}, \tag{11.15}$$

where l_b is the length of the bunch. This shift, δQ_{sc}, is called the Laslett tune shift. We have derived this for the case of a cylindrically symmetric beam, which is close to the case in a machine such as the SSC[2]; in this case the horizontal shift should be the same as the vertical shift,

$$\delta Q_V = \delta Q_H = \delta Q_{SC}. \tag{11.16}$$

In the case of flat beams, such as exist in electron synchrotrons, it is possible to demonstrate[1] that the term $4\sigma_V^2/\beta_V$ should be replaced by $2\sigma_H(\sigma_V + \sigma_H)/\beta_V$, since the beam is larger horizontally. The vertical tune shift then becomes

$$\delta Q_V = -\frac{\beta_V N r_0}{2\pi B_f \sigma_V (\sigma_H + \sigma_V) \beta^2 \gamma}. \tag{11.17}$$

Eqs. (11.13 and 11.17) clearly demonstrate that the space-charge tune shift is effective at low energy, where the term $\beta^3 \gamma^2$ can be very small, particularly for intense and bright beams.

There is some disparity in the accelerator community as to how large a tune shift is acceptable. Some believe that δQ_V should be less than 0.2; although this is not necessary if the effect were truly a tune shift from a linear force, since the shift could be tracked by adjusting the quadrupole currents as the beam is injected into a low energy ring. Furthermore, with the nonlinear effects, as the oscillations grow to large amplitudes, the actual frequency shifts away from the resonance. This is in fact done in the Fermilab booster and also in the AGS at Brookhaven where tune shifts greater than 0.7 are seen. A more serious problem is the increase of the tune-spread due to the nonlinear effects.

11.2 Luminosity and collider rings

For a long period the high energy events took place in a fixed target, hit by the particles of a single beam. Since the important quantity for probing small distances is the energy in the center of mass, it is obvious that much smaller distances can be probed by colliding two beams head-on than by using a single beam of twice the energy. Some of the first machines to collide beams were done with head-on collisions of e^+e^- (Adone[3], ACO[4], and VEPP-2[Ref. 5]) and almost head-on collisions of protons in the ISR[6] at CERN.

Nevertheless the fixed target is made of condensed matter (liquid hydrogen, beryllium, copper, etc.); the number of interaction centers is of the order of Avogadro's number per cubic centimeter. This is about five orders of magnitude greater than the density in a typical beam.

The interaction rate for a fixed target experiment is

$$R = \sigma \dot{N} l n, \tag{11.18}$$

where σ is the interaction cross section, $\dot{N}$ is the number of particles incident on the target per second, l is the length of the target, and n is the number density of the target. Remembering the definition of luminosity from Chapter 1, we may define the luminosity for a fixed target experiment as

$$\mathcal{L} = \frac{R}{\sigma} = \dot{N} l n = \dot{N} l \rho \frac{N_A}{M}, \tag{11.19}$$

with N_A being Avogadro's number and M being the molecular weight. For a liquid hydrogen target ($\rho = 0.07$ g/cm^3) and an intensity of 10^{13} particles per second, this gives a luminosity of about 2×10^{35} cm^{-2}s^{-1}.

The simplest example of collider consists of a ring where a single bunch of particles and a single bunch of antiparticles circulate around the ring in opposite directions. The beams cross every half revolution at two opposite points on the ring. We may treat one beam as the target with a density

$$n = \frac{N_1}{A_{int} l}, \tag{11.20}$$

where A_{int} is the transverse area of the beam crossing, l is the length of the "target" bunch which consists of N_1 particles. The incident current at a single interaction point is then

$$\dot{N} = f N_2, \tag{11.21}$$

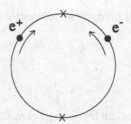

Figure. 11.2 A collider ring with a single bunch in each beam will have two different beam crossings at opposite sides of the ring.

where N_2 is the number of particles in the bunch of the beam, and f is the revolution frequency $\omega_s/2\pi$. The luminosity is then

$$\mathcal{L} = f\frac{N_1 N_2}{A_{\text{int}}}, \tag{11.22}$$

or

$$\mathcal{L} = \frac{fN^2}{A_{\text{int}}}, \tag{11.23}$$

for bunches with identical numbers of particles N. For such a collider typical luminosities are 10^{30} up to 10^{32} cm^{-2}s^{-1}.

11.3 Transverse effects: beam-beam interaction

For a collider with a single ring, as described in the previous section, with N_b bunches per beam, there will be $2N_b$ crossings where the beams overlap almost completely. This is demonstrated in Fig. 11.2 for $N_b = 1$. In these regions there is again a space-charge force arising from one beam which acts on the particles of the other beam. We will assume that one beam is much stronger than the other, and will be unaffected by the weaker beam. This force is similar to the force of Eq. (11.1):

$$F(r) = q(E_r + \beta c B_\theta). \tag{11.24}$$

This is a focusing force for oppositely charged beams, but a defocusing force for beams of the same sign charge as in the intersecting rings of the ISR, or the SSC. The radial force is similar to Eq. (11.7), but with a factor of $(1 + \beta^2)$ instead of $(1 - \beta^2)$:

$$F_\perp(y) = \frac{Nq^2}{2\pi\epsilon_0 l}(1+\beta^2)\frac{1 - \exp\left(-\frac{y^2}{2\sigma_V^2}\right)}{y}. \tag{11.25}$$

For ultrarelativistic particles with small amplitudes,

$$F_\perp(y) \simeq \frac{Nq^2}{2\pi\epsilon_0 l \sigma_V^2}y. \tag{11.26}$$

This may be treated as a small transverse momentum kick with a strength of

$$\Delta p_\perp = F_\perp \delta t, \tag{11.27}$$

where

$$\delta t = \frac{l}{2c}, \tag{11.28}$$

is the length of time that a particle from one beam takes to pass through the other beam. The angular kick is then

$$\Delta y' = \frac{F_\perp \delta t}{p} = \frac{N r_0}{\gamma \sigma_V^2} y. \tag{11.29}$$

Following the treatment for chromaticity in § 6.9, we may write the effect on the one turn transfer matrix as

$$\begin{pmatrix} \cos\mu + \alpha\sin\mu & \beta\sin\mu \\ \gamma\sin\mu & \cos\mu - \alpha\sin\mu \end{pmatrix} = \begin{pmatrix} 1 & 0 \\ \Delta k & 1 \end{pmatrix} \times$$
$$\begin{pmatrix} \cos\mu_0 + \alpha_0\sin\mu_0 & \beta_0\sin\mu_0 \\ \gamma_0\sin\mu_0 & \cos\mu_0 - \alpha_0\sin\mu_0 \end{pmatrix}, \tag{11.30}$$

and expanding the new phase advance as $\mu = \mu_0 + \delta\mu$, for small $\delta\mu$ gives

$$\cos\mu_0 - \sin\mu_0\,\delta\mu = \cos\mu_0 + \frac{1}{2}\beta_0\Delta k\,\sin\mu_0, \tag{11.31}$$

or

$$\delta\mu = -\frac{1}{2}\beta_0\,\Delta k = -\frac{\beta_V^* N r_0}{2\gamma\sigma_V^2}, \tag{11.32}$$

for the change in the phase due to a single beam crossing with a beta-function value of β_V^* at the interaction point. (The γ is the Lorentz factor and not a Twiss parameter.) To calculate the tune shift, we must sum over all the crossings and divide by 2π, to get

$$\Delta Q_{bb} = -\frac{N_{IP} N r_0 \beta_V^*}{4\pi\gamma\sigma_V^2}, \tag{11.33}$$

for the beam-beam tune shift due to like sign beams with N_{IP} collision points in the ring. For oppositely charged beams, Eq. (11.33) has the opposite sign. Since $\sigma_V = \sqrt{\epsilon_V^{rms}\beta_V^*}$, we see that the beam-beam tune shift is independent of β_V^* and only depends on the bunch charge and the emittance for round beams. As with the Laslett tune shift, a more careful calculation would replace the σ_V^2 in the denominator by $\sigma_V(\sigma_V + \sigma_H)/2$ for flattened beams to get[9]

$$\Delta Q_{bb} = -\frac{N_{IP} N r_0 \beta_V^*}{2\pi\gamma\sigma_V(\sigma_V + \sigma_H)}, \tag{11.34}$$

At first glance, this linear tune shift[10] appears to be like the Laslett tune shift, and we might expect to have stability for $|\delta Q_{bb}|$ as large as 0.25. However, the nonlinear terms are much more significant in the beam-beam case and the maximum allowed tune shift per collision point has been measured to be much smaller: ≤ 0.005 in the case of p-p collisions[10] in the ISR, and $\lesssim 0.03$ for e^+e^- collisions in electron storage rings[11,12] where the radiation damping helps.

Why does this occur? The question is still open, in spite of several trials to give a satisfactory answer. Definitely, the sums of nonlinear terms give rise to overlapping resonances. In fact, bunched beams introduce azimuthal dependence with terms like $y^{2n-1}\cos 2k\theta$ which drive the resonances

$$Q_V = \frac{2k}{2n-1+1} = \frac{k}{n}. \tag{11.35}$$

Coupled motion should be considered in addition to the single-resonance approach. Even in the single-resonance model, the superposition of high-order resonances yields[13] stochasticity, in a way similar[14] to the transition from laminar flow to turbulence for a viscous incompressible fluid.

The conservation of emittance, stated by Liouville's theorem, can be interpreted as an incompressibility principle for the phase-space surfaces. Fig. 11.3a shows in a sketchy way how a beam succeeds in hitting a physical obstruction, while preserving its phase-plane area. This is an example of *filamentation* which is driven by the nonlinear terms of the beam-beam interaction. Fig. 11.3b illustrates turbulence or breaking of the phase-plane surface with broadening of the beam and area conservation.

Another model has been suggested[15], based upon the multiple Coulomb scattering in the beam-beam collisions. In this case the particles would behave like a gas which is transversely heated at the expense of the longitudinal energy.

All these models deal with the hypothesis that one beam is much stronger than the other. In most real colliders the two beams have about the same intensity, so that both will be affected by the beam-beam force.

Indeed, the increasing amplitude of the single particle trajectory means that the corresponding bunch swells in the same manner. Then both interacting beams become broader and broader, bringing the particles against any physical obstacle, giving rise to a double beam-loss. This *strong-strong picture*[16] explains why the Gaussian-like space-charge force is still effective at quite large distances: the broadening of both beams means the growth of σ_V and σ_H.

An empirical cure consists of minimizing the vertical beta-function at the interaction regions, i. e., of implementing the low-beta sections, discussed in Chapter 6.

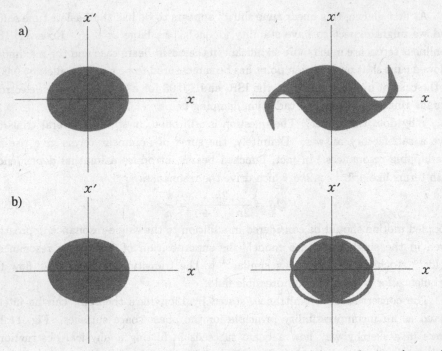

Figure. 11.3 Space charge can cause the simple distribution on the left to evolve into that on the right for a) phase-plane filamentation, b) phase-plane turbulence.

In fact, by inserting $A_{\text{int}} \simeq \pi(2\sigma_V)^2$ for a round beam into Eq. (11.23), gives a luminosity

$$\mathcal{L} \le \frac{fN\gamma}{2r_0\beta_V^*}|\Delta Q_{bb}|. \tag{11.36}$$

Eq. (11.36) clearly demonstrates that the lower the local β_V^* is, the higher the luminosity is. However, if the beta-function is made too small (less than the bunch length), the luminosity will be decreased, since the transverse dimension of the beam will grow too quickly with z, and the overlap integral of the beam densities will be decreased.

11.4 Longitudinal effects

The major longitudinal effect derived from the space charge is a bunch dilation, caused by the electrostatic repulsion. First consider a uniform cylindrical beam of radius a and length l_b in the lab. Also assume that the beam is moving inside a conducting circular beam pipe of radius b, as shown in Fig. 11.4. In the rest system

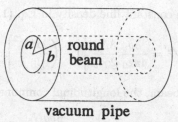

vacuum pipe

Figure. 11.4 A round beam of radius a inside a cylindrical vacuum pipe of radius b.

of the beam, the charge density may be written as

$$\rho^* = \frac{Nq}{\gamma l_b \pi a^2}, \tag{11.37}$$

where N is the number of particles uniformly distributed in a cylinder of radius a and length l_b in the lab. Here the asterisk refers to a quantity in the rest system of the beam. The γ in the denominator clearly reflects the fact that longitudinal lengths in the rest system are dilated, $s^* = \gamma s$. From Gauss's law it is trivial to obtain the component of the electric field which is perpendicular to the motion of the beam:

$$E_\perp^*(r < a) = \frac{Nqr}{2\pi\epsilon_0 \gamma l_b a^2},$$
$$E_\perp^*(r > a) = \frac{Nq}{2\pi\epsilon_0 \gamma l_b r}. \tag{11.38}$$

The voltage from the center of the beam out to the beam pipe of radius b is then

$$V^* = \int_0^b E^* \, dr = \frac{Nq}{4\pi\epsilon_0 \gamma l_b} g, \quad \text{where} \tag{11.39}$$

$$g = 1 + 2\ln\frac{b}{a}. \tag{11.40}$$

For a more general charge distribution which varies along the longitudinal direction, we may replace the constant line-density given by $qN/\gamma l_b$ by some other function. For simplicity, let us assume that the longitudinal distribution in the lab is parabolic:

$$N(z) = \frac{3N_b}{2l_b}\left[1 - \left(\frac{2z}{l_b}\right)^2\right], \tag{11.41}$$

having written the longitudinal distance from the center of the bunch as z. In the rest system of the beam, this becomes

$$N^*(z^*) = \frac{3N_b}{2\gamma l_b}\left[1 - \left(\frac{2z^*}{\gamma l_b}\right)^2\right], \tag{11.42}$$

which, when replacing the constant line density in Eq. (11.39), gives

$$V^* = \frac{qg}{4\pi\epsilon_0}\frac{3N_{\mathrm{b}}}{2\gamma l_{\mathrm{b}}}\left[1 - \left(\frac{2z^*}{\gamma l_{\mathrm{b}}}\right)^2\right]. \qquad (11.43)$$

In the rest system of the beam, the longitudinal component of the electric field is

$$E^*(z^*) = -\frac{\partial V^*}{\partial z^*} = \frac{3gqN_{\mathrm{b}}}{\pi\epsilon_0\gamma^3 l_{\mathrm{b}}^3}\,z^*. \qquad (11.44)$$

To transform the field to the lab, we only need to transform the reference coordinate z^* back to z, since the longitudinal component of an electric field is unchanged by a Lorentz transformation:

$$E(z) = \frac{3gqN_{\mathrm{b}}}{\pi\epsilon_0\gamma^2 l_{\mathrm{b}}^3}\,z. \qquad (11.45)$$

The longitudinal component of force on a single particle of the beam is then

$$F_\parallel = \frac{3}{\pi}\,g\,\frac{N_{\mathrm{b}}q^2}{\epsilon_0 l_{\mathrm{b}}^3\gamma^2}\,z. \qquad (11.46)$$

If we assume that the bunch is much smaller than the rf bucket, the beam will be symmetric about the synchronous particle, and the force on a particle with relative phase φ may be written as

$$F_\parallel(\varphi) = -\frac{3gN_{\mathrm{b}}q^2}{\pi\epsilon_0 l_{\mathrm{b}}^3\gamma^2}\frac{R}{h}\varphi, \qquad (11.47)$$

since

$$\varphi = -\frac{h}{R}z, \qquad (11.48)$$

where z is the particle's position relative to the synchronous particle, and $2\pi R$ is the circumference of the machine. (A particle at the position z is ahead of the synchronous particle and arrives at the rf cavity early, thus having a negative phase.) The energy increment in one revolution for this particle due to the space-charge force is

$$\delta U_\parallel = \int_0^{2\pi R} F_\parallel\,ds = -\frac{6gN_{\mathrm{b}}q^2R^2}{\epsilon_0 l_{\mathrm{b}}^3 h\gamma^2}\varphi. \qquad (11.49)$$

Recalling the derivation of the energy oscillation equation in § 7.4, the equation of motion for W Eq. (7.25) must be modified to

$$\frac{dW}{dt} = \frac{qV}{2\pi h}(\sin\phi_s - \sin\phi) + \frac{3gN_{\mathrm{b}}q^2R^2}{\pi\epsilon_0 l_{\mathrm{b}}^3 h^2\gamma^2}\varphi. \qquad (11.50)$$

Differentiating Eq. (7.28) and combining it with Eq. (11.50) yields

$$\ddot{\varphi} + \frac{\Omega_s^2}{\cos\phi_s}(\sin\phi - \sin\phi_s) - \kappa\varphi = 0, \tag{11.51}$$

for the phase oscillations, where

$$\Omega_s = \omega_s \sqrt{\frac{h\eta_{tr}\cos\phi_s}{2\pi\beta^2\gamma}\frac{eV}{mc^2}}, \quad \text{and} \tag{11.52}$$

$$\kappa = \frac{12gN_b r_0\eta_{tr}c^2}{l_b^3\gamma^3}. \tag{11.53}$$

The small amplitude formula for φ can then be written as

$$\ddot{\varphi} + (\Omega_s^2 - \kappa)\varphi = 0. \tag{11.54}$$

Usually $\kappa \ll \Omega_s^2$, so that Eq. (11.54) reduces to Eq. (7.31). However at transition crossing the rf power must be turned off, since the phase of the rf voltage must be switched from ϕ_s below transition to $\pi - \phi_s$ above transition, as mentioned in Chapter 7. During this time interval the *phase focusing* disappears; thus the coefficient $(\Omega_s^2 - \kappa)$ will result in the difference between a second order infinitesimal quantity, $\Omega_s^2 \propto V\eta_{tr}$, and a first order one, $\kappa \propto \eta_{tr}$. The result is that near transition $\Omega_s^2 < |\kappa|$, and the longitudinal space-charge force will dominate[18], giving rise to an unwanted bunch-lengthening below transition and a low frequency oscillation above transition. The matching of these two behaviors, can cause an increase in momentum spread.

Another phenomenon which plays a dangerous role during transition crossing is the momentum-spread degradation related to the beam-environment microwave interaction. Anyway, both drawbacks can be cured by making the crossing of transition faster, via the implementation of a jump[19] in the value of the momentum compaction by means of pulsing certain quadrupole magnets.

References for Chapter 11

[1] L. J. Laslett, BNL Report 7534 (1963).

[2] *Conceptual Design of the Superconducting Super Collider*, SSC Central Design Group, SSC-SR-2020, March 1989.

[3] F. Amman et al, Lett. Nuovo Cimento 1, 729 (1969) and "Single and Two Beams Operation in Adone", the proceedings of *VII Int. Conf. on High Energy Accelerators*, Yerevan, Vol. 2, p. 9 (1969).

[4] J. E. Augustin et al., "Development and Beam Studies on ACO", the proceedings of *VII Int. Conf. on High Energy Accelerators*, Yerevan, Vol. 2, p. 9 (1969).

[5] V. L. Auslander et al., "Reconstruction of the Colliding $e^{\pm}$ Beams Device VEPP-2" the proceedings of *VII Int. Conf. on High Energy Accelerators*, Yerevan, Vol. 2, p. 9 (1969).

[6] K. Johnsen, Nucl. Inst. and Meth., 108, 205 (1973).

[7] J. Gareyte, "The CERN $p - \bar{p}$ Complex", the proceedings of *XI Int. Conf. on High Energy Accelerators*, p. 79, Geneva (1980).

[8] F. Amman and D. Ritson, "Space-charge Effects in $e - e$ and $e - e^{+}$ Colliding or Crossing Beam Rings", the proceedings of *III Int. Conf. on High Energy Accelerators*, p. 471, Brookhaven (1961).

[9] E. Keil, "Beam-Beam Dynamics", CERN 95-06 Vol. I, 539 (1995).

[10] B. Zotter, "Experimental Investigation of the Beam-Beam Limit of Proton Beams", the proceedings of *X Int. Conf. on High Energy Accelerators*, V. 2, p. 23, Protvino (1977).

[11] L. A. Schick and D. L. Rubin, "Beam Beam Performance as a Function of β_V^* at CESR", 1991 Particle Accelerator Conference, San Francisco.

[12] D. L. Rubin and L. A. Schick, "Single Interaction Point Operation of CESR", 1991 Particle Accelerator Conference, San Francisco.

[13] B. V. Chirikov, Phys. Rep., 52, 263 (1979).

[14] L. C. Teng, IEEE Trans. Nucl. Sci., NS-20, 843 (1973).

[15] M. Conte, Nucl. Inst. and Meth., 188, 269 (1981).

[16] A. W. Chao, "Beam-Beam Instability", AIP Conf. Proc. No. 127, 201 (1985).

[17] T. K. Khoe and R. L. Martin, IEEE Trans. on Nucl. Sci., NS-24, 1025 (1977).

[18] A. Sørenssen, Particle Accel., 6, 141 (1975).

[19] W. Hardt, the proceedings of *IX Int. Conf. on High Energy Accelerators*, p. 434, Stanford (1974).

12

How to Baffle Liouville

We have seen in Chapter 8 that synchrotron radiation provides a mechanism to decrease the volume of phase space occupied by the beam. In fact, the electromagnetic waves emitted by the electrons on curved trajectories, combined with the restoring action of the rf cavities, can induce a reduction of the normalized emittances. At first this emission was considered an obstacle, since more rf power was needed for the acceleration. Later the requirement of having high luminosities in e^+e^- collider rings made people appreciate synchrotron radiation, particularly for its capacity in reducing the sizes of both beams. Furthermore, rings dedicated to producing synchrotron radiation are now used as intense sources of photons for various areas of research. More recently, interest has arisen in the possible reduction of emittances of heavier particle beams, particularly protons and antiprotons.

This phase space compression has always been called *beam-cooling*, due to the similarity with ordinary thermodynamics, where molecules of a cooler gas occupy a smaller region of phase space, than do warmer particles.

Quickly, we shall give a sketchy look at the various problems:

1. cooling[1,3] of freshly-born antiprotons to be used in p$\bar{\text{p}}$ colliders[4];
2. cooling[5] of both p and $\bar{\text{p}}$ beams in a collider ring, aiming at circumventing beam-beam interaction limitations;
3. cooling[6] in a $\bar{\text{p}}$–storage ring, where a gas jet target has been installed.

The first point is the main topic of the chapter and includes the two main cooling techniques: electron cooling[1], and stochastic cooling[3]. With regard to the second point, we may assert that the diffusive processes, responsible for the unexpected beam loss in the beam-beam interaction, may be counteracted by a cooling mechanism which should play the same role as the synchrotron radiation in e^+e^- colliders. In the last point, we have simply to cool down the $\bar{\text{p}}$-beam which was heated by the multiple Coulomb scattering[7] in the jet of gas.

12.1 Beam temperatures

By beam temperature[7,8], we mean the average kinetic energy of particles as measured in the center of mass of the beam. The temperature is separated into longitudinal and transverse components.

Consider a bunch of particles whose center of mass is moving along the z-axis with velocity given by $\beta_0 c$. Likewise, define the Lorentz factor $\gamma_0 = (1 - \beta_0^2)^{-1/2}$. Quantities in the center of mass will be denoted with an asterisk.

Since the transverse components of momenta are Lorentz invariant,

$$\beta_\perp^* = \frac{p_\perp^* c}{U^*} = \frac{p_\perp c}{U/\gamma_0} = \gamma_0 \beta_\perp. \tag{12.1}$$

The momentum of the center of mass in the center of mass system is zero, so that $U = \gamma_0 U^*$. Then the *transverse temperature*, $T_\perp$, may be defined by

$$k_B T_\perp = \left\langle \frac{1}{2}mc^2 \beta_\perp^{*2} \right\rangle = \left\langle \frac{1}{2}mc^2 \gamma_0^2 \beta_\perp^2 \right\rangle, \tag{12.2}$$

where k_B is the Boltzmann constant. Writing $\beta_\perp \simeq \beta_0 p_\perp / p$ gives

$$k_B T_\perp = \left\langle \frac{1}{2}mc^2 (\beta_0 \gamma_0)^2 \frac{p_\perp^2}{p^2} \right\rangle. \tag{12.3}$$

Moreover,

$$\langle p_\perp^2 \rangle = \langle p_x^2 + p_y^2 \rangle = p^2 (\sigma_{x'}^2 + \sigma_{y'}^2), \tag{12.4}$$

since it makes sense to speak about rms quantities when dealing with large statistics. By introducing the 90% emittances, we obtain

$$\langle p_\perp^2 \rangle = \frac{p^2}{4.6} \left(\frac{\varepsilon_H}{\beta_H} + \frac{\varepsilon_V}{\beta_V} \right), \tag{12.5}$$

so

$$k_B T_\perp = \frac{1}{9.2} mc^2 (\beta_0 \gamma_0)^2 \left(\frac{\varepsilon_H}{\beta_H} + \frac{\varepsilon_V}{\beta_V} \right). \tag{12.6}$$

Using normalized emittances, this becomes

$$k_B T_\perp = \frac{1}{9.2} mc^2 \beta_0 \gamma_0 \left(\frac{\varepsilon_H^*}{\beta_H} + \frac{\varepsilon_V^*}{\beta_V} \right); \tag{12.7}$$

thus the ratio $k_B T_\perp / \beta_0 \gamma_0$ is an invariant.

Similarly, the *longitudinal temperature*, $T_\parallel$, is defined by

$$\frac{1}{2} k_B T_\parallel = \left\langle \frac{1}{2}mc^2 \beta_\parallel^{*2} \right\rangle. \tag{12.8}$$

Boosting to the laboratory system,

$$\beta_{\parallel}^* = \frac{\beta_{\parallel} - \beta_0}{1 - \beta_{\parallel}\beta_0}. \tag{12.9}$$

Defining

$$\Delta\beta = \beta_{\parallel} - \beta_0, \tag{12.10}$$

and

$$\Delta\gamma = \gamma - \gamma_0 \tag{12.11}$$

gives

$$\beta_{\parallel}^* = \frac{\Delta\beta}{\gamma_0^{-2} - \gamma_0^{-3}\Delta\gamma} \simeq \gamma_0^2 \Delta\beta. \tag{12.12}$$

It has been assumed that the energy spread is not excessively large, $\Delta\gamma/\gamma_0 \lesssim 10^{-3}$. Then the longitudinal temperature may be got from

$$\frac{1}{2}k_B T_{\parallel} = \frac{1}{2}mc^2 \left\langle \left(\frac{\Delta\gamma}{\beta_0\gamma_0}\right)^2 \right\rangle = \frac{1}{2}mc^2\beta_0^2 \left\langle \left(\frac{\Delta p}{p}\right)^2 \right\rangle, \tag{12.13}$$

since $\Delta U/U = \beta_0^2 \Delta p/p$. Combining Eqs. (12.6 and 12.13) for three degrees of freedom,

$$\frac{3}{2}k_B T = \frac{mc^2}{2}(\beta_0\gamma_0)^2 \left[\frac{\varepsilon_H}{4.6\beta_H} + \frac{\varepsilon_V}{4.6\beta_V} + \frac{1}{\gamma_0^2}\left(\frac{\sigma_p}{p}\right)^2 \right]. \tag{12.14}$$

Notice how Eq. (12.14) clearly demonstrates that, if the beam temperature is somehow reduced, the transverse emittances decrease together with the momentum spread.

It may be useful to write both the transverse and longitudinal kinetic energies in the moving frame as

$$W_{\perp}^* = \frac{1}{2}\beta_0^2\gamma_0^2 mc^2(\langle x'^2 \rangle + \langle y'^2 \rangle), \quad \text{and} \tag{12.15}$$

$$W_{\parallel}^* = \beta_0^2 mc^2 \left(\frac{\sigma_p}{p}\right)^2. \tag{12.16}$$

12.2 Electron cooling

Electron cooling resembles the cooling of a hot gas by mixing it with a cooler one. The most relevant difference is that, in our example, we deal with gases made of charged particles, thus having to face a typical problem of plasma physics.

Consider a storage ring filled with antiprotons for the hot gas. A beam of electrons, with the same speed of the antiprotons, is made to travel parallel to the heavier particles along a straight section of the ring. In the moving frame we have randomly moving ions and electrons, the former with higher kinetic energy than the latter.

The way ions and free electrons interact is not much different than the process through which ions transfer their energy to the bound electrons of condensed matter atoms. In fact, even though the interaction between charged particles is governed by Coulomb scattering, i. e., by an inherently elastic phenomenon, a strong inelastic behavior is given by the displacement and/or the stripping of the electrons from their atomic orbits in the case of ion-matter interactions, and by the removal of the electrons in the other case.

Equilibrium is reached when the average kinetic energies of both beams coincide in the moving frame, that is,

$$\left\langle \frac{1}{2} m_e \beta^{*2} c^2 \theta_e^{*2} \right\rangle = \left\langle \frac{1}{2} m_p \beta^{*2} c^2 \theta_{\bar{p}}^{*2} \right\rangle, \qquad (12.17)$$

with the angular divergences, θ_e^* and $\theta_{\bar{p}}^*$ related by

$$\theta_{\bar{p}}^* = \sqrt{\frac{m_e}{m_p}} \theta_e^*. \qquad (12.18)$$

For electron cooling the electron beam is of very high quality so that in center of mass frame the electrons have extremely small velocities. When an antiproton of velocity $v_{\bar{p}}^*$ passes an electron at rest with an impact parameter r, the momentum transfer to the electron is approximately

$$\Delta \vec{p}_e^* = \int_{-\infty}^{\infty} \vec{F}^* \, dt^* = -\frac{q^2}{4\pi\epsilon_0} \int_{-\infty}^{\infty} \frac{r\hat{r} + z^* \hat{z}}{(r^2 + z^{*2})^{3/2}} \, dt^*. \qquad (12.19)$$

As simplifying approximations, assume:
1. The antiproton does not appreciably change direction since its mass is much larger than the electron's. This is reasonable if we assume a uniform density of electrons.

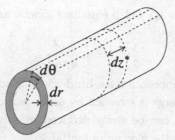

Figure. 12.1 A thin shell of electrons.

2. The electron remains stationary as the antiproton passes it, so that the electron receives an impulse $\Delta \vec{p}_e^*$.

By symmetry the longitudinal component is zero when we integrate over a long time, and the net component is just

$$
\begin{aligned}
p_e^* &= -\frac{q^2 r}{4\pi\epsilon_0} \int_{-\infty}^{\infty} (r^2 + z^{*2})^{-3/2} \frac{dt^*}{dz^*} dz^* \\
&= -\frac{q^2}{2\pi\epsilon_0 v_{\bar{p}}^*} \frac{1}{r}.
\end{aligned} \tag{12.20}
$$

The final kinetic energy of the electron is then

$$
W_e^* = \frac{p_e^{*2}}{2m_e} = \frac{q^2 r_e c^2}{2\pi\epsilon_0 v_{\bar{p}}^{*2} r^2}, \tag{12.21}
$$

and the change in the antiproton's kinetic energy is

$$
\Delta W_{\bar{p}}^* = -W_e^* = -\frac{q^2 r_e c^2}{2\pi\epsilon_0 v_{\bar{p}}^{*2} r^2}. \tag{12.22}
$$

The amount of energy lost by the antiproton due to the electrons in a thin shell of length dz^*, radius r, and thickness dr will then be

$$
-dU^* = W_e^* n_e^* 2\pi r\, dr\, dz^*, \tag{12.23}
$$

where n_e^* is the number density of electrons in the electron beam. This can be integrated to yield the retarding force on the proton,

$$
F^* = \frac{dU^*}{dz^*} \simeq -\frac{q^2 r_e c^2 n_e^*}{\epsilon_0 v_{\bar{p}}^{*2}} \int_{l_L}^{\lambda_D} \frac{dr}{r}, \tag{12.24}
$$

where the Landau distance l_L, which equalizes kinetic and electrostatic energy according to

$$\frac{1}{2}m_e v^{*2} = \frac{q^2}{4\pi\epsilon_0 l_L}, \tag{12.25}$$

is the distance of closest approach for a head-on collision. The Debye length λ_D is the range at which the charge is screened by other particles in the beam.

The screening length can be easily deduced by solving Poisson's equation in cylindrical coordinates for the electric potential Φ,

$$\nabla^2\Phi = \frac{1}{r}\frac{d^2}{dr^2}(r\Phi) = -\frac{\rho_e + \rho_{\bar{p}}}{\epsilon_0} \simeq \frac{qn_e^*}{\epsilon_0}, \tag{12.26}$$

since the electron beam is typically more dense than the antiproton beam. If the normal electron distribution is slightly altered by a thermal condition

$$n^*(k_B T) = n_{e0}^* \exp\left(\frac{-q\Phi}{k_B T}\right), \tag{12.27}$$

in lowest order

$$n_e^* = n_{e0}^* - n^*(k_B T) \simeq n_{e0}^* \frac{q\Phi}{k_B T}. \tag{12.28}$$

(Notice how the density decreases as the temperature increases.) By substitution into Poisson's equation,

$$\frac{d^2(r\Phi)}{dr^2} = \frac{n_{e0}^* q^2}{\epsilon_0 k_B T} r\Phi, \tag{12.29}$$

with the solution

$$\Phi = \frac{q^2}{4\epsilon_0 r} e^{-r/\lambda_D}, \tag{12.30}$$

where the Debye length is defined by

$$\lambda_D = \sqrt{\frac{\epsilon_0 k_B T}{n_{e0}^* q^2}}. \tag{12.31}$$

Evaluating the integral in Eq. (12.24) gives

$$m_p \frac{dv_{\bar{p}}^*}{dt^*} = -\frac{q^2 r_e c^2 n_e^*}{\epsilon_0 v_{\bar{p}}^{*2}} \ln\left(\frac{\lambda_D}{l_L}\right), \tag{12.32}$$

and the deceleration of the proton is

$$\frac{dv_{\bar{p}}^*}{dt^*} = -\frac{4\pi r_e r_p c^4 n_e^* \ln(\Lambda)}{v_{\bar{p}}^{*2}}, \tag{12.33}$$

with $\Lambda = \lambda_D/l_L$, and the classical radius of the antiproton

$$r_p = \frac{q^2}{4\pi\epsilon_0 m_p c^2}. \tag{12.34.}$$

Eq. (12.33) may be rearranged and integrated to estimate the stopping time τ^*;

$$\int_{v_{\bar{p}}^*}^{0} v_{\bar{p}}^{*2} dv_{\bar{p}}^* = -\int_{0}^{\tau^*} [4\pi r_e r_p c^4 n_e^* \ln(\Lambda)] dt^*; \tag{12.35}$$

then

$$\tau^* = \frac{v_{\bar{p}}^{*3}}{12\pi r_e r_p c^4 n_e^* \ln(\Lambda)}. \tag{12.36}$$

A more pessimistic estimate would be to assume the deceleration remains constant rather than increasing as the proton loses energy:

$$\tau^* = -\frac{v_{\bar{p}}^*}{dv_{\bar{p}}^*/dt^*} = \frac{v_{\bar{p}}^{*3}}{4\pi r_e r_p c^4 n_e^* \ln(\Lambda)}. \tag{12.37}$$

In reality the electrons are moving since they have a finite temperature, and a more careful calculation could be made.

A Maxwellian distribution of velocities can be written as

$$f(v^*) = \left(\frac{m}{2\pi k_B T}\right)^{\frac{1}{2}} \exp\left(-\frac{mv^{*2}}{2k_B T}\right). \tag{12.38}$$

By integrating the first and second moments, it is easy to show that

$$\langle v^{*2} \rangle = \frac{3k_B T}{m}, \quad \text{and} \tag{12.39}$$

$$\langle v^* \rangle = \sqrt{\frac{8k_B T}{\pi m}} = \sqrt{\frac{8}{3\pi} \langle v^2 \rangle}. \tag{12.40}$$

Approximating $v_{\bar{p}}^{*3}$ by $\langle v_{\bar{p}}^{*2} \rangle^{3/2}$ and substituting into Eq. (12.37) combined with Eq. (12.14) yields

$$\tau^* = \frac{\beta_0{}^3 \gamma_0{}^3}{4\pi r_e r_p c n_e^* \ln(\Lambda)} \left[\frac{\varepsilon_{\mathrm{H}}}{4.6\beta_{\mathrm{H}}} + \frac{\varepsilon_{\mathrm{V}}}{4.6\beta_{\mathrm{V}}} + \frac{1}{\gamma_0{}^2}\left(\frac{\sigma_p}{p}\right)^2\right]^{\frac{3}{2}}. \tag{12.41}$$

This damping time may be transformed to the laboratory frame by correcting for time dilation,

$$\tau = \gamma_0 \tau^*, \tag{12.42}$$

and for the Lorentz contraction of the number density,

$$n_e^* = \frac{n_e}{\gamma_0} = \frac{J_e}{q\beta_0 c\gamma_0};\tag{12.43}$$

where J_e is the current density of the electron beam. The result is

$$\tau = \frac{q\beta_0{}^4\gamma_0{}^5}{4\pi r_e r_p \eta J_e \ln(\Lambda)} \left[\frac{\varepsilon_{\mathrm{H}}}{4.6\beta_{\mathrm{H}}} + \frac{\varepsilon_{\mathrm{V}}}{4.6\beta_{\mathrm{V}}} + \frac{1}{\gamma_0{}^2}\left(\frac{\sigma_p}{p}\right)^2 \right]^{\frac{3}{2}},\tag{12.44}$$

where the factor η is the ratio of the cooling section to the rings circumference. More correctly, the ratio $\Lambda = \lambda_D/l_L$ should be modified by the Lorentz transformation; however, to first order, the effect should cancel in the ratio. In any case, $\gamma_0 \lesssim 2$ typically in most cooling systems so that with the logarithm the correction is negligible.

If the antiproton temperature is low enough, then

$$\langle v_{\bar p}^2 \rangle \simeq \langle v_e^2 \rangle,\tag{12.45}$$

so that, recalling Eq. (12.39),

$$v_{\bar p}^{*3} \sim v_e^{*3} \sim \left(\frac{3k_B T_e}{m_e c^2}\right)^{\frac{3}{2}} c^3.\tag{12.46}$$

Near this equilibrium condition, the rate for damping of individual antiprotons is then

$$\frac{1}{\tau_{\mathrm{cool}}^*} \simeq \frac{4\pi}{3\sqrt{3}} r_e r_p c n_e^* \ln(\Lambda) \left(\frac{k_B T_e}{m_e c^2}\right)^{-\frac{3}{2}},\tag{12.47}$$

or in the laboratory system

$$\frac{1}{\tau_{\mathrm{cool}}} \simeq \frac{4\pi}{3\sqrt{3}} \frac{r_e r_p \eta J_e \ln(\Lambda)}{q\beta_0 \gamma_0{}^2} \left(\frac{k_B T_e}{m_e c^2}\right)^{-\frac{3}{2}};\tag{12.48}$$

however, at equilibrium an antiproton which is "cooler" than the electrons will be heated up at a rate which is equal but opposite to Eq. (12.48). For a beam of antiprotons which is slightly warmer than the electrons by an amount ΔT, the damping rate may be approximated by

$$\frac{1}{\tau} \simeq \frac{4\pi}{3\sqrt{3}} \frac{r_e r_p \eta J_e \ln(\Lambda)}{q\beta_0 \gamma_0{}^2} \left[\left(\frac{k_B(T_e + \Delta T)}{m_e c^2}\right)^{-\frac{3}{2}} - \left(\frac{k_B T_e}{m_e c^2}\right)^{-\frac{3}{2}} \right] - \frac{1}{\tau_{\mathrm{ext}}}\tag{12.49}$$

where $1/\tau_{\mathrm{ext}}$ is a heating rate which comes from some external source of heat, such as scattering from residual beam gas.

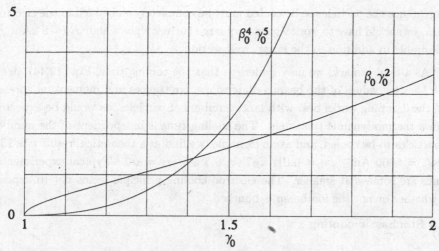

Figure. 12.2 Plots of the two kinematic factors $\beta_0 \gamma_0^2$ and $\beta_0^4 \gamma_0^5$.

To gain a better insight into the realm of electron cooling compare the two kinematic factors $\beta_0 \gamma_0^2$ and $\beta_0^4 \gamma_0^5$ of Eqs. (12.44 and 12.49). As clearly shown in Fig. 12.2, the two functions are equal when $\beta_0 \gamma_0 = 1$, i. e., when $\gamma_0 = \sqrt{2}$ and $\beta_0 = \sqrt{2}/2$; elsewhere they exhibit sharply different behaviors which may be emphasized as follows:

$\gamma_0 < \sqrt{2}$:

$\beta_0 \gamma_0^2$ decreases slowly and is thus of little interest;

$\beta_0^4 \gamma_0^5$ decreases dramatically, giving very efficient cooling.

$\gamma_0 > \sqrt{2}$:

$\beta_0^4 \gamma_0^5$ rapidly becomes too large, and the damping is too slow;

$\beta_0 \gamma_0^2$ increases but may still be useful for intermediate energy antiprotons.

In fact, for a gas-jet target or p̄p collisions, electrons keep the beam cool by taking away the "heat" arising from multiple Coulomb scattering of the antiprotons interacting with either the gas or the opposing beam.

Oddly, electron cooling is not very appropriate for reducing the phase-space volume of freshly created antiprotons; in fact the p̄-beam from the production target has a wide momentum spread and large angular divergences, i. e., huge emittances. While the former disadvantage can be somewhat overwhelmed by setting the electron cooling device in a dispersion-free straight section, there is little to do about the drawback of large transverse emittances other than accept a broad antiproton beam. Since the particles to be cooled must be completely embedded in the electron beam, we should have to work with a very large current since a high J_e ($\sim 5 \text{ kAm}^{-2}$) is desirable in addition to the large cross section.

As a final remark, we may underline that the cooling time, Eq. (12.44), deals with the rms values of the beam's dimensions, emittances and momentum spread, and the damping works best with large numbers of particles, as would be expected from a thermodynamic treatment. The cooling time is independent of the number of particles to be cooled, and as an example, we find as a theoretical result $\tau \simeq 15$ s for $J_e = 5000 \text{ Am}^{-2}$, $\eta = 0.01$, $k_B T = 0.5$ eV, $\gamma_0 = \sqrt{2}$. Typical experimental values are somewhat smaller. The electron cooling principle seems not to depend on whether or not the ion beam is bunched.

12.3 Stochastic cooling

In electron cooling two charged gases are mixed, one hot and one cold; this is a macroscopic thermodynamic process which works well for a large number of particles. In stochastic cooling, on the other hand, fluctuations in the average motions and positions of particles are sensed as the beam passes by a high speed detector and a signal is sent to a correction element which nudges the errant group of charges toward the core of the beam, very much like Maxwell's demon; this works best with a small number of particles and with electronics having a very high bandwidth.

As shown in Chapter 3, the electromagnetic fields in an accelerator cannot compress the volume of phase space that a beam occupies. At first the idea of stochastic cooling seems to somehow violate Liouville's theorem, but it must be realized that the beam is made of individual particles with a lot of empty space surrounding each particle. (The quantum mechanical volume in phase space of a single particle is on the order of $\hbar^3$, which is very much smaller than the average volume occupied by a single particle in a typical beam.) With a sparse beam, the "holes" and particles may be rearranged so that some empty space migrates to the outside of the beam. As the number of particles increases, the detector and corrector must act faster and faster, until the beam becomes a blur, and the fluctuations are undetectable. In the high statistics limit, where thermodynamics works best, Maxwell's demon cannot function because kicks from the corrector become less efficient, redirecting some particles from the core of the beam while attempting a correction. As the corrector works harder, more power is required to drive it, and more energy ends up dissipated in the beam, thus producing a competing mechanism

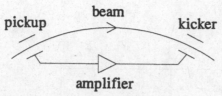

Figure. 12.3 Simple system for stochastic cooling.

which stops the cooling when too many particles are present.

The basic setup, shown in Fig. 12.3, consists of picking up a signal, proportional to the position of the center of mass of part of the beam, and sending it through an amplifier to a correction kicker somewhat farther along the accelerator. It is important that this be done before the synchrotron and betatron oscillations scramble the beam; in other words, the kick must be applied while the group of particles is coherent.

The microscopic system to be cooled is mainly a set of harmonic oscillators, representing either a charged particle beam[11-14] in a dedicated ring, or a collection[15] of charged particles in a Penning trap. Also a system of computerized traffic lights could be considered as a cooling system in a traffic jam of vehicles. Here we will deal mainly with stochastic cooling of a beam in a circular accelerator; that is, of a set of three systems of harmonic oscillators: horizontal and vertical betatron oscillations plus energy oscillations.

The fluctuations are typically detected by either capacitive pickups or a small inductive transformer. The corrector can be either a pulsed bending magnet or a correcting rf gap. Signals can be sent along a chord of the ring with a shorter path length than the arc of the design orbit; this helps offset the slower than light speed of the amplifier, cables, and rise times of the pickup and kicker.

The usual betatron oscillation equations can be approximated by

$$\frac{d^2z}{dt^2} + \omega_s^2 Q^2 z = 0, \tag{12.50}$$

where z and Q are respectively either the horizontal or vertical position and tune.

Let us assume that the correcting kick $qE_\perp f(t)$ starts at a certain instant t^* and lasts a time interval equal to T_s (see Fig. 12.4), which corresponds to the duration of that share of the beam sampled by the pickup. Due to this correcting force, the betatron equation becomes

$$\frac{d^2z}{dt^2} + \omega_s^2 Q^2 z = \frac{qE_\perp}{\gamma m} f(t). \tag{12.51}$$

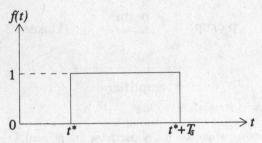

Figure. 12.4 Ideal time structure of the general correcting kick.

Synchrotron oscillations can be treated in a similar way. Recalling § 7.4, the equation energy oscillations for small amplitudes may be written as

$$\frac{d^2u}{dt^2} + \Omega_s^2 u = 0, \tag{12.52}$$

where

$$u = \delta U = U - U_s = -\omega_{\rm rf} W, \tag{12.53}$$

as in Eq. (7.24). The incremental energy increase per turn from the normal rf cavities is

$$\delta u = qV(\sin\phi - \sin\phi_s)$$
$$\simeq qV\cos\phi_s\,\varphi. \tag{12.54}$$

A correcting kick for longitudinal oscillations could be written as $-qE_{\parallel}gf(t)$, where g is the gap length of the rf kicker, and $f(t)$ is again a square pulse, as shown in Fig. 12.4. The oscillation equation must then be changed to

$$\frac{d^2u}{dt^2} + \Omega_s^2 u = \Omega_s^2 qE_{\parallel}gf(t). \tag{12.55}$$

The dispersive part of the horizontal motion,

$$x_p = \eta\frac{\Delta p}{p} = \eta\frac{u}{\beta^2\gamma mc^2}, \tag{12.56}$$

may be solved for u,

$$u = \beta^2\gamma mc^2\frac{x_p}{\eta}, \tag{12.57}$$

and inserted into Eq. (12.55), yielding

$$\frac{d^2x_p}{dt^2} + \Omega_s^2 x_p = \frac{\Omega_s^2 qE_{\parallel}\eta g}{\beta^2\gamma mc^2}f(t), \tag{12.58}$$

which is an equation similar to Eq. (12.51).

Nevertheless, the actual behavior of synchrotron oscillations is slightly different. As shall be seen, the stochastic cooling works better with unbunched or coasting beams, and we may consider the energy oscillations arising from the correcting rf cavity. In broad terms, it may be stated that we observe a small sample of the circulating particles undergoing synchrotron oscillations superimposed on the whole of the unbunched beam.

Having reduced the longitudinal case to the same form as the transverse case, the generic problem can be solved by using Eq. (12.51) for all three dimensions. Considering the kth particle of the sample, we have just before the kick at $t = t^*$,

$$z_k(t^*) = A_k \sin(\omega_k t^* + \phi_k), \quad \text{and} \tag{12.59}$$

$$\frac{dz_k}{dt}(t^*) = \omega_k A_k \cos(\omega_k t^* + \phi_k), \tag{12.60}$$

respectively for the position and velocity of the particle, where the A_k and ϕ_k are determined by initial conditions. The oscillation frequency ω_k is the frequency of the particle's mode of oscillation, i. e., the betatron frequency for transverse motion, and the synchrotron frequency for longitudinal motion. Just after the kick at $t = t^* + T_s$,

$$z_k(t^* + T_s) = A_k' \sin[\omega_k(t^* + T_s) + \phi_k'] = z_k(t^*), \quad \text{and} \tag{12.61}$$

$$\frac{dz_k}{dt}(t^* + T_s) = \omega_k A_k' \cos[\omega_k(t^* + T_s) + \phi_k'] = \frac{dz_k}{dt}(t^*) + \Delta v, \tag{12.62}$$

where the velocity impulse is

$$\Delta v \simeq \begin{cases} \frac{qE_\perp}{\gamma m} T_s & \text{for transverse betatron oscillations,} \\ \Omega_s^2 \eta \frac{qE_\parallel g}{\beta^2 \gamma m c^2} T_s & \text{for longitudinal synchrotron oscillations.} \end{cases} \tag{12.63}$$

If the correcting field is proportional to ω_k, and also to the amplitude A_{cm} of oscillations of the sample's center of mass, we may set

$$E = -\lambda A_{\text{cm}} \omega_k, \quad \text{and} \tag{12.64}$$

$$\frac{\Delta v}{\omega_k} = -G A_{\text{cm}}, \tag{12.65}$$

where λ and G are proportionality constants. Then combining Eqs. (12.61 and 12.62),

$$A_k'^2 = A_k^2 + G^2 A_{\text{cm}}^2 - 2G A_{\text{cm}} A_k \cos(\omega_k t^* + \phi_k). \tag{12.66}$$

12.4 Center of mass of a system of oscillators

Consider the definition of the center of mass of a system of N_s oscillators,

$$z_{\text{cm}} = \frac{1}{N_s} \sum_{k=1}^{N_s} A_k \sin(\omega_k t + \phi_k) = \frac{1}{N_s} \sum_{k=1}^{N_s} A_k \sin \theta_k, \tag{12.67}$$

with $\theta_k = \omega_k t + \phi_k$. The first thought would be to consider it as null, since these oscillators are completely uncorrelated.

Yet if a portion of the beam like our sample tends to assume some off-equilibrium position, its centroid will oscillate with an average amplitude different from zero;

$$z_{\text{cm}} = A_{\text{cm}} \sin(\omega_{\text{cm}} t + \phi_{\text{cm}}) = A_{\text{cm}} \sin \theta_{\text{cm}}, \tag{12.68}$$

where $\theta_{\text{cm}} = \omega_{\text{cm}} t + \phi_{\text{cm}}$. By identifying Eq. (12.67) with Eq. (12.68),

$$\sin \theta_{\text{cm}} = \frac{1}{N_s A_{\text{cm}}} \sum_{k=1}^{N_s} A_k \sin \theta_k, \tag{12.69}$$

where N_s is now the number of particles constituting the sample.

If one considers all the terms of the summation as vertical components of some vectors, the corresponding horizontal components will yield

$$\cos \theta_{\text{cm}} = \frac{1}{N_s A_{\text{cm}}} \sum_{k=1}^{N_s} A_k \cos \theta_k. \tag{12.70}$$

Then squaring and adding Eqs. (12.69 and 12.70) gives

$$A_{\text{cm}}^2 = \frac{1}{N_s^2} \sum_{k=1}^{N_s} A_k^2 + \frac{1}{N_s^2} \sum_{k=1}^{N_s} \sum_{\substack{j=1 \\ j \neq k}}^{N_s} A_k A_j \cos(\theta_k - \theta_j). \tag{12.71}$$

Bearing in mind the definition of an average over N_s items, this is

$$A_{\text{cm}}^2 = \frac{1}{N_s} \langle A^2 \rangle + \frac{1}{N_s^2} \sum_{k=1}^{N_s} \sum_{\substack{j=1 \\ j \neq k}}^{N_s} A_k A_j \cos(\theta_k - \theta_j). \tag{12.72}$$

The velocity of the center of mass may be obtained by differentiating z_{cm}, so

$$v_{\text{cm}}(t) \simeq \omega_{\text{cm}} A_{\text{cm}} \cos(\omega_{\text{cm}} t + \phi_{\text{cm}}), \tag{12.73}$$

having assumed that ω_{cm} and ϕ_{cm} are slowly varying functions of time. By a suitable choice of t^*, which can be made by adjusting the length of cables between the pickup and corrector, we may ensure that the correction takes place when v_{cm} reaches its maximum value;

$$v_{cm}(t^*) = \omega_{cm} A_{cm}. \qquad (12.74)$$

For the best effect, transverse oscillations should be measured at a maximum of the β-function and corrected 90° later in betatron phase; longitudinal fluctuations should be measured at a maximum of the η-function and corrected where $\eta = 0$, since $Q_s \ll 1$. When v_{cm} is a maximum, θ_{cm} is an integral multiple of 2π, so

$$\cos(\omega_k t^* + \phi_k) = \cos[(\omega_k t^* + \phi_k) - (\omega_{cm} t^* + \phi_{cm})] \simeq \cos(\phi_k - \phi_{cm}). \qquad (12.75)$$

Here we have assumed that the amplitude of the fluctuations are not excessively large and that the nonlinearities of the ring are reasonable; ω_k should be about the same as ω_{cm}.

Averaging Eq. (12.66) over the sample produces

$$\langle A'^2 \rangle = \langle A^2 \rangle + G^2 A_{cm}^2$$
$$- \frac{2G A_{cm}}{N_s} \left[\cos \phi_{cm} \sum_{k=1}^{N_s} A_k \cos \phi_k + \sin \phi_{cm} \sum_{k=1}^{N_s} A_k \sin \phi_k \right]. \qquad (12.76)$$

Combining Eqs. (12.69 and 12.70) via Euler's formula,

$$e^{i\omega_{cm}t} e^{i\phi_{cm}} = \frac{1}{N_s A_{cm}} \sum_{k=1}^{N_s} A_k e^{i\omega_k t} e^{i\phi_k}$$

$$\simeq \frac{e^{i\omega_{cm}t}}{N_s A_{cm}} \sum_{k=1}^{N_s} A_k e^{i\phi_k}, \qquad (12.77)$$

so that

$$A_{cm} \cos \phi_{cm} = \frac{1}{N_s} \sum_{k=1}^{N_s} A_k \cos \phi_k, \quad \text{and} \qquad (12.78)$$

$$A_{cm} \sin \phi_{cm} = \frac{1}{N_s} \sum_{k=1}^{N_s} A_k \sin \phi_k, \qquad (12.79)$$

which may be inserted into Eq. (12.76) to give

$$\langle A'^2 \rangle = \langle A^2 \rangle + (G^2 - 2G)A_{\rm cm}^2. \tag{12.80}$$

Forming a difference equation over one complete revolution of period τ_s, we have

$$\frac{d\langle A^2 \rangle}{dt} \simeq \frac{\langle A'^2 \rangle - \langle A^2 \rangle}{\tau_s} = -\frac{2G - G^2}{\tau_s} A_{\rm cm}^2, \tag{12.81}$$

or since $A_{\rm cm}^2 \simeq \langle A^2 \rangle / N_s$,

$$\frac{d\langle A^2 \rangle}{dt} \simeq -\frac{2G - G^2}{N_s \tau_s} \langle A^2 \rangle. \tag{12.82}$$

Notice that N_s is the number of particles in the sample of duration T_s; these quantities are linked to the revolution period τ_s and the total number, N, of particles in the ring by the simple relation

$$\frac{N_s}{T_s} = \frac{N}{\tau_s}. \tag{12.83}$$

From Nyquist's theorem, the required bandwidth of the amplifier for samples of length T_s is

$$W = \frac{1}{2T_s}. \tag{12.84}$$

By examining Eq. (12.82), the exponential cooling rate may now be written as

$$\frac{1}{\tau_{\rm cool}} = \frac{1}{NT_s}(2G - G^2) = \frac{2W}{N}(2G - G^2). \tag{12.85}$$

For $G = 1$ the cooling time is

$$\tau_{\rm cool} = \frac{N}{2W}. \tag{12.86}$$

The factor $2G$ arises from damping of the coherent motion, and G^2 is caused by inherent heating between the different particles.

12.5 Noise

The cooling time of Eq. (12.86) does not account for the noise due to the electronics of the cooling system. We may describe this effect as an additional signal entering the amplifier with a random phase (see Fig. 12.5). Since this noise can be conceived as a blur which overlaps the center of mass amplitude, it may be described as a sinusoidal signal with frequency ω_n, arbitrary phase ϕ_n, and amplitude A_n, depending on the quality of electronic apparatus. Even though the noise will have a broad spectrum, only frequencies close to $\omega_{\rm cm}$ can give a nonzero contribution to the average, so we may use

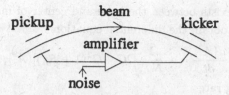

Figure. 12.5 Addition of the electronic noise to the correction signal.

$$\omega_n \simeq \omega_{cm}. \tag{12.87}$$

Eq. (12.68) must now be amended to

$$z_{cm} = A_{cm} \sin(\omega_{cm}t + \phi_{cm}) + A_n \sin(\omega_{cm}t + \phi_n), \tag{12.88}$$

which can be transformed into

$$z_{cm} = A_T \sin(\omega_{cm}t + \phi_{cm} + \Delta\phi_n), \tag{12.89}$$

if

$$A_n \sin(\phi_n - \phi_{cm}) = A_T \sin \Delta\phi_n, \quad \text{and} \tag{12.90}$$

$$A_{cm} + A_n \cos(\phi_n - \phi_{cm}) = A_T \cos \Delta\phi_n. \tag{12.91}$$

Eliminating $\Delta\phi_n$ from the last pair of equations yields

$$A_T^2 = A_{cm}^2 + A_n^2 + 2A_n A_{cm} \cos(\phi_n - \phi_{cm}). \tag{12.92}$$

Similarly, Eq. (12.66) may be revised to

$$A_k'^2 = A_k^2 + G^2 A_T^2 - 2G A_T A_k \cos(\phi_k - \phi_{cm} - \Delta\phi_n). \tag{12.93}$$

After substitution of Eq. (12.92), the last equation becomes

$$A_k'^2 = A_k^2 + G^2 A_{cm}^2 - 2G A_{cm} A_k \cos(\phi_k - \phi_{cm} - \Delta\phi_n)$$
$$- G^2 A_n^2 + 2G^2 A_n A_{cm} \cos(\phi_n - \phi_{cm}). \tag{12.94}$$

Again averaging over N_s particles gives

$$\langle A'^2 \rangle = \langle A^2 \rangle + (G^2 - 2G)A_{cm}^2 + G^2 A_n^2, \tag{12.95}$$

since no correlation exists between the noise and center of mass oscillations. Proceeding as above, we obtain the differential equation

$$\frac{d\langle A^2 \rangle}{dt} = -\frac{1}{NT_s}[2G - G^2(1 + R_n)]\langle A^2 \rangle, \tag{12.96}$$

which has the cooling rate

$$\frac{1}{\tau_{\text{cool}}} = \frac{2W}{N}[2G - G^2(1 + R_n)], \tag{12.97}$$

where

$$R_n = \frac{\langle A_n^2 \rangle}{\langle A^2 \rangle} N_s \tag{12.98}$$

is the reciprocal of the signal to noise ratio with $\langle A^2 \rangle$ averaged over the particles in the sample, and $\langle A_n^2 \rangle$ averaged over one period of revolution.

12.6 Mixing

So far we have assumed a straightforward correction for a random sample of particles running from the pickup to the corrector. The large momentum spread, commonly existing when cooling is required, means that the velocities differ somewhat if the particles are not ultrarelativistic, and the periods of revolution vary for $\eta_{\text{tr}} \neq 0$. This variation of times tends to mix or randomize the particles at some amount of time after a fluctuation has been sensed. The cooling system acts on the statistical average of the sample; if there were no mixing, after the first correction the group of particles would have a centroid of zero, and no more cooling would occur. With mixing, the particles' phases again become random, so that a typical measurement would be $\sqrt{\langle z^2 \rangle / N_s}$ rather than zero.

A complete correction of the offending particles may not occur, since the signal from the pickup may be diluted by other particles in this time slice. Additionally, the corrector will kick all particles which are in the interval of t^* to $t^* + T_s$, and may excite some particles which were already cooled. Indeed, this correction may ultimately take place over several turns because of slow mixing.

Thus the cooling result must again be revised[17]:

$$\frac{1}{\tau_{\text{cool}}} = \frac{2W}{N}[2G - G^2(M + R_n)], \tag{12.99}$$

where $M \geq 1$ is the number of turns over which the correction is applied. For a given signal to noise ratio and a given mixing, the minimum possible value of τ_{cool} occurs when

$$G = \frac{1}{R_n + M}, \tag{12.100}$$

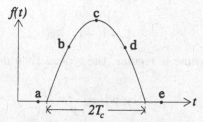

Figure. 12.6 Time dependence on how the cooling system response reacts to momentum spread. For five test particles sensed at the same time, the time of arrival at the kicker will vary with momenta. Particle "c" has the nominal momentum and receives the maximum correction. For $\eta_{tr} < 0$, the higher momentum particles "d" and "e" arrive late. Particles "a" and "b", with low momentum, arrive early. The out of time particles "b" and "d" receive a smaller kick than the nominal particle. If the momentum deviation is large enough, the particles ("a" and "e") will miss the pulse altogether.

giving

$$(\tau_{\text{cool}})_{\min} = \tau_0 = \frac{N}{2W}(R_n + M). \tag{12.101}$$

To obtain the best results, it is desirable to have M as close to one as possible; then the cooling is optimum with the largest velocity impulse Δv for a given A_{cm}^2 (see Eq. 12.65). A quickly cooled sample will no longer be seen by the pickup, leaving room for the correct recording of the next samples. Furthermore, a bad mixing would allow particles already cooled to be coherently reheated by being kicked with a pulse meant for a different sample of hot particles.

On the other hand, a very good mixing is not so appropriate from the point of view of a realistic analysis of the pulse $f(t)$. Off-momentum particles will tend to slip away from their self-induced correction signals. Since the electronics cannot have an infinite bandwidth, the correcting pulse will not actually be a square pulse, but will have finite rising and falling times. The real hardware will be adjusted for the nominal particle (with $\Delta p/p = 0$), so that particles ahead and behind this optimum one will get somewhat smaller kicks, receiving only a partial correction.

A more realistic approximation for the pulse is a parabolic shape,

$$f(t) = 1 - \left(\frac{\Delta t}{T_c}\right)^2, \tag{12.102}$$

where Δt is the deviation in time of flight of a given particle from the nominal value between the pickup and kicker, and $2T_c$ is the length of the parabolic pulse (see Fig. 12.6). The circumference L of the ring may be split into two pieces: the distance l_1 from the pickup to the corrector, and the distance l_2 back around to the pickup, so that

$$L = l_1 + l_2. \tag{12.103}$$

Assuming that the machine is regular, the typical time deviation from pickup to kicker is approximately

$$(\Delta t)_{\text{PK}} \simeq -\eta_{\text{tr}} \frac{\sigma_p}{p} \tau_s \frac{l_1}{L}, \tag{12.104}$$

for a single turn.

The coherent effect of the pulse $f(t)$ must be reduced; on average the kicks are smaller for the new parabolic form than for the old square form. This roughly amounts to letting the term of Eq. (12.99) which is linear in G be replaced by $G(1 - \tilde{M}^{-2})$ with

$$\tilde{M} \simeq \frac{T_c}{(\Delta t)_{\text{PK}}}. \tag{12.105}$$

For a reasonably regular ring

$$\frac{\tilde{M}}{M} \simeq \frac{l_2}{l_1}. \tag{12.106}$$

Revising the cooling rate yet again,

$$\frac{1}{\tau_{\text{cool}}} = \frac{2W}{N} \left[2G \left(1 - \frac{1}{\tilde{M}^2} \right) - G^2 (M + R_n) \right]. \tag{12.107}$$

In principle, a ring could be designed in such a way as to have almost no contribution to the phase slip factor η_{tr} from the pickup to the kicker, and a large contribution from the kicker to the pickup. Such a machine could minimize the mixing between detection of the signal and the correction, but would maximize the randomization after the kick before the particles reappear at the pickup. This type of machine is rather complicated and has not been built.

The stochastic cooling method deals separately with the three degrees of freedom; the horizontal, vertical, and longitudinal amplitudes can be reduced independently of each other. It should be noted that we have found the cooling rate for the square of the amplitude $1/\tau_{(A^2)}$, and that the cooling rate for the amplitude is

$$\frac{1}{\tau_{(A)}} = \frac{1}{2} \frac{1}{\tau_{(A^2)}}. \tag{12.108}$$

Since the sample can be conceived as a self-forming fluctuation in the particle distribution in phase space, the best situation is to have a beam as smeared as possible, i. e., a coasting beam. Nevertheless, bunched beams can still be handled, provided that one is prepared to accept a cooling time scaled by the inverse of the bunching factor. For an unbunched beam synchrotron oscillations are not present, except for those generated by the kicker itself. A more rigorous treatment requires the use of the Fokker-Planck equation[17].

Stochastic cooling does not depend on energy, apart from the obvious requirement of using stronger fields in the kicker. Instead the number of particles to be cooled is very critical; even with optimum conditions ($R_n = 0$, $M = 1$, $\tilde{M} = \infty$), the minimum cooling time ($\tau_{(A)}$) is

$$\tau = \frac{N}{W} \sim \frac{1}{2}\text{day} \tag{12.109}$$

for $W = 250$ MHz, and $N = 10^{13}$. In spite of this rather long exponential decay time, stochastic cooling is unparalleledly useful for increasing the phase space density of a beam emerging from a production target, as occurs in the creation of antiprotons. In fact, no drawbacks arise from the size of the beam to be cooled.

As in all physical processes, there are inherent limitations. In the electron cooling, the space charge of the beam was the final obstacle; with stochastic cooling, the noise is responsible for the actual limit. Of course both methods are not comparable to radiation damping of electrons, which is bounded by quantum limitations; but the mass ratio between ions and electrons makes superfluous any further comment.

Problems

12–1 Invent a new method of beating Liouville and win the Nobel prize.

References for Chapter 12

[1] G. I. Budker, Atomnaya Energiya, 22, 346 (1967).

[2] G. Y. Budker et al., Part. Accel., 7, 197 (1976).

[3] S. van der Meer, "Stochastic Damping of Betatron Oscillations", Internal Report CERN/ISR PO 72-31.

[4] C. Rubbia, P. McIntyre, and D. Cline, "Producing Massive Neutral Intermediate Vector Bosons with Existing Accelerators", Proc. Int. Neutrino Conf., Achen, 1967, p. 683, Vieweg Verlag, Braunschweig (1977).

[5] P. Dal Piaz, "Charmonium and Other Onia at Minimum Energy", KfK Report 2836, H. Poth Ed., p. 111, Karlsruhe (1979).

[6] M. Macrì, "Gas Jet Targets", CERN 84-15, p. 469 (1984).

[7] J. D. Lawson, "Particle Beams and Plasmas", CERN 76-09 (1976).

[8] J. D. Lawson, The Physics of Charged-Particle Beams, Clarendon Press, Oxford (1977).

[9] H. Poth, "Electron Cooling", p. 534, CERN 87-3 (1987)

[10] E. H. Kennard, Kinetic Theory of Gases, McGraw-Hill, New York (1938).

[11] Design Study Team, "Design Study of a Proton–Antiproton Colliding Beam Facility", CERN/PS/AA 78-3 (1978).

[12] Design Study Team, "Design Study of a Antiproton Collector for the Antiproton Accumulator (ACOL)", CERN 83-10 (1983).

[13] R. E. Shafer, "Overview of the Fermilab Antiproton Source", Proceedings of the 12th International Conference on High-Energy Accelerators, p. 24, Batavia (1983).

[14] J. D. McCarthy, "Operational Experience with Tevatron I Antiproton Source", 1987 IEEE Particle Accelerator Conference, p. 1388, Washington, D. C. (1987).

[15] N. Beverini et al., Phys. Rev. 38A, 107 (1988).

[16] A. M. Rosie, Information and Communication Theory, Van Nostrand Co., London (1973)

[17] D. Möhl, "Stochastic Cooling for Beginners", p. 97, CERN 84-15.

13

Spin Dynamics

13.1 Introduction

The topic of spin or *intrinsic angular momentum* is a very rich field in physics. When dealing with the microscopic aspects of fundamental particles, we frequently find that a classical description is not sufficient and that quantum mechanics must be applied, especially where spin is concerned. From the standpoint of accelerator physics, the primary goal is to produce polarized beams of particles to study the fundamental interactions of high energy and nuclear physics.[1] Examples of this are high precision mass measurements of mesons[2] such as the Υ measuring and understanding $g-2$ of muons[3,4]; learning how the protons spin is distributed through its constituents[1]: the intrinsic spins of valence quarks, sea quarks, gluons, and orbital angular momentum these components. Additionally, beams of various isotopes have been studied to measure their spin and magnetic moments.[5,6]

For treating the spin of particles in the beam, quantum effects (such the Sokolov-Ternov effect[7,8] in which spin flip transitions from synchrotron radiation can lead to a build up of polarization of an electron beam in a storage ring) sometimes need to be taken into account, but frequently a simple semiclassical model can be sufficient to understand the spin dynamics of the beam particularly for heavy particles like the proton. Even when doing quantum mechanics, it can be useful to have a simple semiclassical model of the spin.

Since in an introductory chapter like this we cannot hope to cover even all the basics of spin dynamics, we recommend the excellent review article by Mane, Shatunov, and Yokoya[9] for further study beyond our modest introduction.

13.1.1 Simple nonrelativistic model of a spinning body

Consider a slowly (nonrelativistically) spinning object with its center of mass at rest and with rigid distributions of mass density ρ_m and electric density ρ_e. The magnetic dipole moment of the distribution of moving charges is given by

$$\vec{\mu} = \frac{1}{2} \int \vec{r} \times \vec{j}(\vec{r}) \, d^3r = \frac{1}{2} \int \vec{r} \times \rho_e(\vec{r}) \vec{v}(\vec{r}) \, d^3r, \qquad (13.1)$$

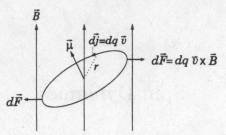

Figure. 13.1 The torque on a circular loop of current j in a uniform external magnetic field may be calculated as $\vec{T} = \oint \vec{r} \times (\vec{v} \times \vec{B}) \, dq = \vec{\mu} \times \vec{B}$, where the magnetic moment is perpendicular to the plane of the loop and has a magnitude $\mu = j\pi r^2 = \frac{1}{2} q r v$ where j is the current in the loop and q is the total charge. If the mass is spread uniformly around the loop and rotates with the same velocity as the charge, then the angular momentum of the loop is $L = r m v$. The ratio of magnetic moment to angular momentum is then $\mu/L = \frac{1}{2} e/m$. The potential energy of the loop is given by $-\vec{\mu} \cdot \vec{B}$.

and the angular momentum from the rotating mass is

$$\vec{S} = \vec{J} = \int \vec{r} \times \rho_m(\vec{r}) \vec{v}(\vec{r}) \, d^3 r. \tag{13.2}$$

It is easily shown that for a fixed center of mass, these integrals for the angular momentum and dipole moment are independent of the choice of origin $\vec{r}$, and so we find that the spin $\vec{S}$ is identical to the angular momentum $\vec{J}$ in this system, i. e. the "orbital" angular momentum of the center of mass is zero. Respectively, the total mass and charge are given by

$$m = \int \rho_m \, d^3 r, \quad \text{and} \quad q = \int \rho_e \, d^3 r. \tag{13.3}$$

In a uniform external magnetic field, there will be a torque

$$\vec{T} = \frac{d\vec{S}}{dt} = \vec{\mu} \times \vec{B} \tag{13.4}$$

on the magnetic moment (as shown in Fig. 13.1) which tends to align the magnetic dipole with the external field.

If we now assume that the shape of the two distributions is the same so that $\rho_m/m = \rho_e/q$, then we find that $\vec{\mu}$ and $\vec{S}$ point in the same direction, and the the ratio of the magnetic dipole moment to the spin is just one half of the charge to mass ratio:

$$\frac{\mu}{S} = \frac{1}{2} \frac{q}{m}. \tag{13.5}$$

For our semiclassical model of beam particles, we may assume that the charge and mass distributions have cylindrically symmetry about a common axis of rotation in the particle's center of mass, and that in this rest system the magnitude of the spin is a fixed quantized value (some integral multiple of $\hbar/2$).* In this model the magnetic dipole moment will also have a fixed value, so that the previous ratio of dipole moment to spin must be rewritten as

$$\frac{\mu}{S} = g \, \frac{q}{2m}, \tag{13.6}$$

where g is sometimes called the gyromagnetic ratio. In the relativistic Dirac theory for the electron (and muon) one finds $g=2$; however there are small radiative corrections which can be calculated with quantum electrodynamics plus hadronic corrections leading to a slight anomalous magnetic moment factor which agrees with the experimentally measured value[10] of

$$a_e = \frac{g-2}{2} = 1.15965218111 \times 10^{-3} \pm 7.4 \times 10^{-13} \tag{13.7}$$

for the electron. A proton on the other hand has a very large anomalous moment with*

$$G = a_p = \frac{g-2}{2} = 1.792847356 \pm 0.000000023. \tag{13.8}$$

Even though protons and electrons are both spin-$\frac{1}{2}$ fermions this large anomaly implies that a proton is not a structureless Dirac particle and must have a more complicated structure than an electron which seems to be more like a point singularity.

For understanding the spin dynamics of a beam of protons, we may consider each proton to be a charged particle with spin and a magnetic dipole moment as illustrated in Fig. (13.2).

When we look in more detail at an electron, the simple-minded idea of a rigidly rotating body becomes unworkable. For example, if we try to model an electron by a rotating thin hoop of mass at radius r, then even for relativistic speeds, this

---

* Since very precise measurements with spin may be made in accelerators, we list some of the physical constants in this chapter with high precision, such as the reduced Planck's constant $\hbar = 6.582119 \times 10^{-16}$ eVs.

* While it is common to use the symbol a for the anomalous factor for electrons and muons, G is frequently used for protons and other hadrons.

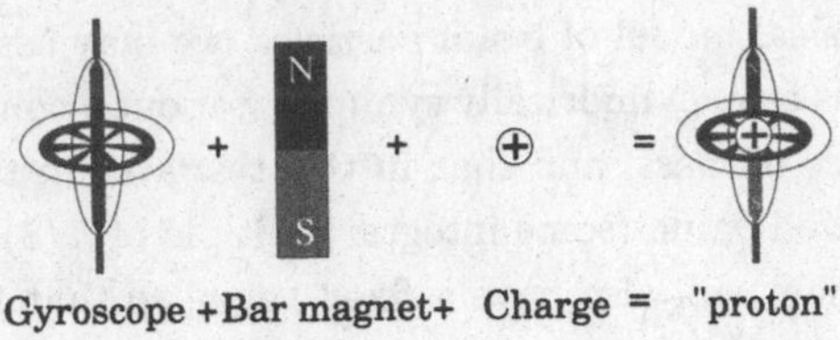

Gyroscope + Bar magnet + Charge = "proton"

	Magnetic	
Spin	Dipole	
	Moment	

Figure. 13.2 From the standpoint of accelerator physics we may consider a simplified semiclassical model of the proton as consisting of an angular momentum, a magnetic dipole, and an electric charge all of which are quantized to fixed magnitudes and with the intrinsic angular momentum and magnetic dipole being rigidly aligned together.

radius would have to be unreasonably large to account for the magnitude of the spin:

$$S = \frac{\hbar}{2} = r \times mc, \tag{13.9}$$

or solving for r we have

$$r = \frac{\hbar}{2mc} \simeq 2 \times 10^{-13} \text{ m}, \tag{13.10}$$

which is more than 100 times larger than the "classical radius" of the electron given in Eq. (8.11).

13.2 Magnetic moments of particles in the rest system

As discussed in §13.1.1 the intrinsic magnetic moment of an elementary particle* is

$$\vec{\mu} = g \frac{q}{2m} \vec{S}, \tag{13.11}$$

where q is usually set equal to the elementary charge $e^{\pm} = \pm 1.602 \times 10^{-19}$ C, with the sign + (positrons and protons) and − (electrons and antiprotons) which make $\vec{\mu}$ and $\vec{S}$ either parallel or antiparallel. In our semiclassical description for these particles, the spin is of constant magnitude quantized in units of $\hbar/2$. In quantum mechanics[11] the square of the magnitude of the spin operator $\vec{S}$ is calculated as

$$\langle\psi|S^2|\psi\rangle = \bar{s}(\bar{s} + 1)\hbar, \tag{13.12}$$

---

* Even though protons, neutrons and other such hadronic particles are now known to have a composite structure containing containing quarks and gluons, they are still frequently referred to as "elementary" particles.

Table I. Spin parameters of several particles.

Particle	s	m [MeV/c^2]	μ/μ_N	$a = (g-2)/2$
e^-	$\frac{1}{2}$	0.510999	1838.282	1.15965×10^{-3}
μ^-	$\frac{1}{2}$	105.658	-8.890597	1.165921
n	$\frac{1}{2}$	939.565	-1.9130427	$(q = 0)$
p	$\frac{1}{2}$	938.272	2.792847	1.792847
$^2_1\text{H}^+$	1	1875.613	0.857438	-0.142987
$^3_1\text{H}^+$	$\frac{1}{2}$	2808.921	2.972962	7.918171
$^3_2\text{He}^{+2}$	$\frac{1}{2}$	2808.391	-2.127498	-4.183963

where $\tilde{s}$ is the spin quantum number of the particle. However in our semiclassical model our spin vector corresponds to the expectation value of the quantum mechanical spin operator which rotates like a normal 3-vector in the rest system of the particle. For example, using the usual Pauli spin matrices (see §13.5) with a spin-up spinor for a spin-$\frac{1}{2}$ particle with quantization axis along z, we have the spin operator $\vec{S} = \vec{\sigma}\hbar/2$ which gives the expectation values

$$S_x = \frac{\hbar}{2}\langle\psi|\sigma_x|\psi\rangle = \frac{\hbar}{2}\begin{pmatrix} 1 & 0 \end{pmatrix}\begin{pmatrix} 0 & -i \\ i & 0 \end{pmatrix}\begin{pmatrix} 1 \\ 0 \end{pmatrix} = 0, \qquad (13.13)$$

$$S_y = \frac{\hbar}{2}\langle\psi|\sigma_y|\psi\rangle = \frac{\hbar}{2}\begin{pmatrix} 1 & 0 \end{pmatrix}\begin{pmatrix} 1 & 0 \\ 0 & -1 \end{pmatrix}\begin{pmatrix} 1 \\ 0 \end{pmatrix} = \frac{\hbar}{2}, \qquad (13.13)$$

$$S_z = \frac{\hbar}{2}\langle\psi|\sigma_z|\psi\rangle = \frac{\hbar}{2}\begin{pmatrix} 1 & 0 \end{pmatrix}\begin{pmatrix} 0 & 1 \\ 1 & 0 \end{pmatrix}\begin{pmatrix} 1 \\ 0 \end{pmatrix} = 0, \qquad (13.13)$$

so that the magnitude of the expectation of the spin vector is just $\hbar/2$. Here we have permuted the Pauli matrices from the usual order:

$$\sigma_x = \begin{pmatrix} 0 & -i \\ i & 0 \end{pmatrix}, \quad \sigma_y = \begin{pmatrix} 1 & 0 \\ 0 & -1 \end{pmatrix}, \quad \text{and} \quad \sigma_z = \begin{pmatrix} 0 & 1 \\ 1 & 0 \end{pmatrix}, \qquad (13.14)$$

since in a flat accelerator ring the stable spin direction is usually vertical rather than along the direction of the beam.

Table I lists[6,10] the masses, spins, magnetic moments, and anomalous factors for several particles. The magnetic moments are given in terms of the nuclear magneton

$$\mu_N = \frac{|e|\hbar}{2m_p} = 3.152451 \times 10^{-8} \text{ eV T}^{-1}. \qquad (13.15)$$

For comparison, the Bohr magneton is almost 1840 times larger

$$\mu_B = \frac{|e|\hbar}{2m_e} = 5.788382 \times 10^{-5} \text{ eV T}^{-1}. \qquad (13.16)$$

We will defer discussing the further ramifications of the quantized nature and treatment of spinors until §13.6, and continue with the semiclassical model where we define the spin vector of a particle at rest as the expectation value of the spin operator, i. e. our spin vector is simply $\vec{S}^*$ which will be generalized to a 4-vector in §13.5.

For an ensemble of N identical particles, the polarization vector is defined as

$$\vec{P} = \frac{1}{N} \sum_{j=1}^{N} \frac{\vec{S}_j}{|\vec{S}|} = \frac{1}{N} \sum_{j=1}^{N} \frac{\vec{S}_j}{\hbar s} \tag{13.17}$$

with s being the spin quantum number as given in Table I. The polarization is relative to a particular axis or unit vector $\hat{n}$ such as the vertical direction $(\hat{n} = \hat{y})$ is

$$P = \hat{n} \cdot \vec{P}. \tag{13.18}$$

If the $N_\uparrow$ particles with spin up and $N_\downarrow$ particles with spin down with the total being $N = N_\uparrow + N_\downarrow$, then the polarization will be

$$P = \hat{y} \cdot \vec{P} = \frac{\hat{y}}{N} \left[\sum_{j=1}^{N^\uparrow} \hat{y} - \sum_{j=1}^{N^\downarrow} \hat{y} \right] = \frac{N_\uparrow - N_\downarrow}{N_\uparrow + N_\downarrow}. \tag{13.19}$$

Polarization of the beam is generally measured by some kind of spin dependent scattering, from a fixed target, another beam of particles or even photons from a laser. When the beam is polarized there is a net angular momentum vector, and we could expect to measure an asymmetry in the distribution of scattering angles in the plane perpendicular to polarization. If the polarization vector is vertical then the measured projection of polarization would be something like

$$P = \frac{1}{A} \frac{N_L - N_R}{N_L + N_R}, \tag{13.20}$$

where N_L and N_R are the number of scattered events to the left and right, respectively and A, called the *analyzing power*, is a normalization to have $P = 1$ for a 100% polarized beam. (Sometimes more complicated expressions[9] for calculating the measured polarization in order to removed systematic errors, but for this introduction these last two equations are sufficient.)

13.3 Basic Thomas precession

Thomas noted that successive Lorentz boosts in different directions produces an extra rotation of the rest system coordinates. He used this effect to explain[14] the experimental value of $g \simeq 2$ measured by Uhlenbeck and Goudsmit[13]. For relativistic spins, this effect must be included in the spin precession equations. In order to derive this effect, we introduce the covariant notation of Lorentz boosts of 4-vectors.

Using a covariant notation with the flat Minkowski metric tensor (west coast convention) we have*

$$g_{\mu\nu} = g^{\mu\nu} : \begin{pmatrix} 1 & 0 & 0 & 0 \\ 0 & -1 & 0 & 0 \\ 0 & 0 & -1 & 0 \\ 0 & 0 & 0 & -1 \end{pmatrix}, \tag{13.21}$$

where the indices μ and ν are numbered from 0 to 3 with the 0-index corresponding to the time-like coordinates and the indices 1, 2, and 3 to the space-like components. Contravariant vectors x^α may be transformed into covariant vectors by lowering the index

$$x_\mu = \sum_{\nu=0}^{3} g_{\mu\nu} x^\nu = g_{\mu\nu} x^\nu, \tag{13.22}$$

where the last expression uses the Einstein summation convention with an implied sum over repeated indices. A covariant index may also be raised to a contravariant index via $g^{\mu\nu}$

$$x^\mu = g^{\mu\nu} x_\nu. \tag{13.23}$$

It is also worth noting for the flat Minkowski metric when we write the metric tensor with one covariant index and the other as a contravariant index that we obtain the identity matrix (equivalent to Kronecker's delta)

$$g^\mu{}_\nu = g^{\mu\kappa} g_{\kappa\nu} : \begin{pmatrix} 1 & 0 & 0 & 0 \\ 0 & 1 & 0 & 0 \\ 0 & 0 & 1 & 0 \\ 0 & 0 & 0 & 1 \end{pmatrix} : g_{\mu\kappa} g^{\kappa\nu} = g_\mu{}^\nu. \tag{13.24}$$

---

* Here, when we define a vector or rank-2 tensor in explicit matrix form, we will place the name of the vector/tensor with Greek indices in the appropriate covariant (lower) and contravariant (upper) locations followed by a colon to the left of the equation.

For a particle with four-momentum p^μ, the invariant rest mass m may be obtained from

$$(mc)^2 = p_\mu p^\mu = p^\mu g_{\mu\nu} p^\nu = (p^0)^2 - \vec{p} \cdot \vec{p}. \qquad (13.25)$$

If we boost* by a velocity $\vec{v} = \vec{\beta} c$, we may write the proper velocity of the boost as the contravariant vector

$$\beta^\mu : \begin{pmatrix} \beta^0 \\ \beta^1 \\ \beta^2 \\ \beta^3 \end{pmatrix} = \begin{pmatrix} \gamma \\ \gamma\beta_x \\ \gamma\beta_y \\ \gamma\beta_z \end{pmatrix}. \qquad (13.26)$$

Here we have used the usual normalization for proper velocity by dividing the spatial velocity by the speed of light c, so that

$$\beta_\mu \beta^\mu = \gamma^2(1 - \vec{\beta} \cdot \vec{\beta}) = 1. \qquad (13.27)$$

The corresponding Lorentz boost may be written as a matrix in terms of the proper velocity (see Problem 13-3) as

$$\Lambda(\vec{\beta})^\mu{}_\nu : \quad \Lambda(\vec{\beta}) = \begin{pmatrix} \beta^0 & \beta^1 & \beta^2 & \beta^3 \\ \beta^1 & 1 + \frac{\beta^1\beta^1}{1+\beta^0} & \frac{\beta^1\beta^2}{1+\beta^0} & \frac{\beta^1\beta^3}{1+\beta^0} \\ \beta^2 & \frac{\beta^2\beta^1}{1+\beta^0} & 1 + \frac{\beta^2\beta^2}{1+\beta^0} & \frac{\beta^2\beta^3}{1+\beta^0} \\ \beta^3 & \frac{\beta^3\beta^1}{1+\beta^0} & \frac{\beta^3\beta^2}{1+\beta^0} & 1 + \frac{\beta^3\beta^3}{1+\beta^0} \end{pmatrix}. \qquad (13.28)$$

Consider a particle instantaneously at rest† at time $t' = 0$ in an inertial system Σ' with its origin located at the particle at time $t' = 0$, and that Σ' is moving with velocity $\vec{\beta} = \beta\hat{z}$ relative to the inertial lab system Σ_{lab}. Assume also that the z-axes of frames Σ' and Σ_{lab} are parallel and that the origins of frames Σ' and Σ_{lab} coincide at time $t = t' = 0$.

If the particle is accelerated over an infinitesimal time Δt so that it changes velocity by an amount

$$c\,\delta\vec{\beta} = \vec{a}'\Delta t = c \begin{pmatrix} 0 \\ \delta\beta \sin\theta \\ \delta\beta \cos\theta \end{pmatrix}, \qquad (13.29)$$

---

* This section follows the flow of Jackson's derivation given in Ref. 12.

† In this section and the following section we use primes to refer to instantaneous rest system and not as derivatives with respect to the longitudinal coordinate of the design orbit.

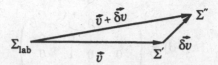

Figure. 13.3 Relationship between the lab frame Σ_{lab} and the two instantaneous rest frames Σ' at time $t' = 0$ and Σ'' at $t' + \Delta t'$.

then new instantaneous rest frame Σ'' of the particle will have the velocity

$$(\vec{\beta} + \vec{\delta\beta})c = c \begin{pmatrix} 0 \\ \delta\beta \sin\theta \\ \beta + \delta\beta \cos\theta \end{pmatrix}, \tag{13.30}$$

relative to the lab frame Σ_{lab}. From Fig. 13.3, we see that Lorentz transformations of a four-vector V^μ from the laboratory frame Σ_{lab} to frame Σ'' may be constructed in two ways:

1. by a direct boost: $(V'')^\mu = \Lambda[-(\vec{\beta} + \vec{\delta\beta})]^\mu{}_\nu V^\nu$, or
2. by successive boosts: $(V'')^\mu = \Lambda(-\vec{\beta})^\mu{}_\kappa \Lambda(-\vec{\delta\beta})^\kappa{}_\nu V^\nu$.

Since Lorentz boosts do not necessarily commute*, these two paths produce different Lorentz transformations. The net transformation $M^\mu{}_\kappa$ from Σ' to Σ'' must be

$$(V'')^\mu = M^\mu{}_\kappa (V')^\kappa = \Lambda[-(\vec{\beta} + \vec{\delta\beta})]^\mu{}_\nu \Lambda(\vec{\beta})^\nu{}_\kappa (V')^\kappa, \tag{13.31}$$

since $\Lambda(\vec{\beta}) = [\Lambda(-\vec{\beta})]^{-1}$.

For the direct boost, the proper velocity of Σ'' relative to Σ_{lab} is

$$\begin{pmatrix} \gamma(1 + \gamma^2\beta\,\delta\beta \cos\theta) \\ 0 \\ \gamma\,\delta\beta \sin\theta \\ \gamma(\beta + \gamma^2\,\delta\beta \cos\theta) \end{pmatrix} + \mathcal{O}(\delta\beta^2). \tag{13.32}$$

The boost from Σ_{lab} to Σ'' is then, with a bit of algebra,

$$\Lambda[-(\vec{\beta} + \vec{\delta\beta})] = \Lambda(-\vec{\beta}) + \Delta_1\,\delta\beta + \mathcal{O}(\delta\beta^2), \tag{13.33}$$

---

* A boost can be thought of as a hyperbolic rotation[12] in a four-dimensional plane containing both the time axis and a spatial axis that is aligned along the direction of the three-dimensional boost velocity. (In higher dimensions than three, we should think of rotations occurring in a plane rather than about some axis.) Just as in three spatial dimensions where rotations about different axes do not commute, in the four-dimensional Minkowski space, the rotations of the Lorentz group in nonparallel planes in general do not commute.

where

$$\Lambda(-\vec{\beta}) = \begin{pmatrix} \gamma & 0 & 0 & -\gamma\beta \\ 0 & 1 & 0 & 0 \\ 0 & 0 & 1 & 0 \\ -\gamma\beta & 0 & 0 & \gamma \end{pmatrix}, \tag{13.34}$$

and

$$\Delta_1 = \begin{pmatrix} \gamma^3\beta\cos\theta & 0 & -\gamma\sin\theta & -\gamma^3\cos\theta \\ 0 & 0 & 0 & 0 \\ -\gamma\sin\theta & 0 & 0 & \frac{\gamma^2\beta\sin\theta}{\gamma+1} \\ -\gamma^3\cos\theta & 0 & \frac{\gamma^2\beta\sin\theta}{\gamma+1} & \gamma^3\beta\cos\theta \end{pmatrix}. \tag{13.35}$$

The Lorentz transformation from Σ' to Σ'' is then

$$\Lambda[-(\vec{\beta} + \delta\vec{\beta})]^\mu{}_\nu \, \Lambda(\vec{\beta})^\nu{}_\kappa = \left[\Lambda(-\vec{\beta}) + \Delta_1\,\delta\beta + \cdots\right]\Lambda(\vec{\beta})$$

$$= \mathbf{I} + \Delta_1\Lambda(\vec{\beta})\,\delta\beta + \mathcal{O}(\delta\beta^2), \tag{13.36}$$

with

$$\Delta_1\Lambda(\vec{\beta}) = \begin{pmatrix} 0 & 0 & -\gamma\sin\theta & -\gamma^2\cos\theta \\ 0 & 0 & 0 & 0 \\ -\gamma\sin\theta & 0 & 0 & -\frac{\gamma^2\beta\sin\theta}{\gamma+1} \\ -\gamma^2\cos\theta & 0 & \frac{\gamma^2\beta\sin\theta}{\gamma+1} & 0 \end{pmatrix}. \tag{13.37}$$

So to first order in $\delta\beta$ the desired transformation is

$$M^\mu{}_\nu : M = \begin{pmatrix} 1 & 0 & -\gamma\,\delta\beta\sin\theta & -\gamma^2\,\delta\beta\cos\theta \\ 0 & 1 & 0 & 0 \\ -\gamma\,\delta\beta\sin\theta & 0 & 1 & -\frac{\gamma^2\beta\,\delta\beta\sin\theta}{\gamma+1} \\ -\gamma^2\,\delta\beta\cos\theta & 0 & \frac{\gamma^2\beta\,\delta\beta\sin\theta}{\gamma+1} & 1 \end{pmatrix} + \mathcal{O}(\delta\beta^2). \tag{13.38}$$

We can write the infinitesimal Lorentz boost by velocity $-\delta\beta$ as

$$\Lambda(-\delta\vec{\beta}') = \begin{pmatrix} 1 & 0 & -\gamma\,\delta\beta\sin\theta & -\gamma^2\,\delta\beta\cos\theta \\ 0 & 1 & 0 & 0 \\ -\gamma\,\delta\beta\sin\theta & 0 & 1 & 0 \\ -\gamma^2\,\delta\beta\cos\theta & 0 & 0 & 1 \end{pmatrix} + \mathcal{O}(\delta\beta^2), \tag{13.39}$$

and a rotation about the x-axis by angle as

$$R_x(\theta) = \begin{pmatrix} 1 & 0 & 0 & 0 \\ 0 & 1 & 0 & 0 \\ 0 & 0 & \cos\theta' & \sin\theta' \\ 0 & 0 & -\sin\theta' & \cos\theta' \end{pmatrix} + \mathcal{O}(\delta\beta^2), \tag{13.40}$$

with

$$\theta' = -\frac{\gamma^2 \beta \sin\theta}{(\gamma+1)} = -\frac{\gamma-1}{\beta}\,\delta\beta\sin\theta. \tag{13.41}$$

Multiplying Eqs. (13.39) and (13.40), we find

$$M^\mu{}_\nu = \Lambda(-\vec{\delta\beta})^\mu{}_\kappa R_y(\theta)^\kappa{}_\nu + \mathcal{O}(\delta\beta^2) = R_y(\theta)^\mu{}_\kappa \Lambda(-\vec{\delta\beta})^\kappa{}_\nu + \mathcal{O}(\delta\beta^2), \tag{13.42}$$

so that to first order in the infinitesimal boost velocity $\delta\beta$ the infinitesimal rotation and boost commute. As with simple three-dimensional spatial rotations we find that adding in the extra hyperbolic rotations of the Lorentz of space-time, infinitesimal rotations (both spherical and hyperbolic) still commute. Thomas[14] was the first person to realize the remarkable fact that there is a rotation of the local coordinate system attached to the particle's noninertial rest frame that moves along with the particle if the particle is accelerated in the lab frame in a direction other than the particle's direction of motion.

We can write this infinitesimal angular rotation of the rest frames as the directed rotation

$$\delta\vec{\theta'} = \frac{\gamma-1}{\beta^2}(\vec{\delta\beta} \times \vec{\beta}). \tag{13.43}$$

Comparing Eq. (13.39) with Eq. (13.28) reveals the change in velocity from rest frame Σ' to rest frame Σ'' to be

$$\delta\vec{v'} = \gamma\delta\vec{v}_\perp + \gamma^2\delta\vec{v}_\parallel = \gamma\left(\delta\vec{v} + \frac{\gamma-1}{v^2}\,(\vec{v}\cdot\delta\vec{v})\vec{v}\right). \tag{13.44}$$

13.4 The Thomas-Frenkel-BMT equation[14-18]

For the precession rotation of the rest spin vector $\vec{S'}$ in Σ', we have

$$\delta\vec{S''} = \delta\vec{S'} + \vec{S'} \times \delta\vec{\theta'}, \tag{13.45}$$

where the second term on the right comes from the rotation of the local rest system. Solving for $\delta\vec{S'}$ gives

$$\delta\vec{S'} = \delta\vec{S''} - \vec{S'} \times \delta\vec{\theta'} = \vec{T'}\delta t' - \vec{S'} \times \delta\vec{\theta'}, \tag{13.46}$$

where $\vec{T'}$ is the instantaneous torque in frame Σ', i. e. from Eq. (13.4):

$$\vec{T'} = \frac{d\vec{S'}}{dt'} = \vec{\mu'} \times \vec{B'} = \frac{gq}{2m}\vec{S'} \times \vec{B'}. \tag{13.47}$$

Eq. (13.46) then becomes*

$$\delta \vec{S}' = -\frac{gq}{2m}(\vec{B}' \times \vec{S}')\delta t' + \frac{(\gamma - 1)}{v^2}(\delta \vec{v} \times \vec{v}) \times \vec{S}'. \qquad (13.48)$$

In boosting from the lab to the rest system Σ' the time between two events is dilated:

$$\delta t' = \frac{\delta t}{\gamma}, \qquad (13.49)$$

so we obtain the differential equation

$$\frac{d\vec{S}'}{dt} = -\frac{gq}{2\gamma m}(\vec{B}' \times \vec{S}') + \frac{(\gamma - 1)}{\gamma v^2}\left(\frac{d\vec{v}}{dt} \times \vec{v}\right) \times \vec{S}'. \qquad (13.50)$$

Then, bearing in mind the Lorentz transformation of the electric and magnetic fields from Eqs. (1.28–31):

$$\vec{B}' = \gamma\left(\vec{B} - \vec{\beta} \times \frac{\vec{E}}{c}\right) - \frac{\gamma^2}{\gamma + 1}\vec{\beta}(\vec{\beta} \cdot \vec{B})$$

$$= \gamma\left(\vec{B} - \vec{\beta} \times \frac{\vec{E}}{c}\right) - (\gamma - 1)\frac{(\vec{v} \cdot \vec{B})}{v^2}\vec{v}, \qquad (13.51)$$

$$\vec{E}' = \gamma(\vec{E} + \vec{\beta} \times \vec{B}c) - \frac{\gamma^2}{\gamma + 1}\vec{\beta}(\vec{\beta} \cdot \vec{E})$$

$$= \gamma(\vec{E} + \vec{v} \times \vec{B}) - (\gamma - 1)\frac{(\vec{v} \cdot \vec{E})}{v^2}\vec{v}, \qquad (13.52)$$

Eq. (13.50) becomes the Thomas-Frenkel-BMT equation[15-17],

$$\frac{d\vec{S}'}{dt} = \vec{\Omega}_s \times \vec{S}'. \qquad (13.53)$$

with

$$\vec{\Omega}_s = \frac{(\gamma - 1)}{v^2}\frac{d\vec{v}}{dt} \times \vec{v} - \frac{gq}{2m}\left[\vec{B} - \frac{(\vec{v} \times \vec{E})}{c^2} - \frac{\gamma - 1}{\gamma}\frac{(\vec{v} \cdot \vec{B})}{v^2}\vec{v}\right], \qquad (13.54)$$

or

$$\vec{\Omega}_s = \vec{\Omega}_{\text{Thomas}} + \vec{\Omega}_{\text{Larmor}}, \qquad (13.55)$$

---

* In this section we return to having $v = \beta c$.

having defined

$$\vec{\Omega}_{\text{Thomas}} = \frac{(\gamma - 1)}{v^2} \frac{d\vec{v}}{dt} \times \vec{v} \tag{13.56}$$

from Eq. (13.43), and

$$\vec{\Omega}_{\text{Larmor}} = -\frac{gq}{m} \left[\vec{B} - \frac{(\vec{v} \times \vec{E})}{c^2} - \frac{\gamma - 1}{\gamma} \frac{(\vec{v} \cdot \vec{B})}{v^2} \vec{v} \right]. \tag{13.57}$$

In order to find a more explicit expression of the Thomas frequency (13.56), we have to evaluate the acceleration in the laboratory system, starting from the the acceleration as represented in the particle's instantaneous rest frame

$$\vec{a}' = \frac{\delta \vec{v}'}{\delta t'} = \gamma^2 \frac{d\vec{v}}{dt} + \frac{\gamma^2 (\gamma - 1)}{v^2} \left(\vec{v} \cdot \frac{d\vec{v}}{dt} \right) \vec{v}, \tag{13.58}$$

having made use of Eqs. (13.44) and (13.49). Since this acceleration is generated by the electric field only, we can write

$$\vec{a}' = \frac{q \vec{E}'}{m} = \frac{q}{m} \left[\gamma(\vec{E} + \vec{v} \times \vec{B}) - (\gamma - 1)\frac{(\vec{v} \cdot \vec{E})}{v^2} \vec{v} \right], \tag{13.59}$$

due to Eq. (13.52). Taking the scalar product of $\vec{v}$ with the previous two equations, we have

$$\gamma^2 \left(\vec{v} \cdot \frac{d\vec{v}}{dt} \right) + \gamma^2 (\gamma - 1) \left(\vec{v} \cdot \frac{d\vec{v}}{dt} \right) = \frac{q}{m} [\gamma(\vec{v} \cdot \vec{E}) - (\gamma - 1)(\vec{v} \cdot \vec{E})], \tag{13.60}$$

or

$$\left(\vec{v} \cdot \frac{d\vec{v}}{dt} \right) = \frac{q}{\gamma^3 m} (\vec{v} \cdot \vec{E}). \tag{13.61}$$

Solving Eq. (13.58) for $d\vec{v}/dt$ and substituting Eqs. (13.59) and (13.61), we eventually obtain

$$\frac{d\vec{v}}{dt} = \frac{q}{\gamma m} (\vec{E} + \vec{v} \times \vec{B}) - \frac{q}{m} \frac{\gamma - 1}{\gamma^3} \frac{\vec{v} \cdot \vec{E}}{v^2} \vec{v} - \frac{q}{m} \frac{\gamma - 1}{\gamma^2} \frac{\vec{v} \cdot \vec{E}}{v^2} \vec{v}, \tag{13.62}$$

or

$$\frac{d\vec{v}}{dt} = \frac{q}{\gamma m} \left[(\vec{E} + \vec{v} \times \vec{B}) - \frac{(\vec{v} \cdot \vec{E})}{c^2} \vec{v} \right], \tag{13.63}$$

since

$$\frac{\gamma-1}{\gamma^3 v^2} + \frac{\gamma-1}{\gamma^2 v^2} = \frac{\gamma-1}{\gamma^2 v^2}\left(\frac{1}{\gamma}+1\right) = \frac{1}{\gamma c^2}\frac{(\gamma-1)(\gamma+1)}{\beta^2\gamma^2} = \frac{1}{\gamma c^2}. \tag{13.64}$$

Therefore the vector product appearing in Eq. (13.56) becomes

$$\frac{d\vec{v}}{dt}\times\vec{v} = \frac{q}{\gamma m}[(\vec{E}\times\vec{v}) + v^2\vec{B} - (\vec{v}\cdot\vec{B})\vec{v}], \tag{13.65}$$

i. e., the Thomas frequency (13.56) can be written as

$$\vec{\Omega}_{\text{Thomas}} = -\frac{q}{m}\frac{\gamma-1}{\gamma}\left[\frac{(\vec{v}\times\vec{E})}{v^2} + \frac{(\vec{v}\cdot\vec{B})}{v^2}\vec{v} - \vec{B}\right]. \tag{13.66}$$

Then summing up both Larmor and Thomas frequencies, and considering the identity

$$\frac{\gamma-1}{\gamma v^2} = \frac{\gamma}{(\gamma+1)c^2}, \tag{13.67}$$

we obtain

$$\vec{\Omega}_s = -\frac{q}{\gamma m}\left[(1+G\gamma)\vec{B} - \frac{\gamma^2}{\gamma+1}\frac{(\vec{v}\cdot\vec{B})}{c^2}\vec{v} - \gamma\left(G+\frac{1}{\gamma+1}\right)\frac{\vec{v}\times\vec{E}}{c^2}\right], \tag{13.68}$$

which is Thomas-Frenkel-BMT precession frequency. Eq. (13.53) then becomes

$$\frac{d\vec{S}'}{dt} = -\frac{q}{\gamma m}\left[(1+G\gamma)\vec{B}_\perp + (1+G)B_\parallel + \gamma\left(G+\frac{1}{\gamma+1}\right)\frac{\vec{E}\times\vec{v}}{c^2}\right]\times\vec{S}', \tag{13.69}$$

where $\vec{B}_\parallel = (\vec{v}\cdot\vec{B})\vec{v}/v^2$ and $\vec{B}_\perp = \vec{B} - \vec{B}_\parallel$ are the respective components of the magnetic field parallel and perpendicular to the particle's velocity.

Take note that, if g were exactly equal to 2, Eq. (13.68) would reduce to

$$\vec{\Omega}_s = -\frac{q}{\gamma m}\left[\vec{B} - \frac{\gamma}{\gamma+1}\frac{(\vec{v}\times\vec{E})}{c^2}\right]. \tag{13.70}$$

It is important to note that in the case with no electric field and no anomaly, a spin which starts parallel to the velocity (i. e. in a pure helicity state) will precess in such a way to remain parallel to the velocity. In other words, the helicity of a pure Dirac particle ($g = 2$) is conserved. In the next section we will show that for a particle with $g = 2$, helicity will remain conserved even for nonzero electric fields.

13.5 Covariant form of spin precession

Even though in §13.1.1 we saw that a simple classical model of a rotating rigid body could not explain how such a large angular momentum can be found in something as small as an electron, we can still gain some insight by continuing with our rotating rigid body model. Although we do not have to stay with a nonrelativistically spinning disk in its rest frame, rather than complicate the following discussion by having to add in the energy from the stresses generated in a rapidly spinning disk as well as the energy stored in the electromagnetic field of a rotating charge distribution, it will be sufficient to consider a slowly spinning disk in which the total energy is dominated by the mass of the disk in the center of mass system.

13.5.1 Relativistic boost of a slowly spinning disk

Following Weinberg[19] we may construct the symmetric energy-momentum tensor

$$T^{\alpha\beta}(x) = T^{\beta\alpha}(x) = \sum_n \frac{p_n^\alpha p_n^\beta}{E_n/c} \delta^3(\vec{x} - \vec{x}_n(t)), \qquad (13.71)$$

for a distribution of point masses where E_n is the energy of the nth mass. For a continuous distribution, the components become densities. The components

$$T^{\alpha 0} = \sum_n p^\alpha \delta^3(\vec{x} - \vec{x}_n) \qquad (13.72)$$

are the density of the four-momentum, so the integral $T^{\alpha 0}$ of over the spatial volume gives the total four-momentum

$$P^\alpha = \int T^{\alpha 0} d^3 x. \qquad (13.73)$$

The nine components

$$T^{ij} = \sum_n p^i v^j \delta^3(\vec{x} - \vec{x}_n), \quad \text{with } i, j \in \{1, 2, 3\} \qquad (13.74)$$

may be thought of as a momentum current density similar to the electric current density being the sum over charge times velocity of each charge. For an isolated system with no external forces total energy and momentum must be conserved, so we find the 4-dimensional divergence of the energy-momentum tensor to be zero

$$\partial_\alpha T^{\alpha\beta} = \frac{\partial T^{\alpha\beta}}{\partial x^\alpha} = 0. \qquad (13.75)$$

Extending the cross product $\vec{r} \times \vec{p}$ of angular momentum to four dimensions, we might expect the angular momentum density to look something like

$$M^{\gamma\alpha\beta} = x^\alpha T^{\beta\gamma} - x^\beta T^{\alpha\gamma}, \tag{13.76}$$

with the conserved 4-divergence

$$\partial_\gamma M^{\alpha\beta\gamma} = 0. \tag{13.77}$$

At a fixed time a total angular momentum tensor may be calculated by integrating over the spatial coordinates:

$$J^{\alpha\beta} = \int M^{0\alpha\beta}\, d^3x = \int (x^\alpha T^{\beta 0} - x^\beta T^{\alpha 0})d^3x. \tag{13.78}$$

To find the part of the angular momentum which is invariant under translations the coordinates may be shifted from x^α to $x^\alpha + r^\alpha$ to obtain a new tensor

$$\bar{M}^{\gamma\alpha\beta} = M^{\gamma\alpha\beta} + r^\alpha T^{\beta\gamma} - r^\beta T^{\alpha\gamma}. \tag{13.79}$$

The new total angular momentum then becomes

$$\begin{aligned}
\bar{J}^{\alpha\beta} &= J^{\alpha\beta} + \int (r^\alpha T^{\beta 0} - r^\beta T^{\alpha 0})d^3x \\
&= J^{\alpha\beta} + r^\alpha P^\beta - r^\beta P^\alpha.
\end{aligned} \tag{13.80}$$

We can make a contraction of $J^{\kappa\lambda}\beta^\mu$ with the antisymmetric Levi-Civita tensor

$$\epsilon^{\kappa\lambda\mu\nu} = \begin{cases} +1 & \text{if } \kappa\lambda\mu\nu \text{ is an even permutation of 0123,} \\ -1 & \text{if } \kappa\lambda\mu\nu \text{ is an odd permutation of 0123,} \\ 0 & \text{otherwise,} \end{cases} \tag{13.81}$$

where

$$\beta^\mu = \frac{P^\mu}{\gamma m c} \tag{13.82}$$

is the proper velocity. Since

$$\tfrac{1}{2}\epsilon_{\kappa\lambda\mu\nu}(r^\lambda P^\mu - r^\mu P^\lambda)\beta^\nu = \tfrac{1}{2}\gamma m c\, \epsilon_{\kappa\lambda\mu\nu}(r^\lambda \beta^\mu - r^\mu \beta^\lambda)\beta^\mu = 0 \tag{13.83}$$

for any translation, we may write a new 4-vector as

$$S^\kappa = \tfrac{1}{2}\epsilon^{\kappa\lambda\mu\nu} J_{\lambda\mu}\, \beta_\nu, \tag{13.84}$$

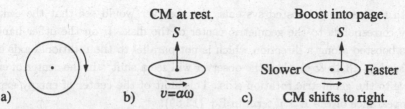

CM at rest.   Boost into page.

Slower ⟨ •• ⟩ Faster

a)   b) $v=a\omega$ c) CM shifts to right.

Figure. 13.4 a) A uniform disk is spinning about its axis as shown with the axis pointing into the page. If the disk is boosted away from an observer along its axis, the transverse velocity components of any piece of the disk remain unchanged. Even though the total energy of the disk is increased by the boost, the increase is uniform across the disk and the center of energy remains centered on the of the disk. b) Consider now a disk spinning with its axis aligned upward with the center of mass being at rest relative to the observer. c) If the disk is then boosted away from the observer and perpendicular to its spin, then the observer will see a higher energy density on right of the disk's geometric center than on the left. Relative to the observer, the right side of the disk is moving faster than the left side of the disk. The effective center of mass, or more correctly center of energy, gets shifted to the right of the geometric center of the disk as the disk is boosted away from the observer.

which is invariant with respect to translations of the observer. When multiplied by γmc, Eq. (13.84) becomes the classical version of the Pauli-Lubański[20] vector

$$W^\kappa = \tfrac{1}{2}\epsilon^{\kappa\lambda\mu\nu} J_{\lambda\mu}\, p_\nu. \tag{13.85}$$

In the rest system Eq. (13.84) becomes

$$S^\mu : \begin{pmatrix} 0 \\ J^{23} \\ J^{31} \\ J^{12} \end{pmatrix}, \tag{13.86}$$

and since this is invariant under translations, then the components must be $J^{23} = S'_x$, $J^{31} = S'_y$, and $J^{12} = S'_z$. When the disk is boosted along the z-axis, the spin 4-vector becomes

$$S^\mu : \begin{pmatrix} \gamma\beta S'_z \\ S'_x \\ S'_y \\ \gamma S'_z \end{pmatrix}. \tag{13.87}$$

If the disk is boosted along its axis of rotation away from an observer, then the observer would see a symmetric distribution of velocities about the rotation axis as shown in Fig. 13.4a. Since the rotational velocities are perpendicular to the direction of the boost, these transverse components of velocity are the same in both the rest and boosted systems; the observer would see that the center of energy corresponds to the geometric center of the disk. If on the other hand the disk is boosted along a direction which is not parallel to the rotational axis of the disk (see Fig. 13.4b & c), then the observer will see a shift[21] in the center of energy relative to the geometric rotation axis. This shift of the center of energy explains the appearance of the $\vec{E} \times \vec{v}$ term in Eq. (13.69).

13.5.2 Covariant formulation of the Thomas-Frenkel-BMT equation

For people who are comfortable with covariant notation and the concept of generalized covariance, we present this alternative derivation.

We want to construct the covariant spin precession equation from the possible four vectors and tensors at hand. The contravariant precession rate is $\frac{dS^\mu}{d\tau}$ where τ is the proper time in the instantaneous rest system of the particle.

We have the proper velocity v^μ, vector potential

$$A^\mu : \begin{pmatrix} \Phi/c \\ A_x \\ A_y \\ A_z \end{pmatrix}, \tag{13.88}$$

the symmetric Minkowski metric tensor $g^{\mu\nu}$, the Levi-Civita antisymmetric tensor $\epsilon^{\kappa\lambda\mu\nu}$, the covariant spin S^μ, and the differential operation ∂^μ (the four dimensional analog to ∇). We must combine these in a covariant way which must also be gauge invariant. Here gauge invariance means that the term must be invariant under the transformation of the vector potential:

$$A^\mu \to A^\mu + \partial^\mu \chi, \tag{13.89}$$

where χ is a scalar function.

Starting with the four vectors we find that

$$S_\alpha \beta^\alpha = 0, \tag{13.90}$$

$$\partial_\alpha \beta^\alpha = 0, \tag{13.91}$$

$$\partial_\alpha S^\alpha = 0, \tag{13.92}$$

and the combinations

$$A^\mu S_\mu \neq (A^\mu + \partial^\mu \chi) S_\mu, \tag{13.93}$$

$$\partial^\mu A_\mu \neq \partial^\mu (A_\mu + \partial_\mu \chi) = \partial^\mu A_\mu + \partial^\mu \partial_\mu \chi, \tag{13.94}$$

are not gauge invariant. However, the combinations

$$A^\mu \beta_\mu = \frac{1}{qc} J_\mu A^\mu, \tag{13.95}$$

$$F^{\mu\nu} = \partial^\mu A^\nu - \partial^\nu A^\mu : \begin{pmatrix} 0 & -E_x/c & -E_y/c & -E_z/c \\ E_x/c & 0 & -B_z & B_y \\ E_y/c & B_z & 0 & -B_x \\ E_z y/c & -B_y & B_x & 0 \end{pmatrix}, \tag{13.96}$$

where $J^\mu = q\beta^\mu c$ is the charge current, and $F^{\mu\nu}$ is the the Minkowski electromagnetic field tensor (sometimes called Faraday tensor) are both gauge invariant. We leave it to the reader to verify that the combination $A^\mu S^\nu - A^\nu S^\mu$ is not gauge invariant as can easily be seen by taking $\chi = x_\alpha x^\alpha$ for the scalar invariant function.

From the gauge invariant electromagnetic tensor, we can also construct the gauge invariant combinations:

$$F^{\mu\nu} S_\nu, \quad S_\kappa F^{\kappa\lambda} \beta_\lambda, \quad {}^*F^{\mu\nu} S_\nu, \quad \text{and} \quad S_\kappa {}^*F^{\kappa\lambda} \beta_\lambda, \tag{13.97}$$

where the dual of the Minkowski electromagnetic field tensor is

$${}^*F^{\kappa\lambda} = \frac{1}{2} \epsilon^{\kappa\lambda\mu\nu} F_{\mu\nu} : \begin{pmatrix} 0 & B_x & B_y & B_z \\ -B_x & 0 & -E_z/c & E_y/c \\ -B_y & E_z/c & 0 & -E_x/c \\ -B_z & -E_y/c & E_x/c & 0 \end{pmatrix}. \tag{13.98}$$

Now we would like to write down a covariant form for the Thomas-Frenkel-BMT equation which must be correct in all inertial frames. To do this we add the three terms in Eq. (13.97) with the arbitrary coefficients a, b, f, and h:

$$\frac{dS^\mu}{d\tau} = aF^{\mu\nu} S_\nu + b(S_\kappa F^{\kappa\lambda} \beta_\lambda)\beta^\mu + f\, {}^*F^{\mu\nu} S_\nu + h(S_\kappa {}^*F^{\kappa\lambda} \beta_\lambda)\beta^\mu. \tag{13.99}$$

The principle of generalized covariance implies that this equation must be true for all inertial reference frames[19]. In the rest system for the spatial coordinates, Eq. (13.99) becomes

$$\frac{d\vec{S}'}{dt'} = a\vec{S}' \times \vec{B}' + \frac{f}{c}\vec{S}' \times \vec{E}' = \vec{\mu}' \times \vec{B}' = \frac{gq}{2m}\vec{S}' \times \vec{B}' \tag{13.100}$$

and so we must have

$$a = \frac{gq}{2m} \quad \text{and} \quad f = 0. \tag{13.101}$$

From Eq. (13.87) we see that

$$S^0 = \vec{\beta} \cdot \vec{S}, \tag{13.102}$$

and the time derivative of S^0 in the rest system is

$$\frac{dS'^0}{dt'} = \vec{S}' \cdot \frac{d\vec{\beta}'}{dt'} = \frac{q}{mc}\vec{S}' \cdot \vec{E}'. \tag{13.103}$$

The time-like component of Eq. (13.99) in the rest system is

$$\frac{dS'^0}{dt'} = \frac{gq}{2mc}\vec{E}' \cdot \vec{S}' - \frac{b}{c}(\vec{S}' \cdot \vec{E}') + (f+h)\vec{S}' \cdot \vec{B}'. \tag{13.104}$$

The equality of the last two equations requires that $f + h = 0$ and

$$b = \frac{g-2}{2}\frac{q}{m}. \tag{13.105}$$

The covariant form of the Thomas-Frenkel-BMT must be

$$\frac{dS^\mu}{d\tau} = \frac{e}{m}\left[F^{\mu\nu} + \frac{g-2}{2}(F^{\mu\nu} + \beta^\mu F^{\nu\kappa}\beta_\kappa)\right] S_\nu. \tag{13.106}$$

The Lorentz force in covariant notation is simply

$$\frac{dp^\mu}{d\tau} = \frac{e}{m}F^{\mu\nu}p_\nu, \tag{13.107}$$

which contains the gauge invariant term in Eq. (13.95). By comparison of the last two equations with $g = 2$, we note that helicity must be conserved for a pure Dirac particle ($g = 2$) with no anomalous magnetic moment.

It should be noted that if the particle has an electric dipole moment aligned with the spin, then f and $h = -f$ will not be equal to zero, and the right-hand side of Eq. (13.100) would have an extra term $\vec{\mathcal{P}}' \times \vec{E}$ for the torque on the electric dipole component $\mathcal{P}'$. The electric dipole moment for an "elementary" particle would be CP-violating as Landau pointed out[22,23] and therefore must be very small.

13.6 Review of nonrelativistic quantum mechanics with emphasis on spin

In a nonrelativistic system, we may write a spinor wave function for a spin-$\tilde{s}$ system as a $2\tilde{s} + 1$-complex vector

$$\Psi = |\Psi\rangle = \begin{pmatrix} \psi_{\tilde{s}} \\ \psi_{\tilde{s}-1} \\ \vdots \\ \psi_{-\tilde{s}} \end{pmatrix}, \tag{13.108}$$

where the ψ_j are complex functions, and we know from quantum mechanics that $\tilde{s}$ must be restricted to nonnegative integer multiples of $\frac{1}{2}$:

$$\tilde{s} \in \left\{ 0, \frac{1}{2}, 1, \frac{3}{2}, 2, \cdots \right\}. \tag{13.109}$$

Such a vector may be thought to have $2 \times (2\tilde{s}+1)$ real parameters; however when we require the spinor to be normalized, then the spinor will only have $2 \times (2\tilde{s}+1) - 1$ free parameters. We define the adjoint of the spinor by taking its hermitian conjugate

$$\Psi^\dagger = \langle \Psi | = (|\Psi\rangle)^\dagger = (|\Psi\rangle)^{*T} = \begin{pmatrix} \psi_{\tilde{s}}^* & \psi_{\tilde{s}-1}^* & \cdots & \psi_{-\tilde{s}}^* \end{pmatrix}. \tag{13.110}$$

By normalized, we mean that the scalar product is 1 where, in the usual quantum mechanical way, we have

$$\Psi^\dagger \Psi = \langle \Psi | \Psi \rangle \equiv \langle \Psi | | \Psi \rangle = \sum_{j=-\tilde{s}}^{\tilde{s}} |\psi_j|^2 = 1. \tag{13.111}$$

The quantum mechanical expectation value of some operator $\mathcal{A}$ is calculated by

$$\langle \mathcal{A} \rangle = \langle \Psi | \mathcal{A} | \Psi \rangle = \int \Psi^\dagger \mathcal{A} \Psi \, d^3x, \tag{13.112}$$

and the wave function Ψ in general consists of the spatial wave function times a spinor or even more generally a summation of complex amplitudes times spatial wave functions times spinors. The wave function must satisfy Schrödenger's equation:

$$i\hbar \frac{\partial |\Psi_s(t)\rangle}{\partial t} = H_s |\Psi_s(t)\rangle, \tag{13.113}$$

where H is the Hamiltonian operator. Here the subscript s indicates that the wave function $|\Psi_s\rangle$ and Hamiltonian H_s are written in the *Schrödinger representation* with time dependence in the wave function. If the Hamiltonian does not explicitly depend on time, then we may easily integrate the wave function to obtain

$$|\Psi_s(t)\rangle = e^{-iH_s t/\hbar} |\Psi_s(0)\rangle. \tag{13.114}$$

Taking the spin Hamiltonian in the Schrödinger representation (see Fig. 13.1)

$$H_s = -\vec{\mu}' \cdot \vec{B}' = -\frac{gq}{2m} \vec{B}' \cdot \vec{S}' = -\vec{\Omega}' \cdot \vec{S}' \tag{13.115}$$

in the rest frame, Eq. (13.114) then gives the expression for rotation of a spinor in a magnetic field

$$|\Psi'(t')\rangle = e^{i(gq/2m\hbar)\vec{B}'\cdot\vec{S}'t'}|\Psi'(0)\rangle. \tag{13.116}$$

With $\vec{B}'$ written in terms of the lab fields and coordinates, Eq. (13.113) becomes

$$\frac{d}{dt'}|\Psi'\rangle = -i\hat{s}(\Omega'\cdot\vec{\sigma})|\Psi'(0)\rangle, \tag{13.117}$$

where

$$\vec{\Omega}' = -\frac{q}{m}[(1+G)\vec{B}'_\perp + (1+G)\vec{B}'_\parallel + 0] = \frac{gq}{2m}\vec{B}', \tag{13.118}$$

since $\gamma = 1$ and $\vec{v} = 0$ in the rest system. In terms of the lab coordinates and fields the time derivative of the spinor may be written as

$$\frac{d}{dt}|\Psi'(t)\rangle = -i\hat{s}(\vec{\Omega}\cdot\vec{\sigma})|\Psi'(0)\rangle, \tag{13.119}$$

In the *Heisenberg representation*, the time dependence of the wave function is transferred to the operators, so we now have

$$|\Psi_h\rangle = |\Psi_s(0)\rangle, \tag{13.120}$$

$$H_h(t) = e^{iH_s t/\hbar}H_s e^{-iH_s t/\hbar}, \tag{13.121}$$

where the subscript h indicates that the wave function and operator entities are written in the Heisenberg representation. From this we can derive Ehrenfest's theorem for the time evolution of an observable:

$$
\begin{aligned}
\frac{d\langle\mathcal{A}\rangle}{dt} &= \frac{d}{dt}\langle\Psi_s(t)|\,\mathcal{A}_s(t)\,|\Psi_s(t)\rangle \\
&= \frac{d}{dt}\langle\Psi_h|\,e^{iH_s t/\hbar}\mathcal{A}_s(t)e^{-iH_s t/\hbar}\,|\Psi_h\rangle \\
&= \frac{i}{\hbar}\langle\Psi_h|\,e^{iH_s t/\hbar}[H,\mathcal{A}_s(t)]e^{-iH_s t/\hbar}\,|\Psi_h\rangle + \langle\Psi_h|\,e^{iH_s t/\hbar}\frac{\partial\mathcal{A}_s(t)}{\partial t}e^{-iH_s t/\hbar}\,|\Psi_h\rangle \\
&= \frac{i}{\hbar}\langle\Psi_s(t)|\,[H_s,\mathcal{A}_s(t)]\,|\Psi_s(t)\rangle + \langle\Psi_s(t)|\,\frac{\partial\mathcal{A}_s(t)}{\partial t}\,|\Psi_s(t)\rangle,
\end{aligned}
\tag{13.122}
$$

or

$$i\hbar\frac{d\langle\mathcal{A}_s\rangle}{dt} = \langle[\mathcal{A}_s(t), H_s]\rangle + i\hbar\left\langle\frac{\partial\mathcal{A}_s}{\partial t}\right\rangle. \tag{13.123}$$

The commutator in quantum mechanics is the analog of the Poisson bracket in classical mechanics, and Eq. (13.123) is the quantum mechanical analog of Liouville's equation (3.103). If the operator $\mathcal{A}$ has no explicit time dependence, then the last term containing the partial derivative with respect to time is zero, and we may simply write this in the Heisenberg representation as

$$i\hbar \frac{dA_h}{dt} = [A_h, H_h]. \tag{13.124}$$

The vector operator for spin (intrinsic angular momentum) may be written as

$$\vec{S} = \hbar \bar{s} \vec{\sigma}, \tag{13.125}$$

where the three components of $\vec{\sigma}$ are a set of group generator matrices in the $(2\bar{s}+1) \times (2\bar{s}+1)$-matrix representation of SU(2). Since spin is an angular momentum, its operator must satisfy the relation

$$\vec{S} \times \vec{S} = [\vec{S}, \vec{S}] = i\hbar \vec{S}. \tag{13.126}$$

For spin-$\frac{1}{2}$, the Pauli matrices are the most common choice for a set of generator matrices:

$$\sigma_x = \begin{pmatrix} 0 & -i \\ i & 0 \end{pmatrix}, \quad \sigma_y = \begin{pmatrix} 1 & 0 \\ 0 & -1 \end{pmatrix}, \quad \text{and} \quad \sigma_z = \begin{pmatrix} 0 & 1 \\ 1 & 0 \end{pmatrix}. \tag{13.127}$$

If we take the Heisenberg representation for the interaction Hamiltonian

$$H_h = -\vec{\mu}' \cdot \vec{B}' = -\frac{gq}{2m} \vec{B}' \cdot \vec{S}', \tag{13.128}$$

Ehrenfest's theorem then gives

$$i\hbar \frac{dS_j'}{dt'} = -\frac{gq}{2m} \sum_{k=1}^{3} B_k'[S_j', S_k'] = -i\hbar \frac{gq}{2m} \sum_{k,l=1}^{3} \epsilon_{jkl} B_k' S_l', \quad \text{or} \tag{13.129}$$

$$\frac{d\vec{S}'}{dt'} = -\frac{gq}{2m} \vec{B}' \times \vec{S}', \tag{13.130}$$

which is just the quantum mechanical expression corresponding to the classical torque on magnetic moment as shown in Fig. 13.1. Eq. (13.116) may now be written as the equation for the rotation of a spinor

$$|\Psi'(\theta)\rangle = e^{-i(\hat{n} \cdot \vec{\sigma})\bar{s}\theta} |\Psi'(0)\rangle, \tag{13.131}$$

where we have taken the unit vector $\hat{n}$ along the direction of the magnetic field ,and the amount of rotation is given by

$$\theta = \frac{gq}{2m} B' \hbar \bar{s} t', \tag{13.132}$$

with $\bar{s}$ being the spin quantum number. Note that with the minus sign in the exponent, Eq. (13.131) produces a right-handed rotation about the vector $\hat{n}$.

Typically we select one of the σ-matrices to be diagonal corresponding to the component of spin along a reference or *quantization axis*. Usually in quantum mechanics texts, this quantization axis is taken as the z-axis; however for accelerator physics, it is more convenient to take the reference direction along the vertical y-axis. In fact the concept of a quantization axis can be misleading when dealing with precessions of an ensemble of particles precessing about different axes as they sample different fields pointing in different directions. For example, a beam has a nonzero emittance, so when the particles on different trajectories pass through a quadrupole or sextupole, they each see magnetic fields of different magnitude and direction. One way to think of this is to simulate a polarized beam referring to a common set of axes for all the particles at any azimuth and track each particle's spin separately. Then when a polarization is measured at the location of the polarimeter target the appropriate averages of the polarization vector and/or spin density matrix can be evaluated by calculating the polarization (spin density matrix) for each particle and then averaging over all the particles in the distribution. For the nth particle we have the expectation of an operator $\mathcal{A}$

$$\langle \mathcal{A} \rangle_n = \langle \Psi_n | \mathcal{A} | \Psi_n \rangle = \text{tr}(\langle \Psi_n | \mathcal{A} | \Psi_n \rangle) = \text{tr}(|\Psi_n\rangle \langle \Psi_n | \mathcal{A}). \qquad (13.133)$$

The spin density matrix for the nth particle is $\rho_n = |\Psi_n\rangle\langle\Psi_n|$. The overall expectation value for $\mathcal{A}$ averaged over the beam of N particles will then be

$$\langle \mathcal{A} \rangle = \frac{1}{N} \sum_{n=1}^{N} \langle \Psi_n | \mathcal{A} | \Psi_n \rangle = \frac{1}{N} \sum_{n=1}^{N} \text{tr}(|\Psi_n\rangle \langle \Psi_n | \mathcal{A}) = \text{tr}\left(\frac{1}{N} \sum_{n=1}^{N} |\Psi_n\rangle \langle \Psi_n | \mathcal{A} \right)$$
$$= \text{tr}(\bar{\rho}\mathcal{A}), \qquad (13.134)$$

where the averaged density matrix

$$\bar{\rho} = \frac{1}{N} \sum_{n=1}^{N} |\Psi_n\rangle\langle\Psi_n|, \qquad (13.135)$$

and we have factored $\mathcal{A}$ out of the summation in the trace since the operator must be independent of the actual particles. It should be noted that the density matrix for spin-$\frac{1}{2}$ does not yield any more information than the spin vector components, but for higher spins, there are more degrees of freedom for each spinor.

For spin-$\frac{1}{2}$, the spinors only have two complex numbers which give four real parameters:

$$|\Psi\rangle = \begin{pmatrix} r_1 e^{i\phi_1} \\ r_2 e^{i\phi_2} \end{pmatrix} = \begin{pmatrix} r_1 \\ r_2 e^{i(\phi_2-\phi_1)} \end{pmatrix} e^{i\phi_1}. \qquad (13.136)$$

The spinor must be normalized $(\sqrt{r_1^2 + r_2^2} = 1)$, so that leaves only three degrees of freedom. Since we are not interested in spin-spin interactions the complex phase may be factored out the top component. This phasor will commute with any spin operator and cancel with its complex conjugate factored from $\langle\Psi|$. This means that the ϕ_1 phase is uninteresting, and only two real parameters remain. In fact for spin-$\frac{1}{2}$, SO(3) rotations of the spin expectation vector $\langle\Psi|\,\vec{\sigma}\,|\Psi\rangle$ can be used instead of the spinor rotation matrices. Some spin codes track with 3-component vectors rather than spinors.

For spin-1, a spinor has six real parameters combined into three complex components. Again, the spinor must be normalized, and an overall phase may be factored to cancel with its complex conjugate, so there will only be four real parameters of interest per particle. In this case, the vector component of the spin $\langle\Psi|\,\vec{S}\,|\Psi\rangle$ will rotate as a usual vector, but there is another part referred to as the tensor components which do not in general rotate the same way as the vector components.

However, the average spin density matrix for an ensemble of spins will have more degrees of freedom than the individual spinor. In fact, for a beam, the $\bar{s}^2$ components of $\bar{\rho}$ could all be zero.

Now even though we are dealing with relativistic particles, we have written Eq. (13.53) with the spin vector in the traveling rest frame. So locally in that instantaneous rest frame, the particle is nonrelativistic and nonrelativistic spinors are sufficient. In the rest frame the spin will rotate about the rest-frame magnetic field vector $\vec{B}'$, but since the local frame will be rotating, an extra Thomas rotation must be added. The construction of spinor rotation matrices will be shown in the next two sections.

13.6.1 Constructing Pauli-like matrices for higher spins

For what follows, let us simplify things by defining a reduced operator for spin $\vec{S} = \mathcal{S}/\hbar$ (equivalent to setting $\hbar = 1$). Since spin or intrinsic angular momentum should be an observable, its operator must be hermitian:

$$\vec{S}^\dagger = \vec{S}. \qquad (13.137)$$

The three components S_j now correspond to generator matrices for the appropriate SU(2) representation. The commutator relations for these matrices are

$$[S_x, S_y] = iS_z, \qquad [S_y, S_z] = iS_x, \qquad \text{and} \quad [S_z, S_x] = iS_y. \qquad (13.138)$$

Taking the scalar product of $\vec{S}$ with itself gives the operator

$$S^2 = \vec{S} \cdot \vec{S} = S_x^2 + S_y^2 + S_z^2, \qquad (13.139)$$

which commutes with each component S_j:

$$
\begin{aligned}
[S^2, S_z] &= [S_x^2 + S_y^2 + S_z^2, S_z] \\
&= [S_x^2 + S_y^2, S_z] \\
&= S_x[S_x, S_z] - [S_z, S_x]S_x + S_y[S_y, S_z] - [S_z, S_y]S_y \\
&= i(-S_x S_y - S_y S_x + S_y S_x + S_x S_y) \\
&= 0, \qquad (13.140)
\end{aligned}
$$

and by cyclic permutation of the indices:

$$[S^2, S_x] = [S^2, S_y] = 0. \qquad (13.141)$$

When the commutator of two quantum mechanical operators is zero, we can simultaneously diagonalize both operators with a common unitary transformation. Since the three spin projection operators do not commute, we can only diagonalize S^2 and one of the projection operators with the same transformation. Selecting S_y as the diagonal matrix, the other two components cannot be simultaneously diagonalized since they do not commute with S_y. It is convenient to define what we will see to be the raising and lowering operators in terms of the two nondiagonalized generator matrices

$$S_\pm = S_z \pm iS_x. \qquad (13.142)$$

Note that these two operators are not hermitian, but rather are hermitian conjugates of each other:

$$S_\pm^\dagger = S_\mp. \qquad (13.143)$$

Since S^2 and S_y commute, we can find simultaneous quantum numbers which correspond to the individual components of the spinor.

We write the usual $2\bar{s} + 1$ orthonormal spinors states in terms of the spin quantum number $\bar{s}$ and the magnetic quantum number of the projection along the reference quantization axis as $|\tilde{s}, m\rangle$. These functions are eigenstates of both the S^2 and S_y operators:

$$S^2|\tilde{s}, m\rangle = \tilde{s}(\tilde{s} + 1)|\tilde{s}, m\rangle, \tag{13.144}$$

$$S_y|\tilde{s}, m\rangle = m|\tilde{s}, m\rangle. \tag{13.145}$$

Note: We have not proven that either $< \psi_m|S^2|\psi_m > = s(s+1)$ or $< \psi_m|S_y|\psi_m > = m$ and refer the reader to their favorite quantum mechanics text for the details (see for example Chapter 2 of Ref. 24).

Since we expect S_+ raise a pure m state to $m + 1$ and S_- to lower the state to $m - 1$ we should expect $S_+ S_-$ and $S_- S_+$ return the same state multiplied by some eigenvalues. Evaluating the operator products gives

$$\begin{aligned} S_+ S_- &= (S_z + iS_x)(S_z - iS_x) = [S_z^2 + S_x^2 + i(S_x S_z - S_z S_x)] \\ &= [S^2 - S_y^2 + S_y], \end{aligned} \tag{13.146}$$

$$\begin{aligned} S_- S_+ &= (S_z - iS_x)(S_z + iS_x) = [S_z^2 + S_x^2 + i(S_z S_x - S_x S_z)] \\ &= [S^2 - S_y^2 - S_y], \end{aligned} \tag{13.147}$$

with the respective eigenvalues $\tilde{s}(\tilde{s}+1) - m^2 + m$ and $\tilde{s}(\tilde{s}+1) - m^2 - m$, respectively. Combining the last two equations gives the commutator

$$\begin{aligned}[S_+, S_-] &= (S^2 - S_y^2 + S_y) - (S^2 - S_y^2 - S_y) \\ &= 2S_y. \end{aligned} \tag{13.148}$$

The operators $S_\pm$ do not commute with the diagonal matrix S_y:

$$\begin{aligned}[S_y, S_\pm] &= [S_y, S_z] \pm i[S_y, S_x] \\ &= iS_x \pm i(-iS_z) \\ &= \pm(S_z \pm iS_x) \\ &= \pm S_\pm. \end{aligned} \tag{13.149}$$

Applying this last commutation relation to the eigenstate $|\tilde{s}, m\rangle$ gives

$$\begin{aligned} S_y S_\pm|s, m\rangle &= S_\pm S_y|\tilde{s}, m\rangle \pm S_\pm|\tilde{s}, m\rangle \\ &= (m \pm 1)S_\pm|\tilde{s}, m\rangle, \end{aligned} \tag{13.150}$$

we find that $S_\pm |\tilde{s}, m\rangle$ are both eigenstates of S_y with the eigenvalues of the states $|\tilde{s}, m \pm 1\rangle$. So we have that S^+ raises the eigenstate one level and S_- lowers it one level. * To find the normalization assume a normalization constant $A_\pm$, so

$$S_\pm |\tilde{s}, m\rangle = A_\pm |\tilde{s}, m \pm 1\rangle, \qquad (13.151)$$

$$\langle \tilde{s}, m | S_\mp S_\pm |\tilde{s}, m\rangle = |A_\pm|^2, \qquad (13.152)$$

which yields

$$A_\pm = \sqrt{(\tilde{s} \mp m)(\tilde{s} \pm m + 1)}. \qquad (13.153)$$

Writing S_x and S_z in terms of $S_\pm$:

$$S_x = \frac{i}{2}(S_- - S_+), \qquad (13.154)$$

$$S_z = \frac{1}{2}(S_+ + S_-). \qquad (13.155)$$

From these we get

$$S_z S_x = \frac{1}{2}(S_- + S_+)\frac{i}{2}(S_- - S_+) = \frac{i}{4}\left(S_z + S_-^2 - S_+^2\right), \qquad (13.156)$$

$$S_x S_z = \frac{i}{2}(S_- - S_+)\frac{1}{2}(S_- + S_+) = \frac{i}{4}\left(-S_z + S_-^2 - S_+^2\right), \qquad (13.157)$$

which can be combined to give

$$\{S_z, S_x\} = \frac{i}{2}\left(S_-^2 - S_+^2\right). \qquad (13.158)$$

Note that Eq. (13.158), which gives zero in the case of spin-$\frac{1}{2}$, does not identically give zero for higher order spins.

---

* If this is not clear to the reader, we admit we have glossed over a number of proofs here in treating this as a quick review rather than a complete treatise. We will proceed along this path in trying to construct actual matrices for the raising and lower operators; at that point it should become obvious that while skipping over details we have managed to construct a set of matrices which obey the previous unproven assumptions.

In matrix form the lowering and raising operators look like:

$$S_- = \begin{pmatrix} 0 & 0 & 0 & \cdots \\ a_1^- & 0 & 0 & \cdots \\ 0 & a_2^- & 0 & \cdots \\ 0 & 0 & a_3^- & \cdots \\ \vdots & \vdots & \vdots & \ddots \end{pmatrix}, \tag{13.159}$$

$$S_+ = \begin{pmatrix} 0 & a_1^+ & 0 & 0 & \cdots \\ 0 & 0 & a_2^+ & 0 & \cdots \\ 0 & 0 & 0 & a_3^+ & \cdots \\ \vdots & \vdots & \vdots & \vdots & \ddots \end{pmatrix}, \tag{13.160}$$

with the maximum number of raisings (lowerings) being $2\tilde{s}$, since $(S_\pm)^{2\tilde{s}+1} = 0$. Using Eqs. (13.152 and 13.153)

$$a_k^- = \sqrt{(2\tilde{s} - k + 1)(k)}, \quad \text{with} \quad m = \tilde{s} - k + 1, \tag{13.161}$$
$$a_k^+ = \sqrt{(k)(2\tilde{s} - k + 1)}, \quad \text{with} \quad m = \tilde{s} - k, \tag{13.162}$$

and we see that $a_k^- = a_k^+$. This now allows us to write the three generator matrices for spin-$\tilde{s}$ as:

$$S_x = \frac{i}{2}(S_+ + S_-) = \frac{i}{2} \begin{pmatrix} 0 & -a_1 & 0 & 0 & \cdots \\ a_1 & 0 & -a_2 & 0 & \cdots \\ 0 & a_2 & 0 & -a_3 & \cdots \\ 0 & 0 & a_3 & 0 & \cdots \\ \vdots & \vdots & \vdots & \vdots & \ddots \end{pmatrix}, \tag{13.163}$$

$$S_y = \begin{pmatrix} s & 0 & 0 & 0 & \cdots & 0 \\ & \tilde{s}-1 & 0 & 0 & \cdots & 0 \\ 0 & 0 & \tilde{s}-2 & 0 & \cdots & 0 \\ 0 & 0 & 0 & \tilde{s}-3 & \cdots & 0 \\ \vdots & \vdots & \vdots & \vdots & \ddots & \vdots \\ 0 & 0 & 0 & 0 & \cdots & -\tilde{s} \end{pmatrix}, \tag{13.173}$$

$$S_z = \frac{1}{2}(S_+ + S_-) = \frac{1}{2} \begin{pmatrix} 0 & a_1 & 0 & 0 & \cdots \\ a_1 & 0 & a_2 & 0 & \cdots \\ 0 & a_2 & 0 & a_3 & \cdots \\ 0 & 0 & a_3 & 0 & \cdots \\ \vdots & \vdots & \vdots & \vdots & \ddots \end{pmatrix}, \tag{13.173}$$

where

$$a_k = \sqrt{k(2\tilde{s} - k + 1)}. \tag{13.174}$$

Table II. Pyramid scheme for calculating values of a_k^2.

s													
1/2							1						
1						2		2					
3/2					3		4		3				
2				4		6		6		4			
5/2			5		8		9		8		5		
3		6		10		12		12		10		6	
7/2	7		12		15		16		15		12		7
$\vdots$	a_1^2		a_2^2		a_3^2		a_4^2		a_5^2		a_6^2		a_7^2

Table II shows a simple pyramid scheme for calculating the square of the a_k with a rather obvious pattern.

It should be obvious that there is a symmetry about the counterdiagonal (aka skew diagonal aka antidiagonal) for S_x and S_z, since

$$a_{2\bar{s}+1-k} = a_k. \tag{13.175}$$

However the diagonalized S_y is antisymmetric about the counterdiagnonal.

Now it is easy to write down a set of spin operator matrices for spin to any order. For spin-$\frac{1}{2}$ we get the Pauli matrices as given in Eq. 13.127. Second order combinations of the Pauli matrices give results which can be written in term of the unit matrix and the Pauli matrices, so operators like $S_j S_k$ give no extra information and can be reduced to first order in S.

For spin-1 we find

$$S_x = \frac{1}{\sqrt{2}} \begin{pmatrix} 0 & -i & 0 \\ i & 0 & -i \\ 0 & i & 0 \end{pmatrix}, \tag{13.176}$$

$$S_y = \begin{pmatrix} 1 & 0 & 0 \\ 0 & 0 & 0 \\ 0 & 0 & -1 \end{pmatrix}, \tag{13.177}$$

$$S_z = \frac{1}{\sqrt{2}} \begin{pmatrix} 0 & 1 & 0 \\ 1 & 0 & 1 \\ 0 & 1 & 0 \end{pmatrix}. \tag{13.178}$$

It is clear by looking at S_y^2 that it cannot be written as sum of constants times the unit-matrix and the three components of the spin operator. Higher order combinations (to second order for spin-1) must be computed to obtain the description of the spin of a particle.

13.7 Rotation matrices for spinors

In this representation, the spinor rotation matrix for a left-handed rotation of angle θ about the y-axis has diagonal components

$$D_y^{\bar{s}}(\theta)_l = 1 + i(\bar{s} - l)\theta - \frac{1}{2!}[(\bar{s} - l)\theta]^2 - i\frac{1}{3!}[(s - l)\theta]^3 + \cdots = e^{i(\bar{s}-l)\theta} \quad (13.179)$$

and putting them together the matrix for the left-handed rotation about the $+\hat{y}$ direction becomes

$$\mathbf{D}_{\hat{y}}^{\bar{s}}(\theta) = \begin{pmatrix} e^{i\bar{s}\theta} & 0 & \cdots & 0 \\ 0 & e^{i(\bar{s}-l)\theta} & \cdots & 0 \\ \vdots & \vdots & \ddots & \vdots \\ 0 & 0 & \cdots & e^{-i\bar{s}\theta} \end{pmatrix}. \quad (13.180)$$

The trace of this matrix is

$$\text{tr}\left[\mathbf{D}_{\hat{y}}^{\bar{s}}(\theta)\right] = \begin{cases} 2\sum_{m=1}^{\bar{s}} \cos(m\theta) & \text{if } 2\bar{s} + 1 \text{ is even (fermions)}; \\ 1 + 2\sum_{m=1}^{\bar{s}} \cos(m\theta) & \text{if } 2\bar{s} + 1 \text{ is odd (bosons)}. \end{cases} \quad (13.181)$$

For spin-$\frac{1}{2}$ the rotation matrices are

$$\mathbf{D}_{\hat{x}}^{\frac{1}{2}}(\theta) = e^{i\sigma_x\theta/2} = \mathbf{I}\cos\frac{\theta}{2} + i\sigma_x\sin\frac{\theta}{2} = \begin{pmatrix} \cos\frac{\theta}{2} & \sin\frac{\theta}{2} \\ -\sin\frac{\theta}{2} & \cos\frac{\theta}{2} \end{pmatrix}, \quad (13.182)$$

$$\mathbf{D}_{\hat{y}}^{\frac{1}{2}}(\theta) = e^{i\sigma_y\theta/2} = \mathbf{I}\cos\frac{\theta}{2} + i\sigma_y\sin\frac{\theta}{2} = \begin{pmatrix} e^{i\theta/2} & 0 \\ 0 & e^{-i\theta/2} \end{pmatrix}, \quad (13.183)$$

$$\mathbf{D}_{\hat{z}}^{\frac{1}{2}}(\theta) = e^{i\sigma_z\theta/2} = \mathbf{I}\cos\frac{\theta}{2} + i\sigma_z\sin\frac{\theta}{2} = \begin{pmatrix} \cos\frac{\theta}{2} & i\sin\frac{\theta}{2} \\ i\sin\frac{\theta}{2} & \cos\frac{\theta}{2} \end{pmatrix}. \quad (13.184)$$

A left-handed rotation of θ about a general unit vector $\hat{n}$ may be written as

$$\mathbf{D}_{\hat{n}}^{\frac{1}{2}}(\theta) = e^{i\hat{n}\cdot\vec{\sigma}\theta/2} = \sum_{j=0}^{\infty} \frac{i^j}{j!}\left[\frac{\theta}{2}\right]^j (\hat{n}\cdot\vec{\sigma})^j, \quad (13.185)$$

but

$$\begin{aligned} (\hat{n}\cdot\vec{\sigma})^2 &= n_x^2\sigma_x^2 + n_y^2\sigma_y^2 + n_z^2\sigma_z^2 \\ &\quad + n_xn_y(\sigma_x\sigma_y + \sigma_y\sigma_x) + n_yn_z(\sigma_y\sigma_z + \sigma_z\sigma_y) + n_zn_x(\sigma_z\sigma_x + \sigma_x\sigma_z) \\ &= (n_x^2 + n_y^2 + n_z^2)\mathbf{I} + 0 \\ &= \mathbf{I}, \end{aligned} \quad (13.186)$$

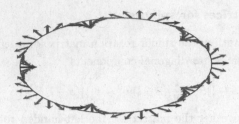

Figure. 13.5 Spin precession on closed orbit for constant vertical bend field with the spin in the horizontal plane. This is plotted for $1 + G\gamma = 6$ with six complete spin rotations per turn in the fixed lab system. In the system rotating with the particle we see 5 cusps corresponding to a spin tune $\nu_{\rm sp} = G\gamma = 5$. The spin and betatron oscillation are plotted for one counterclockwise turn beginning at the middle right side of the plot.

since $\sigma_j \sigma_k = -\sigma_k \sigma_j$ for the Pauli matrices if $j \neq k$. So Eq. (13.185) becomes just

$$
\mathbf{D}_{\hat{n}}^{\frac{1}{2}}(\theta) = e^{i\hat{n}\cdot\vec{\sigma}\theta/2} = \mathbf{I} \sum_{j=1}^{\infty} \frac{(-1)^j}{(2j)!} \left(\frac{\theta}{2}\right)^{2j} + i(\hat{n}\cdot\vec{\sigma}) \sum_{j=0}^{\infty} \frac{(-1)^{2j+1}}{(2j+1)!} \left(\frac{\theta}{2}\right)^{2j+1}
$$

$$
= \mathbf{I} \cos\frac{\theta}{2} + i(\hat{n}\cdot\vec{\sigma}) \sin\frac{\theta}{2}
$$

$$
= \begin{pmatrix} \cos\frac{\theta}{2} + i n_y \sin\frac{\theta}{2} & (n_x + i n_z)\sin\frac{\theta}{2} \\ (-n_x + i n_z)\sin\frac{\theta}{2} & \cos\frac{\theta}{2} - i n_y \sin\frac{\theta}{2} \end{pmatrix}. \tag{13.187}
$$

Similar to Eq. (13.181) with $\bar{s} = 1/2$, note that

$$
\mathrm{tr}\left[\mathbf{D}_{\hat{n}}^{\frac{1}{2}}(\theta)\right] = 2\cos\frac{\theta}{2}. \tag{13.188}
$$

Recalling Eqs. (13.176 to 13.178), the spin-1 rotation matrices about the three axes are

$$
\mathbf{D}_{\hat{x}}^{1}(\theta) = \begin{pmatrix} \frac{1+\cos\theta}{2} & \frac{\sin\theta}{\sqrt{2}} & \frac{1-\cos\theta}{2} \\ -\frac{\sin\theta}{\sqrt{2}} & \cos\theta & \frac{\sin\theta}{\sqrt{2}} \\ \frac{1-\cos\theta}{2} & -\frac{\sin\theta}{\sqrt{2}} & \frac{1+\cos\theta}{2} \end{pmatrix}, \tag{13.189}
$$

$$
\mathbf{D}_{\hat{y}}^{1}(\theta) = \begin{pmatrix} e^{i\theta} & 0 & 0 \\ 0 & 1 & 0 \\ 0 & 0 & e^{-i\theta} \end{pmatrix}, \tag{13.190}
$$

$$
\mathbf{D}_{\hat{z}}^{1}(\theta) = \begin{pmatrix} \frac{\cos\theta+1}{2} & -i\frac{\sin\theta}{\sqrt{2}} & \frac{\cos\theta-1}{2} \\ -i\frac{\sin\theta}{\sqrt{2}} & \cos\theta & -i\frac{\sin\theta}{\sqrt{2}} \\ \frac{\cos\theta-1}{2} & -i\frac{\sin\theta}{\sqrt{2}} & \frac{\cos\theta+1}{2} \end{pmatrix}. \tag{13.191}
$$

13.8 Spin precession on the closed orbit

In a uniform vertical magnetic field, particles can have closed circular orbits in a plane perpendicular to the field. In that case a spin which is in the horizontal plane will precess about the vertical direction with the rate (recalling Eq. (13.69))

$$\frac{d\vec{S}}{dt} = \frac{q}{\gamma m}(1 + G\gamma)\vec{S} \times \vec{B}, \tag{13.192}$$

as shown in Fig. 13.5. Comparing this last equation with the rotation of the orbit

$$\frac{d\vec{p}}{dt} = \frac{q}{\gamma m}\vec{p} \times \vec{B}, \tag{13.193}$$

it is immediately obvious that a horizontal spin in the lab system rotates $1 + G\gamma$ times faster than the velocity. In our usual local coordinate system (see Fig. 2.1) which rotates with the design particle, a horizontal component of spin will precess $G\gamma$ times. This number of precessions in the rotating system is called the spin tune $\nu_{\rm sp}$.

Neither the velocity nor spin are changed in the field free region of a drift, and if the closed orbit passes through the centers of all quadrupole and higher order multipole magnets, then the particle would still experience only the vertical fields of the bending dipoles in a perfectly aligned flat ring. In this case the spin tune will be the same as for the constant vertical field.

If a solenoid with field $B_{\rm sol}$ and length $l_{\rm sol}$ is inserted into a straight section so that its axis is aligned along the closed orbit, then the particle's spin will experience an extra rotation of

$$\phi = \int_0^{l_{\rm sol}/v} \frac{q}{\gamma m}(1 + G)B_{\rm sol}\, dt = (1 + G)\frac{qB_{\rm sol}l_{\rm sol}}{p} \tag{13.194}$$

about the longitudinal axis on every turn. In the rotating coordinate system of the design orbit the full-turn spin rotation matrix for a spin-$\frac{1}{2}$ particle on the closed orbit just before the solenoid will be

$$\begin{aligned}
\mathbf{M} = \mathbf{D}_{\hat{n}}^{\frac{1}{2}}(\mu) &= \mathbf{D}_{\hat{y}}^{\frac{1}{2}}(G\gamma 2\pi)\mathbf{D}_{\hat{z}}^{\frac{1}{2}}(\phi) \\
&= \begin{pmatrix} e^{iG\gamma\pi} & 0 \\ 0 & e^{-iG\gamma\pi} \end{pmatrix} \begin{pmatrix} \cos\frac{\phi}{2} & i\sin\frac{\phi}{2} \\ -i\sin\frac{\phi}{2} & \cos\frac{\phi}{2} \end{pmatrix} \\
&= \begin{pmatrix} e^{iG\gamma\pi}\cos\frac{\phi}{2} & ie^{iG\gamma\pi}\sin\frac{\phi}{2} \\ -ie^{-iG\gamma\pi}\sin\frac{\phi}{2} & e^{-iG\gamma\pi}\cos\frac{\phi}{2} \end{pmatrix}
\end{aligned} \tag{13.195}$$

Recalling Eq. (13.188), the trace of $\mathbf{M}$ gives

$$2\cos\frac{\mu}{2} = 2\cos\left(G\gamma\pi\right)\cos\frac{\phi}{2},\tag{13.196}$$

and comparison of the components yields

$$n_x\sin\frac{\mu}{2} = -\sin(G\gamma\pi)\sin\frac{\phi}{2},\tag{13.197}$$

$$n_y\sin\frac{\mu}{2} = \sin(G\gamma\pi)\cos\frac{\phi}{2},\tag{13.198}$$

$$n_z\sin\frac{\mu}{2} = \cos(G\gamma\pi)\sin\frac{\phi}{2}.\tag{13.199}$$

The fractional part of the spin tune will then be

$$\text{frac}(\nu_{\text{sp}}) = \frac{\mu}{2\pi} = \pm\frac{1}{\pi}\cos^{-1}\left(\cos\left(G\gamma\pi\right)\cos\frac{\phi}{2}\right).\tag{13.200}$$

As with the betatron tunes, the total spin tune cannot be determined from a one-turn matrix and the actual tune may have extra multiple rotations of 2π. The ambiguity of sign is real and reflects the fact that a rotation of angle μ about a vector $\hat{n}$ gives the same result as a rotation of $-\mu$ about the vector $-\hat{n}$.

If $\phi = 0$, then there is no effect from the solenoid, and the closed-orbit spin tune is simply $\nu_{\text{sp}} = G\gamma$. However, if $\phi = \pi$ then the fractional part of the spin tune would be exactly $1/2$ independent of energy. Such a solenoid which rotates the spin by 180° about the longitudinal is an example of a Type I Siberian snake.

A Siberian snake is an insertion device which leaves the orbit unchanged (i. e. behaves like a drift), but produces a net rotation about some axis in the horizontal plane. Type I snakes rotate the spin about the longitudinal z-axis, and Type II snakes rotate the spin about the radial x-axis. In RHIC, the snakes typically rotate about axes which are at $\pm 45°$ to the z-axis in the xz-plane. A full snake rotates the spin by 180°, and partial snakes[28,29] rotate by some smaller angle. The strength of a partial snake is frequently given as a percentage of a full 180° rotation. More generally, an insertion device that leaves the orbit unchanged which rotates any axis is called a spin rotator, so the snakes form a subset of the rotators.

Fig. 13.6 shows how the spin tune varies for different strengths of the partial snake. Notice that at integer values of $G\gamma$, as the snake strength is increased, a gap or stop band opens up in in the curve of the spin tune. When the snake reaches full strength, the spin tune can only have values of an integer plus a half.

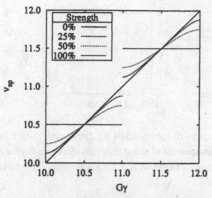

Figure. 13.6 Closed-orbit spin tune ν_{sp} vs $G\gamma$ for a flat ring with one snake.

The stable spin direction $\hat{n}_0$ on the closed orbit just before the snake can be calculated by taking $\sin(\mu/2) \geq 0$ in Eqs.(13.197 to 13.199) and solving for $\hat{n}$ as a function of $G\gamma$. The left plot of Fig. 13.7 shows the how the stable spin direction just before the snake changes as the energy sweeps through integral values of $G\gamma$. Note that $\hat{n}_0$ starts out vertical at the half integer and flips sign as the integer is crossed. This flip of phase is characteristic of a resonance crossing, similar to the phase shift between the voltage and current signals of an LCR circuit as seen when sweeping the frequency across the resonance. We can now ask how the vertical component of the spin changes as the energy is ramped at a rate

$$\alpha = \frac{d(G\gamma)}{d\theta}, \tag{13.201}$$

where θ is the azimuthal angle of the orbit. The middle plot in Fig. 13.7 shows how the vertical component of spin with a 5% snake changes when the energy is ramped at rate of

$$\alpha = \frac{2 \left[\text{units of } G\gamma\right]}{2999 \text{ turns} \times 2\pi} = 1.0614 \times 10^{-4}. \tag{13.202}$$

For the perfectly aligned flat ring the stable spin direction of the closed orbit will be vertical if the snake is turned off ($\phi = 0$). With a nonzero snake right at the resonant frequency, the resonance strength ϵ_{res} can be characterized as the angular tip away from vertical (the stable spin direction with no snake) after one orbit. This is just the rotation angle of the snake $\epsilon_{res} = \phi$.

The middle plot of Fig. 13.7 shows the vertical component of the spin as the energy is ramped from $G\gamma = 10.5$ to 12.5 in 2999 turns for a 5% snake.

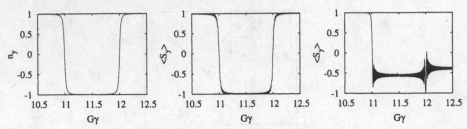

Figure. 13.7 Vertical component of spin tracked for 2999 turns with a linear energy ramp over two units of $G\gamma$. Left: vertical component of the stable spin direction $\hat{n}_0$ for a solenoid snake of strength 5%. Middle: vertical component of spin with a 5% snake. Right: vertical component of spin with a weaker 2% snake.

An experimental set-up[27], consisting of a solenoid, was used at the Indiana University Cooler Ring[30] in order to test the Siberian snake concept and to verify that a polarized proton beam can be accelerated through the depolarizing resonance $a\gamma=2$ (108 MeV) with no polarization losses.

The conservation of the polarization lying parallel to the stable solution, was checked by the collaborations[27,31] with a stored beam sitting on the 108 MeV depolarizing resonance, for a time much longer than the acceleration cycle, which was limited only by the actual beam life-time of about 20 minutes. These measures were performed either with a full snake (180° spin-flip) or a partial snake (60° spin rotation): in both cases the polarization component along the stable solution was fully preserved.

Although of less practical importance, but of more conceptual significance for a deeper understanding of the subtle mechanisms at work, there was an open question on the conservation of spin polarization when particles are injected with their spins transverse to the stable solution: general expectation was that the beam polarization should be fading away; however it was also conceivable that this polarization could survive for an appreciable time, with all the particles spins coherently precessing about the periodic solution.

In a successive experiment, a beam of 108 MeV polarized protons was injected into the IUCF Cooler Ring with their spins perpendicular to the stable solution: after about 10 seconds a residual transverse polarization was measured[31], thus demonstrating the existence of a polarization precession about the stable solution.

Synchrotron oscillations can also cause sidebands of these imperfection resonances if the synchrotron tune is larger than the intrinsic resonance width. Generally in proton rings, the synchrotron tune is smaller than the imperfection width so synchrotron sidebands are not apparent.

13.9 Spin motion in the standard coordinate system

Recalling* the local coordinate system referenced to the design trajectory (see Fig. 2.1) the particle's orbit is described by

$$\vec{r}(s) = (\rho + x)\hat{x} + y\hat{y} + z\vec{s}, \tag{13.203}$$

where $\rho = \rho(s)$ is the bend radius of the design trajectory. The local axes of the design orbit change along the trajectory according to

$$\frac{d\hat{x}}{ds} = \frac{\hat{s}}{\rho}, \quad \frac{d\hat{y}}{ds} = 0, \quad \text{and} \quad \frac{d\hat{s}}{ds} = -\frac{\hat{x}}{\rho}, \tag{13.204}$$

so the velocity is then

$$\vec{v} = \frac{d\vec{r}}{dt} = \frac{d\vec{r}}{ds}\frac{ds}{dt} = \left[x'\hat{x} + y'\hat{y} + \left(1 + \frac{x}{\rho}\right)\hat{s}\right]\frac{ds}{dt}. \tag{13.205}$$

Adding the velocity components in quadrature gives the particle's speed

$$v = \frac{ds}{dt}\sqrt{x'^2 + y'^2 + \left(1 + \frac{x}{\rho}\right)^2} \simeq \frac{ds}{dt}\left(1 + \frac{x}{\rho}\right) + \mathcal{O}(2) \tag{13.206}$$

and we find to first order that the projection of the particles velocity onto the design orbit scales as

$$\frac{ds}{dt} \simeq v\left(1 + \frac{x}{\rho}\right)^{-1}. \tag{13.207}$$

Substituting this back into Eq. (13.205), we have

$$\vec{v} = v\left(1 - \frac{x}{\rho}\right)\left[x'\hat{x} + y'\hat{y} + \left(1 + \frac{x}{\rho}\right)\hat{s}\right]. \tag{13.208}$$

The component of magnetic field parallel to the velocity is

$$\vec{B}_\parallel = (\hat{v} \cdot \vec{B})\hat{v} = \frac{x'B_x + y'B_y + B_s}{\sqrt{1 + x'^2 + y'^2}}\frac{x'\hat{x} + y'\hat{y} + \hat{s}}{\sqrt{1 + x'^2 + y'^2}}$$
$$\simeq x'B_s\hat{x} + y'B_s\hat{y} + (x'B_x + y'B_y)\hat{s} + \mathcal{O}(2). \tag{13.209}$$

---

* In this section we revert to our original usage of symbols: s as the path length along the design trajectory and not the spin quantum number, and a prime ($'$) superscript as a derivative with respect to the coordinate s rather than indicating a rest system value.

Using this, Eq. (13.70) transforms into

$$
\begin{aligned}
\vec{\Omega}_s &= -\frac{q}{\gamma m}[(1+G\gamma)\vec{B} - G(\gamma-1)\vec{B}_\parallel] \\
&= -\frac{q}{\gamma m}\{[(1+G\gamma)B_x - G(\gamma-1)x'B_s]\hat{x} \\
&\quad +[(1+G\gamma)B_y - G(\gamma-1)y'B_s]\hat{y} \\
&\quad +[(1+G\gamma)B_s - G(\gamma-1)(x'B_x + y'B_y)]\hat{s}\} + \mathcal{O}(2).
\end{aligned} \quad (13.210)
$$

On the design trajectory, the instantaneous cyclotron frequency is

$$
\vec{\Omega}_c = -\frac{q}{\gamma_0 m}B_0\hat{y}, \quad (13.211)
$$

where the subscript zeros indicate values for the design particle on its trajectory. and the differential equation for the spinor evolution in the local rest frame of a spin-$\frac{1}{2}$ design particle is then

$$
\begin{aligned}
\frac{d|\Psi\rangle}{ds} &= \frac{d|\Psi\rangle}{dt}\frac{dt}{ds} = -\frac{i}{2}\left(\frac{\rho+x}{\rho v}\vec{\Omega}_s - \vec{\Omega}_c\right)\cdot\vec{\sigma}|\Psi\rangle \\
&= \frac{1}{2}\frac{q}{p_0}\left\{\frac{1+x/\rho}{1+\delta}\left[(1+G\gamma)\vec{B} - G(\gamma-1)\vec{B}_\parallel\right] - B_0\hat{y}\right\}\cdot\vec{\sigma}|\Psi\rangle,
\end{aligned} \quad (13.212)
$$

with the usual definition $\delta = \Delta p/p_0$,

$$
\begin{aligned}
\frac{d|\Psi\rangle}{ds} = \frac{1}{2}\frac{q}{p_0}\frac{\rho+x}{\rho(1+\delta)} &\left[(1+G\gamma)\begin{pmatrix} iB_y & B_x+iB_s \\ -B_x+iB_s & -iB_y \end{pmatrix}\right. \\
&\quad -G(\gamma-1)(x'B_x + y'B_y)\begin{pmatrix} 0 & i \\ i & 0 \end{pmatrix} \\
&\quad \left. -G(\gamma-1)B_s\begin{pmatrix} iy' & x' \\ -x' & -iy' \end{pmatrix}\right]|\Psi\rangle \\
&\quad -\frac{1}{2}\frac{q}{p_0}\begin{pmatrix} iB_0 & 0 \\ 0 & -iB_0 \end{pmatrix}|\Psi\rangle.
\end{aligned} \quad (13.213)
$$

Recalling Eq. (1.17), we can evaluate the factors:

$$
\frac{1+G\gamma}{1+\delta} \simeq (1-\delta)(1+G\gamma_0 + G\gamma_0\beta_0^2\delta) \simeq 1 + G\gamma_0 - \left(1+\frac{G}{\gamma_0}\right)\delta, \quad (13.214)
$$

$$
\frac{G(\gamma-1)}{1+\delta} \simeq (1-\delta)[G(\gamma_0-1) + G\gamma_0\beta_0^2\delta] \simeq G(\gamma_0-1)\left(1+\frac{\delta}{\gamma_0}\right). \quad (13.215)
$$

13.9.1 Sector bend magnets

If we restrict ourselves to considering just simple sector bend magnets with $\vec{B} = B_0 \hat{y}$, then Eq. (13.213) becomes

$$\frac{d|\Psi\rangle}{ds} = \frac{i}{2} \frac{1}{\rho} \left[\left(\frac{1 + G\gamma_0}{1 + \delta} \frac{\rho + x}{\rho} - 1 \right) \sigma_y - \frac{G(\gamma_0 - 1)}{1 + \delta} y' \sigma_s \right] |\Psi\rangle, \quad (13.216)$$

for dipoles with field $\vec{B} = B_0 \hat{y}$. The slope y' remains unchanged throughout the dipole assuming there is no gradient. This can be modified to be explicit in the variable s with the help of the first order orbit equation:

$$x(s) = x_0 \cos \frac{s}{\rho} + x_0' \rho \sin \frac{s}{\rho} + \rho \left(1 - \cos \frac{s}{\rho} \right) \delta, \quad (13.217)$$

for the evolution of the x-coordinate through a sector dipole, since y' remains constant throughout the dipole. If we restrict ourselves to on-momentum particles, then

$$\frac{d|\Psi\rangle}{d\theta} = \rho \frac{d|\Psi\rangle}{ds} = \frac{i}{2} \left\{ \left[G\gamma + (1 + G\gamma) \left(\frac{x_0}{\rho} \cos \theta + x_0' \sin \theta \right) \right] \sigma_y \right.$$
$$\left. - G(\gamma - 1) y' \sigma_s \right\} |\Psi\rangle. \quad (13.218)$$

Ignoring horizontal oscillations, then this just becomes

$$\frac{d|\Psi\rangle}{d\theta} = \frac{i}{2} G \left[\gamma_0 \sigma_y - (\gamma_0 - 1) y' \sigma_s \right] |\Psi\rangle. \quad (13.219)$$

Since y' is constant through the dipole, this can easily be integrated as

$$|\Psi(\theta)\rangle = e^{i[\sigma_y - (1 - 1/\gamma_0) y' \sigma_s] G\gamma_0 \theta/2} |\Psi(0)\rangle$$
$$= \mathbf{I} \cos \frac{\mu}{2} + i(n_y \sigma_y + n_s \sigma_s) \sin \frac{\mu}{2}, \quad (13.220)$$

with

$$\mu(\theta) = G\sqrt{\gamma_0^2 + (\gamma_0 - 1)^2 y'^2} \, \theta = G\gamma_0 \theta + \mathcal{O}(y'^2), \quad (13.221)$$

$$n_x = 0, \quad n_y = \frac{\gamma_0}{\sqrt{\gamma_0 + (\gamma_0 - 1)^2 y'^2}}, \quad \text{and} \quad n_s = \frac{(\gamma_0 - 1) y'}{\sqrt{\gamma_0^2 + (\gamma_0 - 1)^2 y'^2}}, \quad (13.222)$$

where θ is the usual bend angle of the dipole, and we see that there is a slight tilt of the rotation axis in the ys-plane when the trajectory has a vertical slope inside the dipole. Linearizing Eq. (13.220) gives

$$|\Psi(\theta)\rangle \simeq \left\{ \mathbf{I} \cos \frac{G\gamma_0 \theta}{2} + i \left[\sigma_y - \left(1 - \frac{1}{\gamma_0} \right) \sigma_s y' \right] \sin \frac{G\gamma_0 \theta}{2} \right\} |\Psi(0)\rangle; \qquad (13.223)$$

however, one should note that the rotation matrix inside the braces is only unitary for $y' = 0$, and as such is not a good approximation to use for multiturn tracking. For tracking codes, this is easily remedied by using Eq. (13.220) with the normalized unit vector of Eq. (13.222). This effect of the vertical slope in dipoles is rather small compared to that of quadrupoles. If we ignore this small contribution of y' for small amplitude oscillations, then the dipole rotation simplifies to just

$$|\Psi(\theta)\rangle \simeq \mathbf{D}_{\hat{y}}^{\frac{1}{2}}(G\gamma_0 \theta) |\Psi(0)\rangle. \qquad (13.224)$$

13.9.2 Quadrupole magnets

The dominant contribution from vertical oscillations is from the quadrupoles with transverse magnetic fields of the form

$$\vec{B} = B_1(y\hat{x} + x\hat{y}), \qquad (13.225)$$

where the gradient on the design orbit is expressed as $B_1 = (\partial B_y/\partial x)_0$. So the spinor will precess according to

$$\frac{d|\Psi\rangle}{ds} = \frac{i}{2} \frac{1 + G\gamma}{1 + \delta} k\sqrt{x^2 + y^2} \left[\frac{x\sigma_y}{\sqrt{x^2 + y^2}} + \frac{y\sigma_x}{\sqrt{x^2 + y^2}} \right] |\Psi\rangle, \qquad (13.226)$$

in normal quadrupoles with $k = qB_1/p_0$. For purely vertical oscillations with $x = 0$ and $\delta = 0$ this becomes

$$\frac{d|\Psi\rangle}{ds} = \frac{i}{2}(1 + G\gamma_0) ky\sigma_x |\Psi\rangle. \qquad (13.227)$$

Assuming a thin lens approximation with y constant in the quadrupole of length l, integration yields

$$|\Psi(l)\rangle \simeq \mathbf{D}_{\hat{x}}^{\frac{1}{2}}(\theta) |\Psi(0)\rangle = \left(\mathbf{I} \cos \frac{\theta}{2} + \sigma_x \sin \frac{\theta}{2} \right) |\Psi(0)\rangle, \qquad (13.228)$$

with the quadrupole precession angle given by

$$\theta = (1 + G\gamma_0)kly. \qquad (13.229)$$

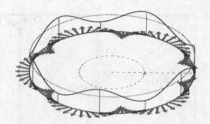

Figure. 13.8   Horizontal spin components plotted for a vertical intrinsic resonance with $G\gamma = Q_V = 5.5$. If the spin precession has the same frequency as the vertical betatron oscillation, then kicks from radial magnetic fields in the quadrupoles will remain in phase with the spin and cause a coherent tipping of the spin away from vertical.

13.10   Vertical betatron oscillations and intrinsic resonances

Let us now consider vertical motion of an on-momentum particle. Combining Eqs. (13.219) and (13.227) gives

$$\frac{d\,|\Psi\rangle}{ds} \simeq \frac{i}{2}\left[\frac{G\gamma}{\rho(s)}\sigma_y - \frac{G}{\rho(s)}(\gamma-1)y'\sigma_s + (1+G\gamma)k(s)y\sigma_x\right]|\Psi\rangle$$

$$= \frac{i}{2}\begin{pmatrix}\frac{G\gamma}{\rho} & \zeta \\ \zeta^* & -\frac{G\gamma}{\rho}\end{pmatrix}|\Psi\rangle, \tag{13.230}$$

where ρ and k are functions of s and

$$\zeta = -(\gamma-1)\frac{Gy'}{\rho} - i(1+G\gamma)ky. \tag{13.231}$$

The vertical orbit around the machine can be described by the quasi-periodic oscillations by

$$y = \sqrt{A\beta_y}\cos(\psi + \psi_0), \tag{13.232}$$

$$y' = -\sqrt{\frac{A}{\beta_y}}[\sin(\psi + \psi_0) + \alpha_y\cos(\psi + \psi_0)], \tag{13.233}$$

for a particle with initial Courant-Snyder invariant A and and betatron phase ψ_0.

When the spin tune matches the vertical betatron tune, then the small precessional kicks about the radial fields from quadrupoles can add up to tip the spin away from the usual vertical direction as shown in Fig. 13.8. This produces another resonant condition when $\nu_{sp} = N + N_y Q_V$, where N and N_y are integers. These spin resonances due to betatron oscillations are called *intrinsic resonances* because they

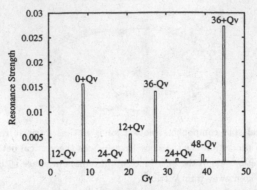

Figure. 13.9 Vertical intrinsic resonance strengths for the AGS with protons, as calculated by Ernest Courant's program DEPOL[34]. These values were calculated for a proton orbit of amplitude corresponding to $2.45\sigma_y$ for a Gaussian beam with normalized 95% emittance of $\pi\epsilon_{95\%}^N = 10\pi \times 10^{-6}$ m and $Q_V = 8.77$. (This calculation was done for a lattice with no partial Siberian snakes.)

exist for even perfectly aligned rings. The strength of these intrinsic resonances is proportional to the amplitude of the oscillation.

Horizontal intrinsic resonances can be driven by horizontal betatron oscillations if the stable spin direction is tilted away from the vertical. Then variations of the precession about the vertical direction can add up to cause tipping of the spins away from the stable direction when $\nu_{sp} = N + N_y Q_H$. Another source of resonance due horizontal oscillations can occur when there is coupling in the ring, so that horizontal oscillations drive vertical oscillations at the horizontal betatron frequency through some part of the ring.

In principle, spin depolarizing resonances can occur when

$$N = N_{sp}\nu_{sp} + N_x Q_H + N_y Q_V + N_{sy} Q_s, \qquad (13.234)$$

where N, N_x, N_y and N_{sy} are integers; however not all these resonance strengths strong enough to depolarize the beam. If the ring's lattice can be described by P identical super cells, i. e. the full turn matrix can be written as

$$\mathbf{M}_{\text{fullturn}} = (\mathbf{M}_{\text{supercell}})^P, \qquad (13.235)$$

then the lattice is said to have a superperiodicity of P. In this case, the stronger intrinsic resonances will occur when N is an integer multiple of P. This just reflects the fact that periodic kicks or pulses which occur closer in time produce higher frequencies (see § 14.2.2), i. e. $f = 1/\tau$.

As an example, the Alternating Gradient Synchrotron (AGS) at BNL has 12 superperiods. Each superperiod consists of 20 dipoles with gradients in a F0F0D0D0 pattern, although the drifts and dipole vary in length depending on location within a superperiod. There are also extra quadrupole and sextupoles place at the same locations in each superperiod to adjust tunes and chromaticities. The stronger intrinsic resonances between $G\gamma = 4.5$ and 46.5 occur for values of $G\gamma$ equal to $0 + Q_V$, $24 - Q_V$ $12 + Q_V$, $36 - Q_V$, $24 + Q_V$, $48 - Q_V$, and $36 + Q_V$ (see Fig. 13.9). In fact it is apparent in the plot that odd multiples of 12 other than $N = 0$ are stronger than the nonzero even multiples. Partial snakes are used to increase the imperfection resonances to give a full spin flip as each integer value of $G\gamma$ is crossed during acceleration.

While solenoids can cause imperfection resonances, for a fixed field strength, the spin precession in a solenoid decreases with energy ($\phi \propto (1 + G)/\gamma$). Vertical misalignments of quadrupoles however cause orbit distortions which must be corrected. Since these misalignments are fixed offsets, the vertical closed orbit will not change with energy, and the contribution to the typical imperfection resonance strengths will tend to increase linearly with energy ($\epsilon_{res} \sim G\gamma$) at high energies.

Since the transverse emittances (Eqs. 5.73 and 5.74) are inversely proportional to momentum, we can expect that at high energy there would be a scaling of the intrinsic resonance strengths which would grow as

$$\epsilon_{res} \propto \frac{G\gamma}{\sqrt{\gamma}} \propto \sqrt{\gamma}. \tag{13.236}$$

Since ϵ_{res} depends on amplitude of the different particles, the amount of tipping away from the stable spin direction $\hat{n}_0$ of the closed orbit will be amplitude dependent. Averaging over the spins of all the particles in the beam, we can expect to see a smaller value of polarization $\vec{P}$ than was injected into the ring.

By installing a full snake in the ring, the fractional part of the spin tune becomes one half, and the stable spin direction outside the snake will be in the horizontal plane. Even though there will be $G\gamma$ precessions of the spin in the ring outside the snake, the snake will rotate the spin by $180°$ about its rotation axis on every turn to produce a net rotation of $180°$ about the snake axis on every turn independent of energy. If $G\gamma$ is too large so there are too many coherent kicks to the spin within one turn, the spins can still depolarize. Even though the snake locks the spin tune at one half, we can consider an effective resonance strength for a single turn, which if too large, will cause too large a tipping away from the stable spin direction of the closed orbit. In this case a second snake on the opposite side of the ring, so that the spin is flipped in each half of the ring. It should be noted that the snake rotation axes must be oriented perpendicular to each other to have a spin tune of one half at all energies (see Problem 13-7). With two snakes on opposite sides of the ring as in RHIC[35], the effective resonance strength between snakes is reduced over that of a full turn. In principle, more snakes need to be added for even higher energies.

13.11 Froissart-Stora formula for a single resonance

To understand the spin flip for a spin-$\frac{1}{2}$ particle while ramping across a resonance more quantitatively, we define the linear ramp of the precession frequency with energy and no solenoid as

$$\tilde{\nu} = G\gamma = \tilde{\nu}_0 + \alpha\theta, \tag{13.237}$$

where

$$\tilde{\nu}_0 = G\gamma_0 \tag{13.238}$$

is the nominal spin tune for the flat ring without perturbing horizontal fields from any solenoids and quadrupoles, and α is the acceleration rate defined by Eq. (13.201).

Treating the solenoid as a perturbation, the resonance occurs when $\tilde{\nu}_0$ is equal to an integer. Taking the azimuthal angle θ as our time-like coordinate, then the Schrödinger equation Eq. (13.119) may be written in the form

$$\frac{d|\Psi\rangle}{d\theta} = -\frac{i}{2}\begin{pmatrix} -\tilde{\nu} & w \\ w^* & \tilde{\nu} \end{pmatrix}|\Psi\rangle, \tag{13.239}$$

where $w(\theta)$ is a small perturbation which describes the instantaneous rotations due to horizontal fields. (Note that the extra minus signs in the matrix are there because we have defined the coordinates with θ increasing as a left-handed rotation about the vertical $\hat{y}$-axis.) This equation may be separated in the usual way as

$$\frac{d|\Psi\rangle}{d\theta} = -i(H_0 + H_1), \tag{13.240}$$

with the unperturbed Hamiltonian

$$H_0 = \frac{1}{2}\begin{pmatrix} -\tilde{\nu} & 0 \\ 0 & \tilde{\nu} \end{pmatrix}, \tag{13.241}$$

and the perturbing Hamiltonian

$$H_1 = \frac{1}{2}\begin{pmatrix} 0 & w \\ w^* & 0 \end{pmatrix}. \tag{13.242}$$

For the unperturbed case, we can transform the spinor to a reference frame precessing with the spin as

$$|\Psi(\theta)\rangle = e^{i\int_0^\theta H_0\theta\, d\theta}|\Psi_r(\theta)\rangle = e^{\frac{i}{2}(\tilde\nu_0\theta+\frac{\alpha}{2}\theta^2)\sigma_y}|\Psi_r\rangle, \tag{13.243}$$

where Ψ_r is the spinor in the Heisenberg interaction representation. Differentiating gives

$$\begin{aligned}
\frac{d|\Psi(\theta)\rangle}{d\theta} &= e^{\frac{i}{2}(\tilde\nu_0\theta+\frac{\alpha}{2}\theta^2)\sigma_y}\left[\frac{i}{2}(\tilde\nu_0+\alpha\theta)\sigma_y|\Psi_r\rangle + \frac{d|\psi_r\rangle}{d\theta}\right] \\
&= \begin{pmatrix} \eta & 0 \\ 0 & \eta^* \end{pmatrix}\left[\frac{i}{2}\tilde\nu\sigma_y|\Psi_r\rangle + \frac{d|\Psi_r\rangle}{d\theta}\right] \\
&= \frac{i}{2}\begin{pmatrix} -\tilde\nu & w \\ w^* & \tilde\nu \end{pmatrix}\begin{pmatrix} \eta & 0 \\ 0 & \eta^* \end{pmatrix}|\Psi_r\rangle
\end{aligned} \tag{13.244}$$

where

$$\eta = e^{\frac{i}{2}(\tilde\nu_0\theta+\frac{\alpha}{2}\theta^2)}. \tag{13.245}$$

Solving for the derivative of $|\Psi_r\rangle$ yields

$$\begin{aligned}
\frac{d|\Psi_r\rangle}{d\theta} &= \frac{i}{2}\begin{pmatrix} \eta^* & 0 \\ 0 & \eta \end{pmatrix}\begin{pmatrix} -\tilde\nu & w \\ w^* & \tilde\nu \end{pmatrix}\begin{pmatrix} \eta & 0 \\ 0 & \eta^* \end{pmatrix}|\Psi_r\rangle - \frac{i}{2}\begin{pmatrix} \tilde\nu & 0 \\ 0 & -\tilde\nu \end{pmatrix}|\Psi_r\rangle \\
&= \frac{i}{2}\begin{pmatrix} 0 & \hat w \\ \hat w^* & 0 \end{pmatrix}|\Psi_r\rangle,
\end{aligned} \tag{13.246}$$

with $\hat w = \eta^{*2}w$. Expand the perturbation in a Fourier series

$$\hat w = \eta^{*2}w = e^{i(\tilde\nu_0\theta+\frac{1}{2}\alpha\theta^2)}\sum_\kappa \epsilon_\kappa e^{i\kappa\theta} = \sum_\kappa \epsilon_\kappa e^{i[(\nu_0-\kappa)\theta+\frac{1}{2}\alpha\theta^2]}. \tag{13.247}$$

Our two-component spinor is

$$|\Psi_r\rangle = \begin{pmatrix} \Psi_r^+ \\ \Psi_r^- \end{pmatrix}, \tag{13.248}$$

with the differential equations in the rotating system:

$$\frac{d\Psi_r^+}{d\theta} = \frac{i}{2}\hat w\Psi_r^-, \tag{13.249}$$

$$\frac{d\Psi_r^-}{d\theta} = \frac{i}{2}\hat w^*\Psi_r^+. \tag{13.250}$$

For a single isolated resonance crossing with $\kappa = \tilde{\nu}_0$

$$\hat{w} = \epsilon_\kappa e^{i\frac{\alpha}{2}\theta^2}.$$ (13.251)

Combining Eqs. (13.249 and 13.250) yields

$$\frac{d^2\Psi_r^+}{d\theta^2} = \frac{i}{2}\left[\frac{d\hat{w}}{d\theta}\Psi_r^- + \hat{w}\frac{d\Psi_r^-}{d\theta}\right] = i\alpha\theta\frac{d\Psi_r^+}{d\theta} + \frac{i}{2}\hat{w}\frac{i}{2}\hat{w}^*\Psi_r^+.$$ (13.252)

After rearranging, we have

$$\frac{d^2\Psi_r^+}{d\theta^2} - i\alpha\theta\frac{d\Psi_r^+}{d\theta} + \frac{|\epsilon_\kappa|^2}{4}\Psi_r^+ = 0,$$ (13.253)

(There equation for Ψ_r^- looks the same except for the sign of the term with the first derivative.) The solution to this equation is a confluent hypergeometric function. Rather than get bogged down into a discussion of the asymptotic behavior of these functions (see Refs. 25 and 32), we present the hand-waving argument by Courant and Ruth[34].

In the vicinity of the resonance the middle term is small an the solution is approximately like a harmonic oscillator. For asymptoticly large (either positive or negative) θ, the differential equation is dominated by the last two terms, since the spinor component must be bounded, i. e. $|\Psi_r^+| \leq 1$, and we have

$$\frac{d\Psi_r^+}{d\theta} \simeq -i\frac{|\epsilon_\kappa|}{4\alpha\theta}\Psi_r^+.$$ (13.254)

Integration gives

$$\frac{\Psi_r^+(\infty)}{\Psi_r^+(-\infty)} \simeq \exp\left(-\frac{i|\epsilon_\kappa|^2}{4\alpha}\int_{-\infty}^{\infty}\frac{d\theta}{\theta}\right) = e^{-\pi|\epsilon_\kappa|^2/2\alpha}.$$ (13.255)

For an initially spin-up spinor, $\Psi_r^+(-\infty) = 1$, and the final component becomes

$$\Psi_r^+(\infty) = e^{-\pi|\epsilon_\kappa|^2/2\alpha}.$$ (13.256)

The expectation of the vertical component of the spin divided by $\hbar/2$, i. e. the single particle polarization becomes

$$P = \hat{y}\cdot\vec{P} = \langle\Psi_r|\sigma_y|\Psi_r\rangle = |\Psi_r^+|^2 - |\Psi_r^-|^2 = 2|\Psi_r^+|^2 - 1.$$ (13.257)

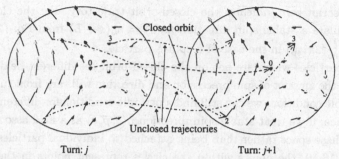

Figure. 13.10 Illustration of the invariant spin field. For the closed orbit $\vec{n}_0(s) = \vec{n}_0(s + L)$, with coordinates $\vec{q}_0(s) = \vec{q}_0(s + L)$ and polarization direction $\vec{P}_0(s) = \vec{P}_0(s + L)$. For other locations in phase space: $\vec{n}(\vec{q}, \vec{p}, s) = \vec{n}(\vec{q}, \vec{p}, s + L)$, even though in general $q(s + L) \neq q(s)$ and $\vec{P}(s + L) \neq \vec{P}(s)$.

Therefore we find that after crossing the resonance the single particle polarization will be

$$\frac{P_f}{P_i} = 2e^{-\pi|\epsilon_\kappa|^2/2\alpha} - 1. \qquad (13.258)$$

If the ramp rate is large compared to the resonance strength ϵ_κ, then the exponential remains close to +1, and the polarization remains essentially unchanged. For example, if the resonance strength is small, say 10^{-5} and we cross the resonance in only five steps, then at most we could only tip the spin by 5×10^{-5} radian. On the other hand if the crossing rate is slow with $\frac{\pi|\epsilon_\kappa|^2}{2\alpha} \gg 1$, then the exponential approaches zero and we get essentially a full spin flip $P_f/P_i = -1$. In the in-between range, the polarization experience only a partial flip or even a decrease without flipping sign. This is exactly the behavior we saw in Fig. 13.7. It is also worth noting that in the vicinity of the resonance (θ near zero) there is also some oscillatory behavior in Fig. 13.7 for the various cases, particularly in the right-hand plot with only partial spin flips.

It is important to realize that there are many subtleties (orbital oscillations in all three planes, lattice structure, misalignments, overlapping spin resonances, acceleration, etc.) which can affect the amount of polarization achieved in an actual accelerator ring. To study these for accelerators like the AGS and RHIC, a comprehensive spin tracking code such as SPINK[36] can be invaluable.

13.12 Invariant spin field

For the usual six dimensional phase space we have a closed orbit with the periodicity condition $\vec{q}_0(s + L) = \vec{q}_0(s)$ where L is the circumference of the ring, and the subscript zero indicates the closed-orbit trajectory. For the closed orbit, we can calculate a stable spin direction $\hat{n}_0(s) = \hat{n}_0(s + L)$ from Eq. (13.187) for $\mathbf{M}_{1\text{turn}} = \mathbf{D}_{\hat{n}_0}^{\frac{1}{2}}(2\pi\nu_{\text{sp}})$ in the case of a spin-$\frac{1}{2}$ particle.

Other particles which are not on the closed orbit see different fields on every turn so that the amount of spin precession and direction will vary from turn to turn. For reasonable conditions we can still find a periodic condition for different locations in phase space, if we treat stable spin direction $\hat{n}(\vec{q}, \vec{p}, s)$ as a field associated with a point in phase space rather than being attached to individual particles[37].

In Fig. 13.10, the points within the circles represent points in the 6d phase space of the beam at a particular azimuth of the ring. The dashed line shows that the closed orbit point "0" maps to itself (dashed line) from one turn (j) to the next ($j + 1$). The arrows pointing out of the page indicate the stable $\hat{n}$-field associated with various points in phase space. If for example, we have only vertical oscillations with tune $Q_V = 1/3$, then a particle will return to the same phase-space point on every third turn ($j \rightarrow j+3$) as indicated by the dash-dotted lines with the ordering: $1 \rightarrow 2$, $2 \rightarrow 3$, $3 \rightarrow 1$. Since the trajectory repeats every three turns, the particle will see exactly the same progression of fields along the trajectory every three turns. So we can calculate a closed 3-orbit matrix for the spin rotation:

$$\mathbf{M}_{3\text{turn}}(\vec{q}_1, \vec{p}_1, s) = \mathbf{M}_{1\text{turn}}(\vec{q}_3, \vec{p}_3, s + 2L)\mathbf{M}_{1\text{turn}}(\vec{q}_2, \vec{p}_2, s + L)\mathbf{M}_{1\text{turn}}(\vec{q}_1, \vec{p}_1, s)$$
$$= \mathbf{D}_{\hat{n}(\vec{q}_1, \vec{p}_1, s)}^{\frac{1}{2}}[6\pi\bar{\nu}_{\text{sp}}(\vec{q}_1, \vec{p}_1, s)] \qquad (13.259)$$

where the spin tune $\bar{\nu}_{\text{sp}}(\vec{q}, \vec{p}, s)$ is averaged over the three turns. More generally, if the spin tune is not on a resonance and the motion is stable, the trajectories will oscillate in all three dimensions, but after some m turns a particle will return within a small enough neighborhood of its initial position so that the m-turn matrix will give a rotation axis which is sufficiently close to $\hat{n}$ for the initial phase space coordinates.

For a given ensemble of particles, the maximum possible polarization cannot be more than the ensemble average of the spin field $\langle \hat{n} \rangle$ averaged over all the particles in the beam, although it may be less than this value.* The individual particles may have spins which are not pointing in the direction of the corresponding $\hat{n}$-vector. Away from spin resonances the spin field vectors will be pointing in roughly the

---

* In the case of a freshly injected beam, the polarization can actually be higher than $\langle \hat{n} \rangle$; however after the beam has circulated for a while, the polarization would then be no larger than $\langle \hat{n} \rangle$.

same direction, but near a resonance they can deviate quite wildly. Even though the polarization average of the ensemble may drop to zero near a resonance, the average may recover a nonzero value later when moving away from the resonance.

13.13 Polarization of electron beams

Recalling the interaction Hamiltonian $H_1 = -\vec{\mu}' \cdot \vec{B}'$ for a spin at rest in a magnetic field, we see that there is a preferred direction for the spin parallel to the magnetic field. This leads to a radiative buildup of polarization in electron storage rings called the Sokolov-Ternov effect. The spin-flip transition rates for relativistic electrons in a homogeneous field can be calculated[7,8] using quantum electrodynamics:

$$W_{\uparrow\downarrow} = \frac{5\sqrt{3}}{16} \frac{e^2 \gamma^5 \hbar}{4\pi\epsilon_0 m_e c^2 |\rho|^3} \left(1 + \frac{8}{5\sqrt{3}}\right), \tag{13.260}$$

$$W_{\uparrow\downarrow} = \frac{5\sqrt{3}}{16} \frac{e^2 \gamma^5 \hbar}{4\pi\epsilon_0 m_e c^2 |\rho|^3} \left(1 - \frac{8}{5\sqrt{3}}\right). \tag{13.261}$$

If we assume that there are no depolarizing effects, then in a flat ring, the equilibrium polarization will be

$$P_{ST} = \frac{W_{\uparrow\downarrow} - W_{\downarrow\uparrow}}{W_{\uparrow\downarrow} + W_{\downarrow\uparrow}} = \frac{8}{5\sqrt{3}} = 0.9238. \tag{13.262}$$

An unpolarized beam will polarize as

$$P(t) = P_{ST} \left[1 - \exp(-t/\tau_{ST})\right], \tag{13.263}$$

where the Sokolov-Ternov polarization rate is given by

$$\tau_{ST}^{-1} = \frac{5\sqrt{3}}{8} \frac{e^2 \gamma^5 \hbar}{4\pi\epsilon_0 m_e^2 c^2} \frac{1}{L} \oint \frac{ds}{|\rho|^3}. \tag{13.264}$$

Table II lists the number of photons and energy radiated per turn, the Sokolov-Ternov polarization time, and ratio of spin-flip photons to total photons for a number of accelerators. For comparison several rings with protons are included to show how fruitless this effect is for polarizing hadrons.

Table III. Sokolov-Ternov polarization rates.

Ring	Particle	Energy [GeV]	N_γ [/turn]	ΔU [loss/turn]	τ_{ST}	$\frac{W_{\uparrow\downarrow}}{f_{rev} N_\gamma}$
CESR	$e^\pm$	5.5	700	-1 MeV	167 min	1×10^{-13}
HERAe	$e^\pm$	27.5	3600	-83 MeV	23 min	1×10^{-12}
LEP	$e^\pm$	45	5800	-120 MeV	300 min	2×10^{-13}
LEP	$e^\pm$	60	7800	-380 MeV	81 min	8×10^{-13}
RHIC	p	100	7	-3 meV	3×10^{14} yr	6×10^{-29}
RHIC	p	250	18	-0.13 eV	3×10^{12} yr	2×10^{-27}
HERAp	p	920	65	-8.5 eV	1×10^{11} yr	3×10^{-26}
Tevatron	p	1000	70	-8.5 eV	2×10^{11} yr	2×10^{-26}
SSC	p	20000	1400	-0.12 MeV	7×10^{7} yr	3×10^{-23}

In contrast to protons, electrons (or positrons) can lose the memory of polarization very quickly due to radiation effects. If the $\hat{n}$-field has a large spread across the beam, then when the momentum changes due to a radiated photon usually without flipping the spin of the electron, a spin which had been aligned along the $\hat{n}$ will find itself tilted away from $\hat{n}$ at the new phase space point after the emission. This hopping of points in phase space from the radiation of quanta can cause a very fast diffusion of the spins resulting in a very rapid depolarization of the electron beam. Spin diffusion is the main cause of depolarization near resonances, not spin-flip radiation.

Derbenev and Kondratenko[26,9] derived more accurate formulae for the equilibrium polarization

$$P_{DK} = \frac{8}{5\sqrt{3}} \frac{\oint \left\langle \frac{1}{|\rho|^3} \hat{b} \cdot \left(\hat{n} - \frac{\partial \hat{n}}{\partial \delta}\right) \right\rangle_s ds}{\oint \left\langle \frac{1}{|\rho|^3} \left[1 - \frac{2}{9}(\hat{n} \cdot \hat{v})^2 + \frac{11}{18}\left(\frac{\partial \hat{n}}{\partial \delta}\right)^2\right] \right\rangle_s ds} \tag{13.265}$$

and polarization rate

$$\frac{1}{\tau_{DK}} = \frac{5\sqrt{3}}{8} \frac{r_e \gamma^5 \hbar}{m_e} \frac{1}{L} \oint \left\langle \frac{1}{|\rho|^3} \left[1 - \frac{2}{9}(\hat{n} \cdot \hat{v})^2 + \frac{11}{18}\left(\frac{\partial \hat{n}}{\partial \delta}\right)^2\right] \right\rangle_s ds \tag{13.266}$$

where $\hat{v}$ is the unit vector tanget to the trajectory, $\hat{b} = \vec{v} \times \dot{\vec{v}}/|\vec{v} \times \dot{\vec{v}}|$ is the unit vector binormal to the trajectory, and the averages are taken over all the particles in the beam and their oscillations in phase-space. As is quite evident, we see that the $\hat{n}$-field and its variation with $\delta = \Delta p/p$ appear in both equations. These reduce to the Sokolov-Ternov values for the flat ring case on the closed orbit when $\hat{b} = \hat{n}$ and there is no variation of $\hat{n}$ with respect to energy.

Problems

13-1   In the center-of-mass rest system, show that the integral in Eq. (13.2) is invariant if the origin is translated by a fixed vector $\vec{a}$. Assume that the rotation is nonrelativistic and for simplicity that the mass distribution has cylindrical symmetry.

13-2   Show that a Lorentz boost by the velocity $\vec{v} = \beta c\,\hat{z}$ has the form

$$\Lambda(\vec{v}) = \begin{pmatrix} \cosh\eta & 0 & 0 & \sinh\eta \\ 0 & 1 & 0 & 0 \\ 0 & 0 & 1 & 0 \\ \sinh\eta & 0 & 0 & \cosh\eta \end{pmatrix},$$

with the hyperbolic rotation angle $\eta = \frac{1}{2}\ln\left(\frac{1+\beta}{1-\beta}\right)$.

13-3   Show that the matrix for a Lorentz boost[33] by the contravariant proper velocity defined in Eq. (13.26) may be written in the form given in Eq. (13.28). Hint: Start with a boost along the z-axis and applying a rotation of the coordinates to the boost matrix: $\mathbf{R}_x(\zeta)\mathbf{R}_y(\xi)\Lambda(v\hat{z})\mathbf{R}_y(\xi)^{-1}\mathbf{R}_x(\zeta)^{-1}$.

13-4   Verify that Eqs. (13.32), (13.35) and (13.39) are correct to first order in $\delta\beta$.

13-5   Starting from the covariant form of the Thomas-Frenkel-BMT equation Eq. (13.106) becomes Eq. (13.69) when we express the position, time, and electromagnetic fields in the fixed lab system and spin in the instantaneous rest system of the particle.

13-6   Consider a flat ring with snakes and a polarized proton beam.

a) Prove that $\mathbf{R}_{\hat{y}}(\theta)\mathbf{R}_{\hat{a}}(\pi)\mathbf{R}_{\hat{y}}(\theta) = \mathbf{R}_{\hat{a}}(\pi)$, where $\hat{a} = \hat{z}\cos\phi + \hat{x}\sin\phi$, i. e. $\hat{a} \perp \hat{y}$.

b) Find the closed-orbit spin tune for a ring with $2\frac{1}{2}$ snakes with the full-turn spin rotation matrix:

$$\mathbf{M} = \mathbf{R}_{\hat{z}}(\pi/2)\mathbf{R}_{\hat{y}}[G\gamma(\pi - \alpha)]\mathbf{R}_{\hat{x}}(\pi)\mathbf{R}_{\hat{y}}(G\gamma\pi)\mathbf{R}_{\hat{z}}(\pi)\mathbf{R}_{\hat{y}}(G\gamma\alpha).$$

c) What is the stable spin direction $\hat{n}_0$ for the matrix in part b?

13-7   Consider a proton ring with two snakes on opposite sides having a one-turn rotation map for the closed orbit as

$$\mathbf{M} = \mathbf{R}_{\hat{y}}(G\gamma\pi)\mathbf{R}_{\hat{a}}(\pi - \delta)\mathbf{R}_{\hat{y}}(G\gamma\pi)\mathbf{R}_{\hat{z}}(\pi)$$

with

$$\hat{a} = \hat{x}\cos\phi + \hat{z}\sin\phi.$$

Find the spin tune ν_{sp} as a function of δ and ϕ. Plot the ν_{sp} for small values of ϕ and δ as the snake is detuned from the optimum settings.

13–8   Expand the exponentials to verify the rotation matrices in Eqs. (13.189, and 13.191) for a spin-1 particle.

References for Chapter 13

[1] Elliot Leader, *Spin in particle physics*, Cambridge U. Press, Cambridge (2001).

[2] W. W. MacKay et al., Phys. Rev. **D29**, 2483 (1984).

[3] J. Bailey *et al.*, Nuclear Physics B 150 (1979) 1.

[4] G. W. Bennett et al., *Phys. Rev.* **D 73**, 072003 (2006).

[5] H. Okuno et al., *Phys. Lett.* bf B354, 41 (1995).

[6] N. J. Stone, *Atomic Data and Nuclear Data Tables*, **90**, 75 (2005).

[7] A. A. Sokolov and I. M. Ternov, *Synchrotron Radiation*, (English translation) Pergamon Press, New York, (1968).

[8] V. B. Berestetskii, E. M. Lifshitz, and L. P. Pitaevskii, *Quantum Electrodynamics*, 2nd ed., Pergamon Press, New York (1982).

[9] S. R. Mane, Yu. M. Shatunov and K. Yokoya, "Spin-polarized charged particle beams in high-energy accelerators", *Rep. Prog. Phys.*, **68**, 1997 (2005).

[10] http://www.nist.gov/srd, National Institute of Standards and Technology [NIST] "CODATA Fundamental Physical Constants", (2007).

[11] L. D. Landau and E. M. Lifshitz, *Quantum Mechanics*, 3rd Ed., Pergamon Press, Oxford (1977).

[12] John David Jackson, *Classical Electrodynamics*, 2nd Ed., John Wiley & Sons, New York, 1975.

[13] G. E. Uhlenbeck and S. A. Goudsmit, **117**, 264 (1926).

[14] L. H. Thomas, Nature, 117 (1926) 514.

[15] L. H. Thomas, Phil. Mag. S. 7, **3**, 1 (1927).

[16] J. Frenkel, Z. Physik, **37**, 243 (1926).

[17] V. Bargman, L. Michel and V. L. Teledgi, Phys. Rev. Letters, **2** (1959) 435.

[18] J. S. Bell, Polarized Particles for Accelerator Physicists, CERN 75-11 (1975).

[19] Steven Weinberg, *Gravitation and Cosmology*, John Wiley and Sons, New York (1972).

[20] J. K. Lubański, *Physica*, **IX**, 310 (1942).

[21] M. H. L. Pryce, *Proc.of the Royal Soc. of London*, Series A Math. and Sci., **195**, 62 (1949).

[22] L. Landau, Nucl. Phys. 3, 127 (1957).

[23] L. I. Schiff, Phys. Rev. 132, 2194 (1963).

[24] M. E. Rose, *Elementary Theory of Angular Momentum*, John Wiley & Sons (1957).

[25] M. Froissart and R. Stora, Nucl. Instr. Meth., **7**, 297 (1960).

[26] Y. Derbenev and A. M. Kondratenko, Sov. Phys. JETP **37**, 968 (1973).

[27] A. D. Krisch et al., Phys. Rev. Lett., **63**, 1137 (1989).

[28] T. Roser, Proc. Int. Workshop on Hadron Fac. Tech., Los Alamos Nat. Lab. Report LA-11130-C, 317 (1987).

[29] H. Huang *et al.*, Phys. Rev. Lett., **73**, 22 (1994).

[30] R. Pollock, Proc. XIII Int. Conf. on High Energy Acc., Novosibirsk, 7-11.8.1986, Vol. I, p. 209.

[31] N. Akchurin *et al.*, Phys. Rev. Lett., **69**, 1753 (1992).

[32] S. Y. Lee, *Spin Dynamics and Snakes in Synchrotrons*, World Scientific Publ. Co., Singapore (1997).

[33] S. Gasiorowicz, *Elementary Particle Physics*, John Wiley & Sons, New York, 1966.

[34] E. D. Courant and R. D. Ruth, "The Acceleration of Polarized Protons in Circular Accelerators", BNL 51270 (1980).

[35] W. W. MacKay et al., *Proceedings of Spin 2004: 16th International Spin Physics Symposium*, 163 (2005).

[36] Alfredo U. Luccio, "Spin Tracking in RHIC (Code SPINK)", in *Trends in Collider Spin Physics*, Eds. Y. Onel, N. Paver, and A. Penzo, World Scientific Publ. Co., Singapore (1997).

[37] D. P. Barber et al., "Electrons are not Protons", in EICA Workshop, Upton NY, BNL-52663 (2002).

14

Position Measurements and Spectra

14.1 Transverse position measurement

One of the most common methods for measuring the transverse position of a bunched beam is to sense the electric field with capacitive pickups. As the bunch travels down the beam pipe, image charges move along the inner surface of the pipe. (Here we are of course assuming that the beam pipe is made of metal.) From Gauss' law it is easy to see that the total charge moving along the pipe must be equal but of opposite sign to the charge in the bunch.

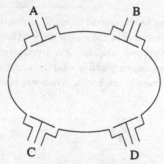

Figure. 14.1 Concept of a dual plane beam-button position monitor. The signals from each button are brought out to the readout electronics via high quality coaxial cables. Here the buttons have been placed as in a synchrotron light source to keep them out of the band of synchrotron radiation in the midplane.

One type monitor employs button shaped electrodes to sense the electric field strength at the surface of the chamber as shown in Fig. (14.1). Buttons are particularly used when the bunches are short as in electron accelerators. The short length of the electrode can limit the amount of induced charge if the bunches are longer than the radius of the button.

When the bunches are longer as in hadron accelerators, longer stripline electrodes are frequently used. Fig. (14.2) shows a shorted stripline monitor with a measured signal from the lower stripline.

An estimate of the charge density on the inner surface of a cylindrical beam pipe of radius b can be made by treating the bunch as a line of charge of density λ at a distance a from the center of the pipe. Ignoring the pipe, a second line of image charge (line density: $-R\lambda$) can be placed parallel to the beam but at a distance A from the origin. We solve for distance A and the ratio R to have an equipotential surface on a cylinder of radius b. The equation for the potential lines of charge is then

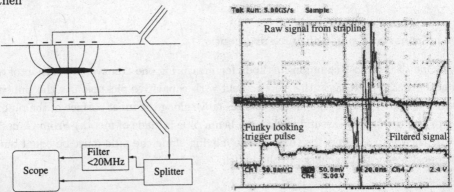

Figure. 14.2 Concept of a stripline beam position monitor. Induced currents on the wall are picked up on coaxial cables by using a small gap at the end of the stripline. The capacitive nature of such a pickup differentiates the bunch current. On the right are measurements of a fully stripped gold bunch passing through a stripline position monitor after being extracted from the AGS. As can be seen, there was quite a bit of structure in the bunch. (This was taken early in the commissioning of the extraction system.)

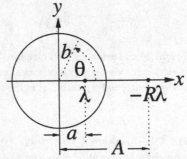

Figure. 14.3 Equipotential surface with line of charge density λ at radius a and an image line of charge of density $-R\lambda$ at radius A.

$$V = \frac{\lambda}{2\pi\epsilon_0}\left[\ln\left(\sqrt{(x-a)^2 + y^2}\right) - R\ln\left(\sqrt{(x-A)^2 + y^2}\right)\right] + V_0. \qquad (14.1)$$

Rearranging terms produces

$$\frac{2\pi\epsilon_0}{\lambda}(V - V_0) = \frac{1}{2}\ln\frac{(x-a)^2 + y^2}{[(x-A)^2 + y^2]^R}. \tag{14.2}$$

After exponentiating, we may define

$$\alpha = e^{\frac{4\pi\epsilon_0}{\lambda}(V-V_0)} = \frac{(x-a)^2 + y^2}{[(x-A)^2 + y^2]^R}. \tag{14.3}$$

In order to have a cylindrical equipotential surface, we must have a quadratic equation, so $R = 1$, and

$$(x-a)^2 + y^2 = \alpha[(x-A)^2 + y^2].$$

Since the potential at $(x, y) = (b, 0)$ must be the same as at $(x, y) = (0, b)$, we can write

$$\alpha = \frac{(b-a)^2}{(A-b)^2} = \frac{a^2 + b^2}{A^2 + b^2}. \tag{14.4}$$

Solving for A gives the nontrivial answer

$$A = \frac{b^2}{a}. \tag{14.5}$$

(The other root is $A = a$ which just cancels the charge at $x = a$.) Transforming to polar coordinates ($x = r\cos\theta$, $y = r\sin\theta$) and evaluating the radial component of electric field at $r = b$ gives

$$E_r(r = b, \theta) = -\frac{\lambda}{2\pi\epsilon_0}\left[\frac{b^2 - a^2}{b(b^2 + a^2 - 2ab\cos\theta)}\right]. \tag{14.6}$$

Since the field is zero inside the conductor, the surface on the inner wall of the beam pipe must be

$$\sigma(\theta) = \epsilon_0 E_\perp = -\frac{\lambda}{2\pi}\left[\frac{b^2 - a^2}{b(b^2 + a^2 - 2ab\cos\theta)}\right]. \tag{14.7}$$

Now the charge per length on a stripline subtending the arc from θ_1 to θ_2 (See Fig. (14.4).) obtained by integrating

$$\frac{dq}{dz} = -\frac{\lambda}{2\pi}\int_{\theta_1}^{\theta_2}\left[\frac{b^2 - a^2}{b(b^2 + a^2 - 2ab\cos\theta)}\right]d\theta$$

$$= \frac{\lambda}{\pi}\tan^{-1}\left[\left(\frac{b+a}{b-a}\right)\tan\frac{\theta}{2}\right]\Bigg|_{\theta_1}^{\theta_2}. \tag{14.8}$$

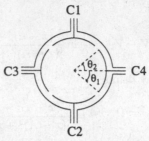

Figure. 14.4 Definition of angles of integration for stripline "C4".

For the opposite plate we can reverse the sign of a. The voltage induced across the coaxial cables will be proportional to the dq/dz on the stripline. To go further requires either a lot more algebra or invocation of a symbolic calculator such as MAPLE[3]. To first order in x/b we get

$$\frac{V_1 - V_2}{V_1 + V_2} = 4\frac{\sin\frac{\alpha}{2}}{\alpha}\frac{x}{b} + O\left(\frac{x^3}{b^3}\right) + O\left(\frac{xy^2}{b^3}\right). \qquad (14.9)$$

where $\alpha = \theta_2 - \theta_1$, and we have replaced a by x. A somewhat more linear formula comes from

$$\ln\frac{V_1}{V_2} = 8\frac{\sin\frac{\alpha}{2}}{\alpha}\frac{x}{b} + O\left(\frac{x^3}{b^3}\right) + O\left(\frac{xy^2}{b^3}\right). \qquad (14.10)$$

As an example, assume that the striplines subtend an angle of $\alpha = 70°$ ($\theta_1 = -35°$, $\theta_2 = 35°$) and a radius for the striplines of $b = 56.5$ mm, then we get to first order from Eq. (14.9): Fig. (14.5) shows typical signals from the four plates of a dual plane stripline with above dimensions.

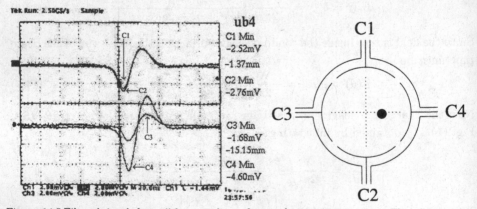

Figure. 14.5 Filtered signals from all four striplines of a two plane position monitor. The relative sizes of the voltages show that the beam was displaced to the right and slightly downward as indicated in the picture on the right. The positions were calculated using Eq. (14.10).

The accuracy of relative position measurements in a circular accelerator with a stripline monitor can be better than 10 μm when the signals are averaged over several turns.

14.2 Coherent and Schottky frequency spectra of beams

As we have seen, there are a number of frequencies associated with the beam – in particular, the synchrotron and betatron tunes. (It is not easy to measure the spin tune directly and so we ignore it here.) If a pickup such as a beam position monitor, (or a wall current monitor which measures the currents in resistors across a ceramic gap in the beam pipe) then coherent oscillations of these modes can be quite apparent. Schottky noise in the beam can also be detected and used to cool the beam stochastically as discussed in Chapter 12. To start with we examine the coherent frequency spectrum of a bunched beam.

14.2.1 A single Gaussian bunch

A Gaussian bunch of total charge q passing a pickup at location s will have a longitudinal profile

$$dQ(s,t) = \frac{q}{\sqrt{2\pi}\sigma_s} e^{\frac{-(s-vt)^2}{2\sigma_s}} ds, \tag{14.11}$$

where σ_s is the rms width of the distribution. The distribution of current passing the pickup is then

$$
\begin{aligned}
i(t) = \frac{dQ}{dt} &= v \frac{dQ}{ds} \\
&= \frac{qv}{\sqrt{2\pi}\sigma_s} e^{\frac{-(s-vt)^2}{2\sigma_s}} \\
&= \frac{q}{\sqrt{2\pi}\sigma_t} e^{\frac{-(t-t_0)^2}{2\sigma_t}},
\end{aligned} \tag{14.12}
$$

where $t_0 = s/v$. For the following discussion, we will place the pickup at $s = 0$, so $t_0 = 0$. The harmonic content of the signal may be found from the Fourier transform[†] of the current:

$$\hat{i}(\omega) = \frac{q}{\sqrt{2\pi}\sigma_t} \int_{-\infty}^{\infty} e^{-j\omega t} e^{-\frac{t^2}{2\sigma_t^2}} dt,$$

[†] Here we have used the engineering convention of $j = \sqrt{-1}$ to minimize confusion with the current i.

$$= \frac{q}{\sqrt{2\pi}\sigma_t} e^{-\frac{\sigma_t^2 \omega^2}{2}} \int_{-\infty}^{\infty} e^{-\frac{(t+j\sigma_t^2\omega)^2}{2\sigma_t^2}} \, dt,$$

$$= q e^{-\frac{\sigma_t^2 \omega^2}{2}}. \tag{14.13}$$

So we find that the Fourier transform of a Gaussian distribution is again Gaussian with an rms width of $\sigma_\omega = 1/\sigma_t$. In the limit of an infinitesimally short bunch the Gaussian distribution becomes

$$i(t) = \lim_{\sigma_t \to 0} \frac{q}{\sqrt{2\pi}\sigma_t} e^{-\frac{t^2}{2\sigma_t}} = q\,\delta(t), \tag{14.14}$$

and the spectral content becomes flat

$$\hat{\imath}(\omega) = q \int_{-\infty}^{\infty} e^{-j\omega t}\,\delta(t)\,dt = q. \tag{14.15}$$

14.2.2 Circulating bunches

The bunched beam current in a circular accelerator of circumference L with N_p equally spaced bunches may be approximated by

$$i(t) = \sum_{n=-\infty}^{\infty} \sum_{m=1}^{N_p} \int_{-\infty}^{\infty} Q_m(t')\,\delta\left(t - t' - \frac{nmL}{N_p v}\right) dt', \tag{14.16}$$

where $Q_m(t') = dq_m/dt'$ is the longitudinal profile of charge in the mth bunch. The frequency spectrum may be obtained from the Fourier transform of the current:

$$\hat{\imath}(\omega) = \sum_{m=1}^{N_p} \sum_{n=-\infty}^{\infty} \int_{-\infty}^{\infty} \int_{-\infty}^{\infty} e^{-j\omega t}\, Q_m(t')\,\delta\left(t - t' - \frac{nmL}{N_p v}\right) dt'\, dt,$$

$$= \sum_{m=1}^{N_p} \sum_{n=-\infty}^{\infty} \hat{Q}_m(\omega) \exp\left(-j\frac{nmL}{N_p v}\omega\right). \tag{14.17}$$

When the bunches are identical this becomes:

$$\hat{\imath}(\omega) = \hat{Q}(\omega) \sum_{n=-\infty}^{\infty} \exp\left(-j\frac{nL}{N_p v}\omega\right) = \frac{Q(\omega)}{N_p \omega_s} \sum_{n=-\infty}^{\infty} \delta\left(\omega - nN_p\omega_s\right) \tag{14.18}$$

with $\omega_o = 2\pi v/L$ being the angular revolution frequency.

If we approximate the bunch shape by a delta function, then the spectrum will have a "comb"-shape as shown in Fig. (14.6).

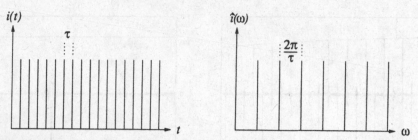

Figure. 14.6 The left plot shows the current distribution for equal δ-function bunches of charge $Q_m(t') = q\delta(t'))$ and spacing $\tau = L/N_p v$. The right plot shows the corresponding Fourier transform.

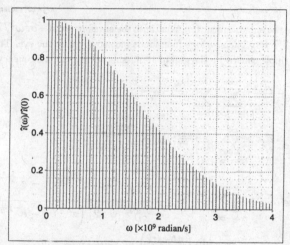

Figure. 14.7 Relative frequency of equal Gaussian bunches with 106.6 ns spacing between bunches.

If we now allow a gap in the number of bunches so that only N_b bunches are placed with the same spacing leaving $N_p - N_b$ holes in the bunch train, we should expect to see additional harmonics of the revolution lines between those of Eq. (14.18). Consider N_b bunches of equal charge q and width σ placed in N_p equally spaced buckets:

$$i(t) = \sum_{n=-\infty}^{\infty} \sum_{m=1}^{N_b} \int_{-\infty}^{\infty} \int_{-\infty}^{\infty} e^{-j\omega t} \frac{q}{\sqrt{2\pi}\sigma} e^{-\frac{t'^2}{2\sigma^2}} \delta\left(t - t' - \frac{nm}{N_p}\tau_s\right) dt' dt. \quad (14.18)$$

The frequency spectrum is then

$$\hat{i}(\omega) = qe^{-\frac{\sigma^2\omega^2}{2}} \sum_{n=-\infty}^{\infty} \sum_{m=1}^{N_b} \int_{-\infty}^{\infty} e^{-j\omega t} \delta\left(t - \frac{nm}{N_p}\tau_s\right) dt$$

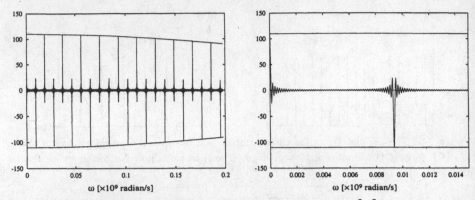

Figure. (14.8) Plot of the enhancement function times the envelope $\exp(-\sigma^2\omega^2/2)$ for $N_p = 120$, $N_b = 110$, $\tau_s = 12.7\ \mu s$. The envelope function is plotted to guide the eye. The enlargement on the right shows the ripple caused by the gap of 10 missing bunches at a frequency of 78 kHz.

$$= qe^{-\frac{\sigma^2\omega^2}{2}} \sum_{n=-\infty}^{\infty} \sum_{m=1}^{N_b} e^{-j\frac{nm\omega\tau_s}{N_p}}$$

$$= qe^{-\frac{\sigma^2\omega^2}{2}} \sum_{n=-\infty}^{\infty} e^{-j\frac{n\omega\tau_s}{N_p}} \frac{1 - e^{-j\frac{nN_b\omega\tau_s}{N_p}}}{1 - e^{-j\frac{n\omega\tau_s}{N_p}}}$$

$$= qe^{-\frac{\sigma^2\omega^2}{2}} \sum_{n=-\infty}^{\infty} e^{-j\frac{n\omega\tau_s}{N_p}} e^{-j\frac{nm\omega\tau_s}{2N_p}} \frac{2j\left(e^{j\frac{nN_b\omega\tau_s}{2N_p}} - e^{-j\frac{nN_b\omega\tau_s}{2N_p}}\right)}{e^{-j\frac{n\omega\tau_s}{2N_p}} \, 2j\left(e^{j\frac{n\omega\tau_s}{2N_p}} - e^{-j\frac{n\omega\tau_s}{2N_p}}\right)}$$

$$= qe^{-\frac{\sigma^2\omega^2}{2}} \sum_{n=-\infty}^{\infty} \frac{\sin\left(\frac{nN_b\omega\tau_s}{2N_p}\right)}{\sin\left(\frac{n\omega\tau_s}{2N_p}\right)} e^{-j\frac{n(N_b+1)\omega\tau_s}{2N_p}}. \tag{14.19}$$

The modulation factor, called the *enhancement* function, is

$$\mathcal{E}_n(\omega) = \frac{\sin\left(\frac{nN_b\tau_s}{2N_p}\omega\right)}{\sin\left(\frac{n\tau_s}{2N_p}\omega\right)}. \tag{14.20}$$

From this we see that having an irregular pattern of bunches will produce more closely spaced lines separated by the revolution frequency $(1/\tau_s)$ which is smaller than the typical bunch frequency (N_p/τ_s). Other enhancement factors can be calculated for bunches of differing intensity or with more gaps between bunch trains.

14.2.3 Momentum spread

So far we have assumed that the particles are all oscillating at the same frequency. In fact any real beam has a nonzero momentum spread, and unless we have an isochronous ring, there will be a spread in the revolution frequencies of the individual particles. In the case of beams bunched by rf cavities, there will be synchrotron oscillations with each individual particle having a varying revolution period.

14.2.4 Longitudinal Schottky spectrum of an unbunched beam

The current of the mth particle can be written as

$$i_m(t) = qf_m \sum_{n=-\infty}^{\infty} e^{jn[\omega_m t + \psi_m]}$$

$$= qf_m \left[1 + 2 \sum_{n=1}^{\infty} \cos[n(\omega_m t + \psi_m)] \right] \tag{14.21}$$

where $f_m = \omega_m/2\pi$ and ψ_m are the respective revolution frequency and phase of the particle.

Averaging over N particles in the beam the rms frequency spread of the nth revolution harmonic of the synchronous particle's frequency f_s will be the absolute value of

$$\text{harmonic bandwidth} = n\sigma_f = nf_s |\eta_{\text{tr}}| \frac{\sigma_p}{p}, \tag{14.22}$$

where we recall that the phase slip factor was defined in Eq. (7.3):

$$\eta_{\text{tr}} = \frac{1}{\gamma^2} - \frac{1}{\gamma_{\text{tr}}^2}. \tag{14.23}$$

The total average current for a large number of particles is

$$\langle i \rangle = \sum_{i=1}^{N} i_m(t) = Nq\langle f \rangle = Nqf_s, \tag{14.24}$$

which is just the dc component of the current. The rms component of current may be found from

$$\sigma_i^2 = \langle (i - \langle i \rangle)^2 \rangle$$

$$= \left\langle \left[\sum_{m=1}^{N} qf_m \left(1 + 2 \sum_{n=1}^{\infty} \cos(n\omega_m t + \psi_m) \right) - Nqf_s \right]^2 \right\rangle$$

$$= 2q^2 f_s^2 N. \tag{14.25}$$

So the rms current component is then

$$\sigma_i = q f_s \sqrt{2N}, \tag{14.26}$$

which is independent of harmonic number. The bandwidth is however proportional to the number n of the revolution harmonic as given by Eq. (14.22) and indicated in Fig. (14.9).

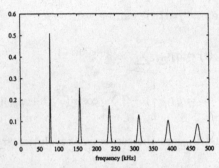

Figure. 14.9 Plot of first six revolution harmonics. (We have exaggerated the momentum spread so that the widths can be seen.) Note that the area under each peak (i.e., peak times width) is constant.

14.2.5 Synchrotron oscillations of a single particle

For simplicity, consider only a single particle which is now undergoing a small synchrotron oscillation. This modulates the arrival time at the detector as

$$t \to t + a \sin(\Omega_s t + \psi), \tag{14.27}$$

where $a \; (\ll \tau_s)$ and ψ are respectively the amplitude and phase of modulation. The current seen by the detector on the nth-turn is then

$$
\begin{aligned}
i_n &= \sum_{n=-\infty}^{\infty} q \, \delta \left(t - n \left[\tau_s + a \sin(\Omega_s n \tau_s + \psi) \right] \right) \\
&\simeq \frac{q}{\tau_s} \sum_{n=-\infty}^{\infty} e^{j n \omega_s [\tau_s + a \sin(n \Omega_s \tau_s + \psi)]},
\end{aligned}
\tag{14.28}
$$

where we have made use of the identity

$$\sum_{-\infty}^{\infty} \delta(t - n\tau) = \frac{1}{\tau} \sum_{-\infty}^{\infty} e^{j \frac{n 2 \pi t}{\tau}}. \tag{14.29}$$

Recalling another identity

$$e^{jz\sin\theta} = \sum_{m=-\infty}^{\infty} J_m(z)\, e^{j\theta}, \qquad (14.30)$$

we may now write

$$i_n = q \sum_{n=-\infty}^{\infty} e^{-jn\omega_s\tau_s} \sum_{m=-\infty}^{\infty} J_m(n\omega_s a)\, e^{-jm(n\omega_s\tau_s+\psi)}. \qquad (14.31)$$

Now each revolution harmonic gets split into a sequence of synchrotron satellites of relative height $J_m(a)$ for the mth satellite. Since $J_m(n\omega_s a)$ decreases with increasing m only the nearest lines are important. As a rule of thumb, lines with $m \gtrsim n\, 2a\omega_s$ are negligible[4].

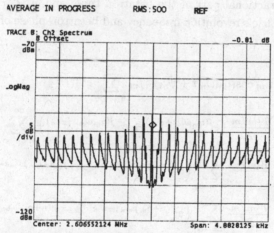

Figure. 14.10 Longitudinal spectrum showing synchrotron sidebands of a high order revolution line in RHIC. The pickup was a small Schottky cavity with a resonant frequency near 2 GHz. The signal was mixed down to the lower range of the spectrum analyzer. The peak in the middle is a revolution line of infinitesimal width, although the spike as measured does not go to infinity due to the finite bandwidth of the analyzer. The characteristic shape of the sidebands, as having a sharp edge away from the revolution line with a sloping fall towards the revolution line, is due to the fact that the synchrotron frequency is a maximum at the center of the bucket, where the particle distribution also peaks, and then falls off toward the edge of the bucket.

14.2.6 Transverse Schottky spectrum of an unbunched beam

Again we first consider a single particle with a betatron oscillation in one plane. Here we must use a transverse pickup which is sensitive to the amplitude a_m of the oscillation. The measured signal is then proportional to the dipole oscillation signal

$$d_m = a_m(t)i_m(t), \tag{14.32}$$

for the mth particle. We immediately see that the signal is proportional to both the amplitude of oscillation and to the current of the bunch, so in addition to the effect of betatron oscillations, transverse spectra may also exhibit aspects of the longitudinal spectra. For the betatron oscillation we have

$$a_m(t) = a_m \cos(q_m \omega_m t + \psi_m), \tag{14.33}$$

where q_m is the fractional part of the betatron tune, and $\omega_m/2\pi$ and ψ_m are respectively the particle's revolution frequency and betatron-phase offset. Eq. (14.32) now becomes

$$
\begin{aligned}
d_m(t) &= a_m \cos(q_m \omega_m t + \psi_m) q f_m \sum_{n=-\infty}^{\infty} e^{jn\omega_m t} \\
&= \frac{q a_m f_m}{2} \left(e^{j(q_m \omega_m t + \psi_m)} + e^{-j(q_m \omega_m t + \psi_m)} \right) \sum_{n=-\infty}^{\infty} e^{jn\omega_m t} \\
&= \frac{q a_m f_m}{2} \sum_{n=-\infty}^{\infty} \left(e^{j[(n+q_m)\omega_m t + \psi_m]} + e^{j[(n-q_m)\omega_m t - \psi_m]} \right) \\
&= \frac{q a_m f_m}{2} \sum_{n=-\infty}^{\infty} \left(e^{j[(n+q_m)\omega_m t + \psi_m]} + e^{-j[(n+q_m)\omega_m t + \psi_m]} \right) \\
&= q a_m f_m \sum_{n=-\infty}^{\infty} \cos[(n+q_m)\omega_m t + \psi_m]. \tag{14.34}
\end{aligned}
$$

So the spectrum will have lines spaced like the revolution harmonics but offset by an amount $q_m f_m$. Since we cannot tell the difference between negative and positive frequencies, the negative frequencies fold over to give lines at $(n \pm q_m)f_m$ as shown in Fig. (14.11).

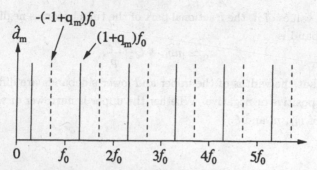

Figure. 14.11 Spectrum of betatron lines. The dashed lines are folded over from negative frequencies by plotting the absolute value $|(n + q)f_s|$. Here we have plotted lines for q < 0.5.

Summing over N particles we get the total average transverse signal

$$\langle d \rangle = 0, \tag{14.35}$$

with the rms spread

$$\sigma_d = \sqrt{\langle (d_m - \langle d_m \rangle)^2 \rangle} = q f_s \sigma_a \sqrt{\frac{N}{2}}, \tag{14.36}$$

where σ_a is the rms betatron amplitude. The Schottky power in each sideband is proportional to σ_d^2 and is again independent of the revolution harmonic number n.

Comparing Eqs. (14.26 & 14.36) we see that the rms betatron amplitude may be obtained from

$$\sigma_a = \frac{2\sigma_d}{\sigma_i}. \tag{14.37}$$

If the detectors are very well calibrated, then this can be used to obtain the transverse emittance.

The momentum spread and betatron tune spread contribute to give a nonzero width to the betatron line. If the betatron tune spread is only due to chromaticity, then

$$\sigma_q = |(n + q)\eta_{tr} + Q\xi| \frac{\sigma_p}{p}, \tag{14.38}$$

where Q is the total betatron tune (including integer part) and ξ is the chromaticity

$$\xi = \frac{p}{Q}\frac{dQ}{dp}. \tag{14.39}$$

Other contributions to the betatron tune spread which are not chromatic should be added in quadrature with Eq. (14.38), since they would be independent of the momentum oscillation.

For large values of n, the fractional part of the tune becomes negligible and the width of the band is

$$\sigma_q \simeq |n\eta_{tr} + Q\xi| \frac{\sigma_p}{p}. \tag{14.40}$$

We now see that the widths of the upper and lower sidebands are different since n can be either positive or negative. Whether the upper is narrower or wider depends on the signs of η_{tr}, n, and ξ.

Problems

14-1   A relativistic ($\beta \simeq 1$) bunched beam travels down the beam pipe shown below with a ring pickup. The radius of the pipe and pickup is $b = 4$ cm, and the length of the pickup is $w = 5$ cm. Ignore longitudinal fields. Plot the shape of a voltage pulse seen on an oscilloscope with a 50Ω termination. Assume that the coaxial cable is also a 50Ω cable.

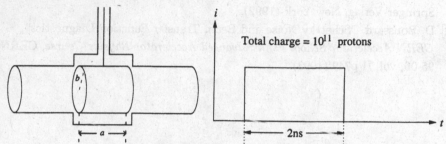

14-2   Given the following Schottky spectrum for RHIC with $^{197}\text{Au}^{79+}$ ions (fully stripped) where only a single storage cavity was powered, calculate the rf voltage in the cavity. Assume the following parameters for RHIC: $\gamma_{tr} = 22.8$, circumference of $L = 3833.845$ m, harmonic number $h = 7 \times 360 = 2520$, and $U_s = 100$ GeV/nucleon at fixed energy.

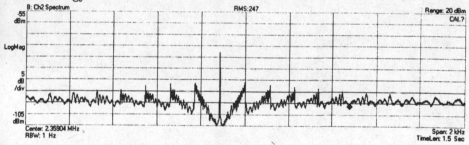

References for Chapter 14

[1] R. E. Shafer, "Beam Position Monitoring", AIP Conf. Proc. 212, p. 26, New York (1990).

[2] R. E. Shafer, "Characteristics of Directional Coupler Beam Position Monitors", *IEEE Trans. on Nucl. Sci.*, Vol. NS-32, #5 p. 1933 (1985).

[3] B. W. Char et al., *Maple V First Leaves: A Tutorial Introduction* Springer-Verlag, New York (1992)

[4] D. Boussard, "Schottky Noise and Beam Transfer Function Diagnostics", *CERN Accelerator School Fifth Advanced Accelerator Physics Course*, CERN 95-06, vol. II p749 (1993).

Appendix A

Confusion of Definitions in Other Sources

In a field such as accelerator physics, which is widespread, both in content and geography, it is impossible to have all definitions of quantities and symbols describing a concept to be defined in an identical manner, by all people. Throughout this book, we have tried to use consistent definitions with few exceptions. In this appendix, we try to give other popular definitions of quantities and relate them to the definitions in this book.

A.1 Coordinates

The local coordinate system (x, y, z) with x pointing in the radial direction, y vertically, and z along the trajectory of the design particle is perhaps more commonly used in North America than in Europe. Another common arrangement[2] is to have z as the vertical coordinate, x as the horizontally transverse coordinate, and s pointing along the design trajectory.

A.2 Emittance

There are several definitions of emittance used throughout the world, and the differences are somewhat confusing. The basic concept is the same: the emittance somehow specifies an area of a region of transverse phase space where the beam is located at some instant in time. As the beam moves through the accelerator, the shape of the beam may change, but the area of the new shape containing the same fraction of beam as before must remain constant, if the conditions of Liouville's theorem are satisfied. The confusion is just how to specify this area. What is the fraction of the total beam, and what is the proportionality constant of area with respect to the emittance?

In this book, we fall prey to the literature and use two different conventions. For linacs and proton rings, we have followed Courant[1] in specifying the emittance as the area of the region in x, x' or y, y'-phase space. For an elliptical region in a linear machine, this means

$$\text{emittance} = \pi \varepsilon. \tag{A.1}$$

Various people use different fractions of the beam. For linacs and proton rings,

335

this fraction is frequently 90%. Some people divide the area by π and call ε the emittance, and confuse the issue even more by writing something like

$$\text{``}\varepsilon = 4\pi \text{ mm mrad''} \tag{A.2}$$

which seems to imply that the area of an ellipse has a factor of π^2 in it! Hey, who said man was a rational animal anyway?

For electron rings, the beam distribution is close to a Gaussian distribution, since the synchrotron radiation keeps populating the tails of the distribution even though particles get scraped out of the tails by the beam pipe and other obstructions. The convention here is to use a fraction of the beam contained within 1σ of the distribution. A more confusing point is that the emittance is defined for electron rings as the area of this region divided by π. In the linear approximation, this means the emittance is just the Courant-Snyder invariant for an ellipse which contains about 39% of the beam (i. e. 1σ for a 2-d Gaussian distribution.) This gives $\sigma_\beta = \sqrt{\epsilon\beta}$ for this standard deviation of the amplitude of the betatron oscillation.

A.3 Tunes: Q or ν

We have used Q_H and Q_V respectively for the horizontal and vertical betatron tunes, and Q_s for the synchrotron tune. Some authors use the Greek letter ν instead of Q and may replace the subscript with the corresponding coordinate. In the case of the betatron tunes, this should not cause any problems, but the synchrotron tune can be confused with another tune called the *spin tune* which is **always** designated by ν_s. The spin tune is the number of times that a particle's spin precesses in the rest system of the synchronous particle during one full revolution of the beam in a circular accelerator. To avoid confusion, we recommend never using ν_s for the synchrotron tune.

A.4 Betatron functions

The betatron functions are sometimes written with the coordinate for a subscript rather than the "H" and "V" we have used.

A.5 Dispersion function

The periodic dispersion function for a cyclic machine is represented by at least four different symbols in the literature: $\eta(s)$, $D(s)$, $X_p(s)$, and α_p. A subscript of the corresponding coordinate is frequently appended, particularly if there is dispersion in both the vertical and horizontal planes. The letter D is used mostly in

Europe; however, this is sometimes done in a confusing manner since the letter D is also sometimes used for the matrix element, $\partial x_2/\partial \delta_1$, for a transformation from $(x_1, x_1', y_1, y_1', z_1, \delta_1)$ to $(x_2, x_2', y_2, y_2', z_2, \delta_2)$. In the U. S. and elsewhere both η and X_p are used for the periodic dispersion function; η typically at electron machines, and X_p and α_p more often at proton machines.

A.6 Chromaticity

In § 6.9 we have defined the chromaticity by the ratio of fractional change in tune to the fractional change in momentum:

$$\xi = \left(\frac{\Delta Q}{Q}\right) / \left(\frac{\Delta p}{p}\right). \tag{A.3}$$

A different definition is the change in tune divided by the fractional change in momentum:

$$\xi = p\frac{\Delta Q}{\Delta p}. \tag{A.4}$$

A.7 Momentum compaction

The term *momentum compaction* bothers some authors since a positive value corresponds to an increase in orbit length as momentum is increased. Sands[3], prefers the term *momentum dilation*, and Russian authors use the, perhaps more reasonable, *orbit expansion coefficient*.[4] Be that as it may, we have opted not to obscure life any more than it already is, and to use the more common English phrase, momentum compaction.

A.8 Multipole numbering

In the U. S. most people label the multipole coefficients b_n and a_n as the respective normal and skew $2(n+1)$–poles. In Europe the convention is to use b_n and a_n for the respective normal and skew $2n$–poles.

A.9 Phase slip

We define the phase slip factor

$$\eta_{tr} = \frac{1}{\gamma^2} - \alpha_p. \tag{A.5}$$

We have used the subscript "tr" to eliminate confusion with the dispersion function η. Other definitions have been made with the opposite sign[5]. In the early work by Courant, Snyder, and Livingston[6] the symbol γ was used rather than η.

A.10 Longitudinal canonical momentum W

The longitudinal canonical momentum coordinate,

$$W = -\delta U/\omega_{\text{rf}}, \tag{A.6}$$

is sometimes defined by dividing by the synchrotron frequency Ω_s rather than ω_{rf}. The transformation from $(-\delta U, t)$ to $(-\delta U/\Omega_s, \varphi)$, however does not preserve the phase space volume and is not a true canonical transformation. The rf frequency is an easily determined and independent input parameter to the accelerator, whereas the synchrotron frequency depends on the structure of the ring as well as ω_{rf}, ϕ_s, and the amplitude of the rf voltage. By dividing by ω_{rf} the definition of W is easily extended to longitudinal motion in a linac.

References for Appendix A

[1] E. D. Courant, "Introduction to Accelerator Physics" (Lecture notes written by R. D. Ruth and W. T. Weng,) *Physics of High Energy Particle Accelerators* AIP Conf. Proc. 87, Ed.: R. A. Carrigan and R. F. Huson, AIP, New York (1982).

[2] K. G. Steffen: High Energy Beam Optics, John Wiley and Sons, London-New York 1965.

[3] M. Sands, "The Physics of Electron Storage Rings", SLAC Report 121 (1970).

[4] A. A. Kolomensky and A. N. Lebedev, *Theory of Cyclic Accelerators*, North-Holland, Amsterdam (1966).

[5] E. D. Courant and H. S. Snyder, *Ann. Physics*, **3**, 1 (1958)

[6] E. D. Courant, M. S. Livingston and H. S. Snyder, Phys. Rev., **88**, 1190 (1952).

Appendix B

Properties of a Generic Optical System

Given a generic optical system, characterized by the matrix

$$\mathbf{M} = \begin{pmatrix} M_{11} & M_{12} \\ M_{21} & M_{22} \end{pmatrix}, \tag{B.1}$$

it is always possible to build an ancillary matrix by multiplying the original matrix by either one or two free-flight matrices:

$$\overline{\mathbf{M}} = \begin{pmatrix} \overline{M}_{11} & \overline{M}_{12} \\ \overline{M}_{21} & \overline{M}_{22} \end{pmatrix} = \begin{pmatrix} 1 & l_1 \\ 0 & 1 \end{pmatrix} \begin{pmatrix} M_{11} & M_{12} \\ M_{21} & M_{22} \end{pmatrix} \begin{pmatrix} 1 & l_2 \\ 0 & 1 \end{pmatrix}$$

$$= \begin{pmatrix} M_{11} + l_2 M_{21} & M_{12} + l_1 M_{11} + l_2 M_{22} + l_1 l_2 M_{21} \\ M_{21} & M_{22} + l_1 M_{21} \end{pmatrix}. \tag{B.2}$$

We shall see that, setting each element of the matrix $\overline{M}$ equal to zero, it is possible to obtain some relevant properties of the generic system.

Setting $\overline{M}_{11} = 0$ gives the second focal distance, $l_2 = -M_{11}/M_{21}$:

$$\begin{pmatrix} x_2 \\ x_2' \end{pmatrix} = \begin{pmatrix} 0 & \overline{M}_{12} \\ \overline{M}_{21} & \overline{M}_{22} \end{pmatrix} \begin{pmatrix} x_1 \\ 0 \end{pmatrix} = \begin{pmatrix} 0 \\ \overline{M}_{21} x_1 \end{pmatrix}, \tag{B.3}$$

since $x_1' = 0$ means we have an incoming parallel beam, and $x_2 = 0$ describes a beam-concentration point (N. B. $\overline{M}_{21} x_1$ represents the converging rays.)

Letting $\overline{M}_{22} = 0$ yields the first focal distance, $l_1 = -M_{22}/M_{21}$:

$$\begin{pmatrix} x_2 \\ x_2' \end{pmatrix} = \begin{pmatrix} \overline{M}_{11} & \overline{M}_{12} \\ \overline{M}_{21} & 0 \end{pmatrix} \begin{pmatrix} 0 \\ x_1' \end{pmatrix} = \begin{pmatrix} \overline{M}_{12} x_1' \\ 0 \end{pmatrix}. \tag{B.4}$$

With $\overline{M}_{12} = 0$ the image-object relation, $M_{12} + l_1 M_{11} + l_2 M_{22} + l_1 l_2 M_{21} = 0$:

$$\begin{pmatrix} x_2 \\ x_2' \end{pmatrix} = \begin{pmatrix} \overline{M}_{11} & 0 \\ \overline{M}_{21} & \overline{M}_{22} \end{pmatrix} \begin{pmatrix} 0 \\ x_1' \end{pmatrix} = \begin{pmatrix} 0 \\ \overline{M}_{22} x_1' \end{pmatrix}. \tag{B.5}$$

A point on the axis ($x_1 = 0, x_1' \neq 0$) is transformed into another point on the axis ($x_2 = 0, x_2' = \overline{M}_{22} x_1'$). For a thin lens $M_{11} = M_{22} = 1$ and $M_{12} = 0$, and we obtain the usual formula from geometric optics;

$$\frac{1}{l_1} + \frac{1}{l_2} = \frac{1}{f} = -M_{21}. \tag{B.6}$$

Having $\overline{M}_{21} = 0$ means that the system is a telescopic or affine one[1] in which the two focal planes are at infinity, and parallel incoming rays leave the other end parallel but magnified or demagnified. Matrices of this last sort are of minor interest in beam dynamics.

References for Appendix B

[1] M. Born and E. Wolf, *Principles of Optics*, Pergamon, New York, 1980.

[2] M. V. Klein, *Optics*, pp. 84-103, John Wiley and Sons, New York 1970.

Appendix C

Generating Functions

A Hamiltonian $H(\vec{x}, \vec{p}; s)$ is related to a Lagrangian L by the equation

$$L(\vec{x}, \vec{x}'; s) = \vec{p} \cdot \vec{x}' - H(\vec{x}, \vec{p}; s). \tag{C.1}$$

The variation of the action integral

$$\delta \int_{s_1}^{s_2} L \, ds = \delta \int_{s_1}^{s_2} (\vec{p} \cdot d\vec{x} - H \, ds) = 0, \tag{C.2}$$

between two fixed endpoints leads to Hamilton's equations of motion

$$\frac{dx_i}{ds} = \frac{\partial H}{\partial p_i} \tag{C.3}$$

$$\frac{dp_i}{ds} = -\frac{\partial H}{\partial x_i}. \tag{C.4}$$

We would like to transform from the coordinate system $(\vec{x}, \vec{p})$ to a new system $(\vec{X}, \vec{P})$ with a new Hamiltonian $K(\vec{X}, \vec{P}; s)$ having the new equations of motion

$$\frac{dX_i}{ds} = \frac{\partial K}{\partial P_i} \quad \text{and} \tag{C.5}$$

$$\frac{dP_i}{ds} = -\frac{\partial K}{\partial X_i}. \tag{C.6}$$

The new Hamiltonian and coordinates should satisfy a similar variation principle as the old system, i. e.

$$\delta \int_{s_1}^{s_2} (\vec{P} \cdot d\vec{X} - K \, ds) = 0. \tag{C.7}$$

Comparison of Eqs. (C.2) and (C.7) means that

$$\int_{s_1}^{s_2} (\vec{p} \cdot \vec{x}' - H) ds - \int_{s_1}^{s_2} (\vec{P} \cdot \vec{X}' - K) ds = \text{a constant}. \tag{C.8}$$

This constant may be written as

$$\text{constant} = \int_{s_1}^{s_2} \frac{dF}{ds} ds, \tag{C.9}$$

341

for some arbitrary function F. Since this must be true for all possible pairs of endpoints, we must have

$$\vec{p} \cdot \vec{x}' - H = \vec{P} \cdot \vec{X}' - K + \frac{dF}{ds}. \tag{C.10}$$

Suppose that $F = F_1(\vec{x}, \vec{X}; s)$, then

$$\frac{dF}{ds} = \sum_i \left(\frac{\partial F_1}{\partial x_i} \frac{dx_i}{ds} + \frac{\partial F_1}{\partial X_i} \frac{dX_i}{ds} \right) + \frac{\partial F_1}{\partial s}. \tag{C.11}$$

By requiring

$$p_i = \frac{\partial F_1}{\partial x_i} \quad \text{and} \tag{C.12}$$

$$-P_i = \frac{\partial F_1}{\partial X_i} \tag{C.13}$$

we find that Eqs. (C.10) and (C.11) combine to give

$$K = H + \frac{\partial F_1}{\partial s}. \tag{C.14}$$

By writing $F = F_2(\vec{x}, \vec{P}; s) - \vec{X} \cdot \vec{P}$, we find that

$$\frac{dF}{ds} = \frac{\partial F_2}{\partial s} + \sum_i \left(\frac{\partial F_2}{\partial x_i} x_i' + \frac{\partial F_2}{\partial P_i} P_i' - P_i X_i' - P_i' X_i \right), \tag{C.15}$$

which combines with Eq. (C.10) to give

$$K = H + \frac{\partial F_2}{\partial s} \quad \text{with} \tag{C.16}$$

$$p_i = \frac{\partial F_2}{\partial x_i}, \quad \text{and} \tag{C.17}$$

$$X_i = \frac{\partial F_2}{\partial P_i}. \tag{C.18}$$

By reversing the roles of the old and new quantities, it is easy to see that the function $F = F_3(\vec{p}, \vec{X}; s)$ transforms the Hamiltonian to

$$K = H + \frac{\partial F_3}{\partial s} \quad \text{provided that} \tag{C.19}$$

$$x_i = -\frac{\partial F_3}{\partial p_i} \quad \text{and} \tag{C.20}$$

$$P_i = -\frac{\partial F_3}{\partial X_i}. \tag{C.21}$$

The generating function

$$F = \vec{x} \cdot \vec{p} - \vec{X} \cdot \vec{P} + F_4(\vec{p}, \vec{P}; s), \qquad (C.22)$$

will produce the Hamiltonian

$$K = H + \frac{\partial F_4}{\partial s}, \quad \text{provided that} \qquad (C.23)$$

$$x_i = -\frac{\partial F_4}{\partial p_i} \quad \text{and} \qquad (C.24)$$

$$X_i = \frac{\partial F_4}{\partial P_i}. \qquad (C.25)$$

References for Appendix C

[1] H. Goldstein, *Classical Mechanics*, Addison-Wesley, Reading, Massachusetts (1981).

Appendix D

Luminosity

D.1 Basic definition of luminosity

The luminosity rate of interactions, at time t, between two colliding beams may be easily understood by treating one beam as a target, with number density $\rho_2(\vec{\mathbf{X}}_2, t)$, and the other as an incident beam, whose flux density is $v\rho_1(\vec{\mathbf{X}}_1, t)$, where

$$v = |\vec{v}_1 - \vec{v}_2| \tag{D.1}$$

is the relative velocity.† The coordinates $\mathbf{X}_2$ and $\mathbf{X}_1$ are points in the six dimensional phase space $(x, x', y, y', s, \delta)$. The interaction rate may then be written as

$$\frac{dN}{dt} = \int \sigma(\vec{\mathbf{X}}_2, \vec{\mathbf{X}}_1)\, \rho_2(\vec{\mathbf{X}}_2, t)\, |\vec{v}_1 - \vec{v}_2|\, \rho_1(\vec{\mathbf{X}}_1, t)\, d^6\mathbf{X}_2\, d^6\mathbf{X}_1, \tag{D.2}$$

where σ is the cross section for single particle interactions.

Of course, for the actual rate seen by an experiment one must fold in the detector acceptance and efficiency for a desired reaction by integrating the differential cross section times acceptance and efficiency over the the phase space of final particles.

For most high energy collisions the probability of interaction is extremely small, unless the particles are very close, e. g.,

$$|\vec{x}_1 - \vec{x}_2| \lesssim 10^{-15}\text{m}. \tag{D.3}$$

This is much less than the size of a typical beam ($\gtrsim 10^{-6}$m). It is therefore reasonable to approximate the cross section by

$$\sigma(\vec{\mathbf{X}}_2, \vec{\mathbf{X}}_1) \simeq \sigma(\vec{p}_2, \vec{p}_1)\, \delta^3(\vec{x}_1 - \vec{x}_2). \tag{D.4}$$

---

† For relativistic colliding beams this relative velocity is the difference of the two velocities $v \simeq |c - (-c)| = 2c$ as calculated in the between the two beams. Perhaps the best way to think of this is that an observer in the rest system of the collision point sees the bunches pass through each other at twice the effective speed as a single bunch.

If the beams are almost monoenergetic, σ may be factored outside the integral and replaced by the total cross section, so that

$$\frac{dN}{dt} \simeq |\vec{v}_1 - \vec{v}_2|\, \sigma(\vec{p}_2, \vec{p}_1) \int \rho_2(\vec{x}, t)\, \rho_1(\vec{x}, t)\, d^3x. \tag{D.5}$$

If the bunch shape does not change appreciably while the opposing bunches overlap, then the time dependence of the ith densities may be written as

$$\rho_i(\vec{x}, t) = \rho_i(\vec{x} - \vec{v}_i t),$$

and the total number of interactions from one bunch crossing is

$$N = |\vec{v}_1 - \vec{v}_2|\, \sigma \int \rho_2(\vec{x} - \vec{v}_2 t)\, \rho_1(\vec{x} - \vec{v}_1 t)\, d^3x\, dt. \tag{D.6}$$

When there are N_b equally spaced bunches per beam in a circular collider whose revolution frequency is f_0, the interaction rate per interaction region is

$$\frac{dN}{dt} = |\vec{v}_1 - \vec{v}_2|\, f_0 N_b \sigma \int \rho_2(\vec{x} - \vec{v}_2 t)\, \rho_1(\vec{x} - \vec{v}_1 t)\, d^3x\, dt. \tag{D.7}$$

The ratio of interaction rate to total cross section is called the instantaneous luminosity

$$\mathcal{L} = |\vec{v}_1 - \vec{v}_2|\, f_0 N_b \int \rho_2(\vec{x} - \vec{v}_2 t)\, \rho_1(\vec{x} - \vec{v}_1 t)\, d^3x\, dt, \tag{D.8}$$

which for high energy collisions in the center of mass system becomes

$$\mathcal{L} = 2c f_0 N_b \int \rho_2(x, y, z - ct)\, \rho_1(x, y, z + ct)\, d^3x\, dt, \tag{D.9}$$

where beam-2 moves in the $+z$ direction. The total or integrated luminosity refers to the instantaneous luminosity integrated over the time of the experiment.

If the bunch densities change shape in the overlap region, then the simplification of Eq. (D.5) is invalid and the luminosity must be calculated from

$$\mathcal{L} = 2c f_0 N_b \int \rho_2(x, y, z, t)\, \rho_1(x, y, z, t)\, d^3x\, dt. \tag{D.10}$$

If the beams cross where the transverse beta-functions both have minima, then the minima of the beta-functions should not be smaller than the bunch length, otherwise too much of the overlap region has a lower density, and the peak luminosity is degraded. The beta-function about an interaction point goes like

$$\beta(s) = \beta^* + \frac{s^2}{\beta^*}, \tag{D.11}$$

with the minimum value of β^* at the interaction point ($s = 0$). For bunches of length l_b, the overlap occurs between $s = \pm l_b/2$. If $l_b = \beta^*$ the longitudinal centers of the bunches cross at $s = 0$ where $\beta = \beta^*$ and the opposite ends of the bunches cross where $\beta = 1.25\beta^*$.

D.2 Gaussian beam distributions

For one transverse horizontal degree of freedom the Courant-Snyder invariant is

$$W_x = (x_\beta \quad x'_\beta) \begin{pmatrix} \gamma_x & \alpha_x \\ \alpha_x & \beta_x \end{pmatrix} \begin{pmatrix} x_\beta \\ x'_\beta \end{pmatrix}$$
$$= \gamma x_\beta^2 + 2\alpha x_\beta x'_\beta + \beta x'^2_\beta, \tag{D.12}$$

where β_x, α_x, and $\gamma_x = (1+\alpha_x^2)/\beta_x$ are the Twiss parameters for betatron motion, and x_β and x'_β are the transverse betatron coordinate and angle in a paraxial phase space. A Gaussian distribution of particles undergoing betatron oscillations for this degree of freedom may be written as

$$f_\beta(x_\beta, x'_\beta) = \frac{N_z}{2\pi\epsilon} e^{-\frac{1}{2}W_x/\epsilon_x}, \tag{D.13}$$

where N is the number of particles and $\pi\epsilon_x$ is the rms horizontal emittance.

Similarly for the longitudinal motion we may write an invariant

$$W_z = (z \quad \delta) \begin{pmatrix} \gamma_z & \alpha_z \\ \alpha_z & \beta_z \end{pmatrix} \begin{pmatrix} z \\ \delta \end{pmatrix}$$
$$= \gamma_z z^2 + 2\alpha_z z\delta + \beta_z \delta^2, \tag{D.14}$$

The corresponding longitudinal distribution function for a Gaussian bunch is

$$f_z(z, \delta) = \frac{N}{2\pi\epsilon_z} e^{-\frac{1}{2}W_z/\epsilon_z}, \tag{D.15}$$

where $z = s - vt$ and $\delta = \Delta p/p$ are respectively the longitudinal and fractional momentum deviations of a particle from the design particle when it passes the coordinate s.

The total transverse coordinates are the sum of the betatron coordinates and the effect of dispersion:

$$X = \begin{pmatrix} x \\ x' \end{pmatrix} = \begin{pmatrix} x_\beta \\ x'_\beta \end{pmatrix} + \begin{pmatrix} 0 & \eta_x \\ 0 & \eta'_x \end{pmatrix} \begin{pmatrix} z \\ \delta \end{pmatrix}, \tag{D.16}$$

where η_x and η'_x are the horizontal closed-orbit dispersion functions. Since the Jacobian of the transformation from coordinates $(x_\beta, x'_\beta, z, \delta)$ to (x, x', z, δ) is one, we may write a combined distribution function as

$$f(x, x', z, \delta) = \frac{N}{(2\pi)^2\epsilon_x\epsilon_z} e^{-\frac{1}{2}[(X-DZ)^T\Xi_\beta(X-DZ)+Z^T\Xi_z Z]}, \tag{D.17}$$

with the definitions:

$$X = \begin{pmatrix} x \\ x' \end{pmatrix}, \quad Z = \begin{pmatrix} z \\ \delta \end{pmatrix}, \quad D = \begin{pmatrix} 0 & \eta_x \\ 0 & \eta'_x \end{pmatrix}, \tag{D.18}$$

$$\Xi_\beta = \frac{1}{\epsilon_x} \begin{pmatrix} \gamma_x & \alpha_x \\ \alpha_x & \beta_x \end{pmatrix}, \quad \text{and} \quad \Xi_z = \frac{1}{\epsilon_z} \begin{pmatrix} \gamma_z & \alpha_z \\ \alpha_z & \beta_z \end{pmatrix}. \tag{D.19}$$

Rearranging terms in the distribution and adding in a similar distribution function for uncoupled vertical betatron motion yields the equation

$$f(x, x', y, y', z, \delta) = \frac{N}{(2\pi)^3 \epsilon_x \epsilon_y \epsilon_z} e^{-\frac{1}{2}\hat{X}^T \Xi \hat{X}}, \tag{D.20}$$

where

$$\hat{X} = \begin{pmatrix} x \\ x' \\ y \\ y' \\ z \\ \delta \end{pmatrix}, \quad \text{and} \tag{D.21}$$

$$\Xi = \begin{pmatrix} \frac{\gamma_x}{\epsilon_x} & \frac{\alpha_x}{\epsilon_x} & 0 & 0 & 0 & -\frac{\gamma_x \eta_x + \alpha_x \eta'_x}{\epsilon_x} \\ \frac{\alpha_x}{\epsilon_x} & \frac{\beta_x}{\epsilon_x} & 0 & 0 & 0 & -\frac{\alpha_x \eta_x + \beta_x \eta'_x}{\epsilon_x} \\ 0 & 0 & \frac{\gamma_y}{\epsilon_y} & \frac{\alpha_y}{\epsilon_y} & 0 & 0 \\ 0 & 0 & \frac{\alpha_y}{\epsilon_y} & \frac{\beta_y}{\epsilon_y} & 0 & 0 \\ 0 & 0 & 0 & 0 & \frac{\gamma_z}{\epsilon_z} & \frac{\alpha_z}{\epsilon_z} \\ -\frac{\gamma_x \eta_x + \alpha_x \eta'_x}{\epsilon_x} & -\frac{\alpha_x \eta_x + \beta_x \eta'_x}{\epsilon_x} & 0 & 0 & \frac{\alpha_z}{\epsilon_z} & \frac{\beta_z}{\epsilon_z} + \frac{H}{\epsilon_x} \end{pmatrix}, \tag{D.22}$$

with the function $H = \gamma_x \eta_x^2 + 2\alpha_x \eta_x \eta'_x + \beta_x \eta'^2_x$. The function H is sometimes mistakenly called the *dispersion invariant* since it is constant through regions of the lattice with no bends (i. e., straight sections). It is however **not** an invariant inside bends and will in general be different from one bend free region to the next. With a bit of algebra one can show that

$$\det(\Xi) = \frac{1}{\epsilon_x^2 \epsilon_y^2 \epsilon_z^2}. \tag{D.23}$$

The beam sigma matrix (Σ) or variance matrix for the distribution is just the inverse of Ξ:

$$\Sigma = \begin{pmatrix} \beta_x \epsilon_x + \eta_x^2 \sigma_\delta^2 & -\alpha_x \epsilon_x + \eta_x \eta'_x \sigma_\delta^2 & 0 & 0 & -\eta_x \alpha_z \epsilon_z & \eta_x \sigma_\delta^2 \\ -\alpha_x \epsilon_x + \eta_x \eta'_x \sigma_\delta^2 & \gamma_x \epsilon_x + \eta'^2_x \sigma_\delta^2 & 0 & 0 & -\eta'_x \alpha_z \epsilon_z & \eta'_x \sigma_\delta^2 \\ 0 & 0 & \beta_y \epsilon_y & -\alpha_y \epsilon_y & 0 & 0 \\ 0 & 0 & -\alpha_y \epsilon_y & \gamma_y \epsilon_y & 0 & 0 \\ -\eta_x \alpha_z \epsilon_z & -\eta'_x \alpha_z \epsilon_z & 0 & 0 & \beta_z \epsilon_z & -\alpha_z \epsilon_z \\ \eta_x \sigma_\delta^2 & \eta'_x \sigma_\delta^2 & 0 & 0 & -\alpha_z \epsilon_z & \sigma_\delta^2 \end{pmatrix}, \tag{D.24}$$

where $\sigma_\delta^2 = \gamma_z \epsilon_z$. Moreover if we consider coupling in all three dimensions, the beam sigma matrix may be written as the usual symmetric variance matrix for a Gaussian distribution:

$$\Sigma = \begin{pmatrix} \sigma_x^2 & \sigma_{xx'} & \sigma_{xy} & \sigma_{xy'} & \sigma_{xz} & \sigma_{x\delta} \\ \sigma_{xx'} & \sigma_{x'}^2 & \sigma_{x'y} & \sigma_{x'y'} & \sigma_{x'z} & \sigma_{x'\delta} \\ \sigma_{xy} & \sigma_{x'y} & \sigma_y^2 & \sigma_{yy'} & \sigma_{yz} & \sigma_{y\delta} \\ \sigma_{xy'} & \sigma_{x'y'} & \sigma_{yy'} & \sigma_{y'}^2 & \sigma_{y'z} & \sigma_{y'\delta} \\ \sigma_{xz} & \sigma_{x'z} & \sigma_{yz} & \sigma_{y'z} & \sigma_z^2 & \sigma_{z\delta} \\ \sigma_{x\delta} & \sigma_{x'\delta} & \sigma_{y\delta} & \sigma_{y'\delta} & \sigma_{z\delta} & \sigma_\delta^2 \end{pmatrix}, \tag{D.25}$$

which has 21 free parameters. The general Gaussian distribution is then given by

$$f(x, x', y, y', z, \delta) = \frac{N\sqrt{|\Xi|}}{(2\pi)^3} e^{-\frac{1}{2}\hat{X}^T \Xi \hat{X}}, \tag{D.26}$$

where $\Xi = \Sigma^{-1}$.

The distribution given by Eq. (D.26) has a simple hyperelliptical shape. For a long beam passing through minimum of the beta-function, it should have a dog-bone shape. So far we have described the particle distribution relative to a longitudinal position s, where the z coordinate is a time-like coordinate specifying how far the particle in question is in advance of the design particle. In order to evaluate the overlap integral of two colliding beams, we need to specify the density function in terms of $(x, x', y, y', s, \delta; t)$, rather than $(x, x', y, y', z, \delta; s)$. For a particle of velocity v, the relation between s, z, and t is $z = vt - s$, and the required transformation between coordinates is defined by

$$\hat{X}(x, x', y, y', z, \delta; t) = M^{-1} X(x, x', y, y', s, \delta; t), \tag{D.27}$$

i. e.,

$$\begin{pmatrix} x \\ x' \\ y \\ y' \\ z \\ \delta \end{pmatrix} = \begin{pmatrix} 1 & -s & 0 & 0 & 0 & 0 \\ 0 & 1 & 0 & 0 & 0 & 0 \\ 0 & 0 & 1 & -s & 0 & 0 \\ 0 & 0 & 0 & 1 & 0 & 0 \\ 0 & 0 & 0 & 0 & 1 & -\frac{\beta s}{\gamma^2} \\ 0 & 0 & 0 & 0 & 0 & 1 \end{pmatrix} \begin{pmatrix} x \\ x' \\ y \\ y' \\ vt - s \\ \delta \end{pmatrix} = \begin{pmatrix} x - x's \\ x' \\ y - y's \\ y' \\ vt - \left(1 + \frac{\beta\delta}{\gamma^2}\right)s \\ \delta \end{pmatrix} \tag{D.28}$$

where $\beta = v/c$ and $\gamma = \sqrt{1 - \beta^2}$ are the usual Lorentz parameters. Substituting Eq. (D.28) into Eq. (D.26) gives

$$f(x, x', y, y', s, \delta; t) = \frac{N\sqrt{|\Xi|}}{(2\pi)^3} e^{-\frac{1}{2}X^T(M^{-1})^T \Xi M^{-1} X}, \tag{D.29}$$

which will now have the correct dog-bone shape at the waist. For high energy bunched beams the $\frac{\beta\delta}{\gamma^2} \ll 1$ and may be ignored. (For RHIC[3,4,5] $\frac{\beta\delta}{\gamma^2} \lesssim 10^{-7}$.) Integrating Eq. (D.29) over the momentum-like coordinates yields the particle density per volume as a function of spatial coordinate and time:

$$\rho(x, y, s; t) = \frac{N\sqrt{|\Xi|}}{(2\pi)^3} \iiint e^{-\frac{1}{2}\mathbf{X}^{\mathrm{T}}(\mathbf{M}^{-1})^{\mathrm{T}}\Xi\mathbf{M}^{-1}\mathbf{X}} \, dx' \, dy' \, d\delta. \qquad (\text{D.30})$$

Let us now consider the limited case of a crossing point located in a straight section with no dispersion or horizontal-vertical coupling; here Ξ of Eq. (D.22) becomes block-diagonal:

$$\Xi = \begin{pmatrix} \frac{\gamma_x}{\epsilon_x} & \frac{\alpha_x}{\epsilon_x} & 0 & 0 & 0 & 0 \\ \frac{\alpha_x}{\epsilon_x} & \frac{\beta_x}{\epsilon_x} & 0 & 0 & 0 & 0 \\ 0 & 0 & \frac{\gamma_y}{\epsilon_y} & \frac{\alpha_y}{\epsilon_y} & 0 & 0 \\ 0 & 0 & \frac{\alpha_y}{\epsilon_y} & \frac{\beta_y}{\epsilon_y} & 0 & 0 \\ 0 & 0 & 0 & 0 & \frac{\gamma_z}{\epsilon_z} & \frac{\alpha_z}{\epsilon_z} \\ 0 & 0 & 0 & 0 & \frac{\alpha_z}{\epsilon_z} & \frac{\beta_z}{\epsilon_z} \end{pmatrix}. \qquad (\text{D.31})$$

With this form, the integration of Eq. (D.30) is fairly simple and gives:

$$\rho(x, y, s; t) = \frac{Ne^{-\frac{x^2}{2\epsilon_x(\beta_x^* - 2\alpha_x^* s + \gamma_x^* s^2)}} e^{-\frac{y^2}{2\epsilon_y(\beta_y^* - 2\alpha_y^* s + \gamma_y^* s^2)}} e^{-\frac{(vt-s)^2}{2\epsilon_z\beta_z^*}}}{\sqrt{(2\pi)^3(\beta_x^* - 2\alpha_x^* s + \gamma_x^* s^2)(\beta_y^* - 2\alpha_y^* s + \gamma_y^* s^2)\beta_z^* \epsilon_x \epsilon_y \epsilon_z}}, \qquad (\text{D.32})$$

where the superscript "*"'s refer to the value at the design crossing point where $s = 0$. It is worthwhile noting that the expressions of transverse Twiss parameters in parentheses are just the evolution of the transverse β-functions along s:

$$\beta_x(s) = \beta_x^* - 2\alpha_x^* s + \gamma_x^* s^2, \qquad \text{and} \qquad \beta_y(s) = \beta_y^* - 2\alpha_y^* s + \gamma_y^* s^2. \qquad (\text{D.33})$$

For two colliding bunches the densities may be written as

$$\rho_1(x, y, s; t) = \frac{N_1 e^{-\frac{x^2}{2\epsilon_{x1}\beta_{x1}(s)}} e^{-\frac{y^2}{2\epsilon_{y1}\beta_{y1}(s)}} e^{-\frac{(s-vt)^2}{2\epsilon_{z1}\beta_{z1}^*}}}{\sqrt{(2\pi)^3\epsilon_{x1}\epsilon_{y1}\epsilon_{z1}\beta_{x1}(s)\beta_{y1}(s)\beta_{z1}^*}}, \qquad (\text{D.34})$$

with the s-axis taken along the trajectory of beam-1 and

$$\rho_2(x, y, s; t) = \frac{N_2 e^{-\frac{(x - h_x - \theta_x s)^2}{2\epsilon_{x2}\beta_{x2}(s)}} e^{-\frac{(y - h_y - \theta_y s)^2}{2\epsilon_{y2}\beta_{y2}(s)}} e^{-\frac{(s+vt+\Delta)^2}{2\epsilon_{z2}\beta_{z2}^*}}}{\sqrt{(2\pi)^3\epsilon_{x2}\epsilon_{y2}\epsilon_{z2}\beta_{x2}(s)\beta_{y2}(s)\beta_{z2}^*}}, \qquad (\text{D.35})$$

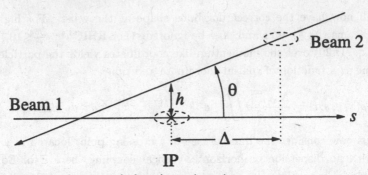

Figure. D.1 Geometry of errors at the beam's crossing point.

the trajectory of beam-2 being offset at $s = 0$ by h_x in the x-direction and h_y in the y-direction. (See Fig. D.1.) Using a small angle approximation, the slopes of the second beam's trajectory in the x and y directions are respectively θ_x and θ_y relative to the first beam's trajectory. The parameter Δ accounts for a mistiming of the bunch crossing; the bunches then cross at $s = \Delta/2$ from the design interaction point. The luminosity of this bunch crossing may be found by integrating the overlap of the two densities:

$$L = 2v \iiiint \rho_1(x, y, s; t)\rho_2(x, y, s; t) \, dx \, dy \, ds \, dt. \tag{D.36}$$

Integrating the x-dependent exponentials over x yields

$$\int_{-\infty}^{\infty} e^{-\frac{(x - h_x - \theta_x s)^2}{2\epsilon_{x1}\beta_{x1}(s)}} e^{-\frac{x^2}{2\epsilon_{x2}\beta_{x2}(s)}} \, dx = \sqrt{\frac{2\pi\epsilon_{x1}\epsilon_{x2}\beta_{x1}(s)\beta_{x2}(s)}{\epsilon_{x1}\beta_{x1}(s) + \epsilon_{x2}\beta_{x2}(s)}} \, e^{-\frac{(h_x + \theta_x s)^2}{2[\epsilon_{x1}\beta_{x1}(s) + \epsilon_{x2}\beta_{x2}(s)]}}. \tag{D.37}$$

A similar integral occurs for the y-dimension. The integral of the t-dependent exponentials is

$$\int_{-\infty}^{\infty} e^{-\frac{(s - vt)^2}{2\epsilon_{z1}\beta_{z1}^*}} e^{-\frac{(s + vt + \Delta)^2}{2\epsilon_{z2}\beta_{z2}^*}} \, v \, dt = \sqrt{\frac{2\pi\epsilon_{z1}\epsilon_{z2}\beta_{z1}^*\beta_{z2}^*}{\epsilon_{z1}\beta_{z1}^* + \epsilon_{z2}\beta_{z2}^*}} \, e^{-\frac{(2s + \Delta)^2}{2(\epsilon_{z1}\beta_{z1}^* + \epsilon_{z2}\beta_{z2}^*)}}. \tag{D.38}$$

Using these results Eq. (D.36) becomes

$$L = \int_{-\infty}^{\infty} \frac{2N_1 N_2 \, e^{-\frac{(h_x + \theta_x s)^2}{2[\epsilon_{x1}\beta_{x1}(s) + \epsilon_{x2}\beta_{x2}(s)]}} e^{-\frac{(h_y + \theta_y s)^2}{2[\epsilon_{y1}\beta_{y1}(s) + \epsilon_{y2}\beta_{y2}(s)]}} e^{-\frac{(2s + \Delta)^2}{2(\epsilon_{z1}\beta_{z1}^* + \epsilon_{z2}\beta_{z2}^*)}}}{\sqrt{(2\pi)^3 [\epsilon_{x1}\beta_{x1}(s) + \epsilon_{x2}\beta_{x2}(s)][\epsilon_{y1}\beta_{y1}(s) + \epsilon_{y2}\beta_{y2}(s)][\epsilon_{z1}\beta_{z1}^* + \epsilon_{z2}\beta_{z2}^*]}} \, ds, \tag{D.39}$$

for a single crossing of two bunches.

For the special case where the colliding beams are collinear with identical beam sizes and shapes at the interaction point, and having $2\sigma_z \lesssim \beta_x^*$, $2\sigma_z \lesssim \beta_y^*$, and $\Delta = 0$, the previous integral simplifies to

$$L = \frac{N_1 N_2}{4\pi \sigma_x \sigma_y}. \tag{D.40}$$

The average instantaneous luminosity for N_b crossing bunches per revolution with N_1 and N_2 particles per bunch respectively for the opposing beams is then

$$\mathcal{L} = \frac{f_0 N_b N_1 N_2}{4\pi \sigma_x \sigma_y}. \tag{D.41}$$

When the beams are round with $\epsilon_{rms} = \epsilon_x^{rms} = \epsilon_y^{rms}$ and $\beta^* = \beta_x^* = \beta_y^*$ and of equal current, then

$$\mathcal{L} = \frac{f_0 N_b N^2 \gamma}{4\pi \beta^* \epsilon_{rms}}. \tag{D.42}$$

In a more realistic situation the different bunches may have random intensities, so that the product $N_b N_1 N_2$ must be replaced by a sum of the products of the colliding bunches' intensities, i. e.,

$$N_b N_1 N_2 \rightarrow \Sigma_{j=1}^{N_b} N_{1j} N_{2j}. \tag{D.43}$$

D.3 Estimation of luminosities in RHIC

Table I lists the RHIC[3,4,5] design parameters for both protons and gold ions. The results of several simulated scans are shown in Figures D.2 to D.9 for both protons and gold ions.

Table I. RHIC parameters.

Parameter	Protons	Gold Ions
$\beta_{x min}$	1.0 m	1.0 m
$\beta_{y min}$	1.0 m	1.0 m
$\epsilon_{x 95\%}^N$	20 μm	10 μm
$\epsilon_{y 95\%}^N$	20 μm	10 μm
σ_z	0.075m	0.2m
N_1	1×10^{11}	1×10^9
N_2	1×10^{11}	1×10^9
N_b	55	55
f_{rev}	78.25 kHz	78.25 kHz
γ	260	108

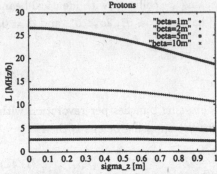

Figure D.2 Luminosity versus bunch length for proton beams at various β^* values.

Figure D.3 Longitudinal scan of luminosity for protons with β^*=1 m.

Fig. D.2 shows the variation of the luminosity with bunch length for β^* values from 1 to 10 m. Here I have assumed that the beams are round and of the same size

$$\beta_{x1}^* = \beta_{y1}^* = \beta_{x2}^* = \beta_{y2}^* \quad \text{and} \quad \epsilon_{x1}^* = \epsilon_{y1}^* = \epsilon_{x2}^* = \epsilon_{y2}^*. \tag{D.44}$$

This demonstrates the usual rule of thumb that the bunch length should be no larger than the value of β^*

Fig. D.5 shows the effect of misalignment of the waist of the the beam. Here I have assumed that the horizontal and vertical waists of both beams are displaced by the same amount but in opposite directions for both rings. This shows that to peak up luminosity, we should measure the optics carefully around the interaction regions. A scan of luminosity versus each waist should be made at some time to ensure that the optics are placing all four waists within about 10 cm of the interaction point.

The pair of beam position monitors (BPM) in each interaction region are separated by about 16.6 m. The beam pipe aperture through the DX and D0 magnets limit the maximum crossing angle to about 1 mr. The alignment of the beams from the BPM's should be better than 200 μm, so that we should be able to dead reckon the crossing angle very well, $\theta_{x,y} \leq 6$ μr. Figs. D.6 and D.7 show that there is very little degradation of luminosity for crossing angles up to several hundred microradians. Clearly the longer bunches of the gold beams are a bit more sensitive to crossing angle.

The rf system can easily phase the beams to cross at the required location within a couple of centimeters. Figs. D.3 and D.9 show that a $\Delta/2 = 20$ cm longitudinal shift of the IP has a marginal effect for both gold and proton beams

Since the BPM system should align the crossing point to a few hundred microns, we should be able to set the beam to collide with at least 30% of the peak luminosity. Even if we miss the crossing point by 0.5 mm, we can perform a two dimensional scan in 100μm steps to get quite close to the peak. Then a finer transverse scan can be performed to reach the peak.

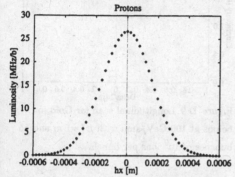

Figure D.4 Scan of h_x for proton beams at 250 GeV with β^*=1 m.

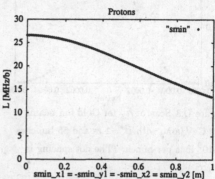

Figure D.5 Longitudinal scan of beam waists for protons at 250 GeV (β^*=1 m).

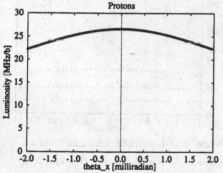

Figure D.6 Scan of θ_x for proton beams at 250 GeV with β^*=1 m.

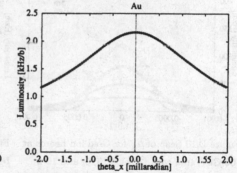

Figure D.7 Scan of θ_x for Gold ion beams at 100 GeV/amu with β^*=1 m and 55 bunches of 10^9 ions per bunch.

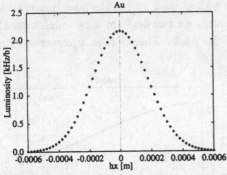

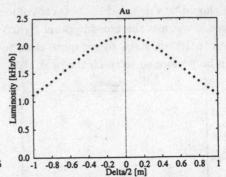

Figure D.8 Scan of h_x for Gold ion beams at 100 GeV/amu with β^*=1 m and 55 bunches of 10^9 ions per bunch. (The dot spacing is 20 μm.)

Figure D.9 Longitudinal scan for Gold ion beams at 100 GeV/amu with β^*=1 m and 55 bunches of 10^9 ions per bunch.

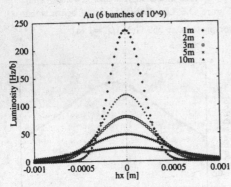

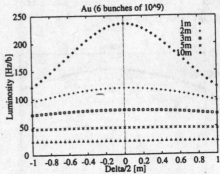

Figure D.10 Scan of h_x for Gold ion beams at 100 GeV/amu with β^*=1, 2, 3, 5, and 10 m and 6 bunches of 10^9 ions per bunch. (The dot spacing is 20 μm.)

Figure D.11 Longitudinal scan for Gold ion beams at 100 GeV/amu with β^*=1, 2, 3, 5, and 10 m and 6 bunches of 10^9 ions per bunch.

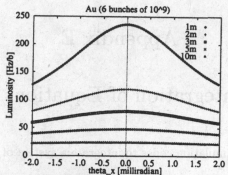

Figure D.12 Scan of θ_x for Gold ion beams at
100 GeV/amu with β^*=1, 2, 3, 5, and 10 m
and 6 bunches of 10^9 ions per bunch.

References for Appendix D

[1] M. Sands, "The Physics of Electron Storage Rings", SLAC Report 121 (1970).

[2] W. W. MacKay, "Luminosity as Calculated from Machine Parameters", C-A/AP/89 (2002).

[3] "RHIC Design Manual", unpublished (1998).

[4] H. Hahn et al., Nucl. Inst. and Meth. **A499**, 245 (2003).

[5] I. Alekseev et al., Nucl. Inst. and Meth. **A499**, 392 (2003).

Appendix E

Leapfrog Integration of Equations of Motion

When we write simulation codes to integrate a system of equations of motion like

$$\frac{d\phi}{dt} = \alpha W \tag{E.1}$$

$$\frac{dW}{dt} = -\beta\phi, \tag{E.2}$$

where α and β are constants, things sometimes go awry. An obvious way to solve this without computers is to write a second order differential equation:

$$\frac{d^2\phi}{dt^2} + \alpha\beta\,\phi = 0, \tag{E.3}$$

with solutions which are sine-like or exponential depending on the sign of $\alpha\beta$. For the case of $\omega^2 = \alpha\beta > 0$ we have

$$\phi(t) = \phi_0 \cos[\omega(t - t_0)], \tag{E.4}$$

which is periodic so that a particle traces out an ellipse in the (ϕ, W)-phase space.

We frequently resort to integrating by stepping through a pair of difference equations. A simple naive approach may run into problems. First I will demonstrate the incorrect method with what might appear at first to be a reasonable choice of difference equations:

$$\phi_{n+1} = \phi_n + \alpha W_n\,\Delta t, \tag{E.5}$$

$$W_{n+1} = W_n - \beta\phi_n\Delta t. \tag{E.6}$$

Writing them in matrix form we have

$$\begin{pmatrix} \phi_{n+1} \\ W_{n+1} \end{pmatrix} = \begin{pmatrix} 1 & \alpha\Delta t \\ -\beta\Delta t & 1 \end{pmatrix} \begin{pmatrix} \phi_n \\ W_n \end{pmatrix} = \mathbf{M} \begin{pmatrix} \phi_n \\ W_n \end{pmatrix}, \tag{E.7}$$

where the

$$\det(\mathbf{M}) = 1 + \alpha\beta\Delta t^2 \neq 1 \tag{E.8}$$

unless either α or β are zero. For $\omega^2 = \alpha\beta > 0$ this would give results with the particle proceeding to larger amplitudes in phase space instead of tracing out an ellipse. The resulting integration is nonsymplectic as demonstrated in Fig. E.1.

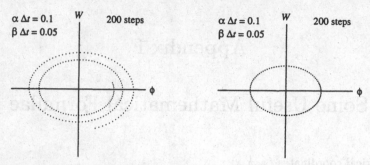

Figure. E.1 The plot on the left shows how the improper method causes a blowup of the oscillation, whereas the plot on the right was tracked with the leap-frog method.

$$\phi_{\frac{1}{2}} \longrightarrow \phi_{1+\frac{1}{2}} \longrightarrow \phi_{2+\frac{1}{2}}$$
$$W_1 \longrightarrow W_2 \longrightarrow W_3$$

Figure. E.2 By integrating the variables so that the time steps hop over each other, we can obtain a symplectic result.

If instead we consider Eqs. (E.1 & E.2) as modeling the longitudinal motion in a circular ring with a single rf cavity, then a convenient time step would correspond to one turn with a long drift followed by a small thin-lens-type energy kick:

$$\begin{pmatrix} \phi_{n+\frac{1}{2}} \\ W_{n+1} \end{pmatrix} = \begin{pmatrix} 1 & 0 \\ -\beta\Delta t & 1 \end{pmatrix} \begin{pmatrix} 1 & \alpha\Delta t \\ 0 & 1 \end{pmatrix} \begin{pmatrix} \phi_{n-\frac{1}{2}} \\ W_n \end{pmatrix}$$

$$= \begin{pmatrix} 1 & \alpha\Delta t \\ -\beta\Delta t & 1 - \alpha\beta\,\Delta t^2 \end{pmatrix} \begin{pmatrix} \phi_{n-\frac{1}{2}} \\ W_n \end{pmatrix}$$

$$= \mathbf{M} \begin{pmatrix} \phi_{n-\frac{1}{2}} \\ W_n \end{pmatrix}, \tag{E.9}$$

with $\det(\mathbf{M}) = 1$. See how the ellipse is preserved in the right plot of Fig. E.1.

This type of two-step integration where one variable is integrated ($\phi_{n-\frac{1}{2}} \to \phi_{n+\frac{1}{2}}$) and then the other is integrated ($W_n \to W_{n+1}$) is referred to as *leap-frog* integration, since the time steps for the two variables are interleaved as indicated by Fig. (E.2).

References for Chapter 14

[1] R. W. Hockney and J. W. Eastwood, *Computer Simulation Using Particles*, Adam Hilger, Bristol (1988).

Appendix F

Some Useful Mathematical Formulae

In cylindrical coordinates (ρ, ϕ, z):

1. $\quad \nabla f = \dfrac{\partial f}{\partial \rho}\hat{\rho} + \dfrac{1}{\rho}\dfrac{\partial f}{\partial \phi}\hat{\phi} + \dfrac{\partial f}{\partial z}\hat{z}$

2. $\quad \nabla \cdot \vec{A} = \dfrac{1}{\rho}\dfrac{\partial}{\partial \rho}(\rho A_\rho) + \dfrac{1}{\rho}\dfrac{\partial A_\phi}{\partial \phi} + \dfrac{\partial A_z}{\partial z}$

3. $\quad \nabla \times \vec{A} = \left(\dfrac{1}{\rho}\dfrac{\partial A_z}{\partial \phi} - \dfrac{\partial A_\phi}{\partial z}\right)\hat{\rho} + \left(\dfrac{\partial A_\rho}{\partial z} - \dfrac{\partial A_z}{\partial \rho}\right)\hat{\phi} + \dfrac{1}{\rho}\left(\dfrac{\partial}{\partial \rho}(\rho A_\phi) - \dfrac{\partial A_\rho}{\partial \phi}\right)\hat{z}$

4. $\quad \nabla^2 f = \dfrac{1}{\rho}\dfrac{\partial}{\partial \rho}\left(\rho\dfrac{\partial f}{\partial \rho}\right) + \dfrac{1}{\rho^2}\dfrac{\partial^2 f}{\partial \phi^2} + \dfrac{\partial^2 f}{\partial z^2}$

In spherical coordinates (r, θ, ϕ):

1. $\quad \nabla f = \dfrac{\partial f}{\partial r}\hat{r} + \dfrac{1}{r}\dfrac{\partial f}{\partial \theta}\hat{\theta} + \dfrac{1}{r\sin\theta}\dfrac{\partial f}{\partial \phi}\hat{\phi}$

2. $\quad \nabla \cdot \vec{A} = \dfrac{1}{r^2}\dfrac{\partial}{\partial r}(r^2 A_r) + \dfrac{1}{r\sin\theta}\dfrac{\partial}{\partial \theta}(A_\theta \sin\theta) + \dfrac{1}{r\sin\theta}\dfrac{\partial A_\phi}{\partial \phi}$

3. $\quad \nabla \times \vec{A} = \dfrac{1}{r\sin\theta}\left[\dfrac{\partial}{\partial \theta}(A_\phi \sin\theta) - \dfrac{\partial A_\theta}{\partial \phi}\right]\hat{r} + \dfrac{1}{r}\left[\dfrac{1}{\sin\theta}\dfrac{\partial A_r}{\partial \phi} - \dfrac{\partial}{\partial r}(rA_\phi)\right]\hat{\theta}$
$\qquad\qquad + \dfrac{1}{r}\left[\dfrac{\partial}{\partial r}(rA_\theta) - \dfrac{\partial A_r}{\partial \theta}\right]\hat{\phi}$

4. $\quad \nabla^2 f = \dfrac{1}{r^2}\dfrac{\partial}{\partial r}\left(r^2\dfrac{\partial f}{\partial r}\right) + \dfrac{1}{r^2\sin\theta}\dfrac{\partial}{\partial \theta}\left(\sin\theta\dfrac{\partial f}{\partial \theta}\right) + \dfrac{1}{r^2\sin^2\theta}\dfrac{\partial^2 f}{\partial \phi^2}$

General vector identities:

1. $\qquad\qquad \vec{A}_\perp = -\hat{e}_\parallel \times (\hat{e}_\parallel \times \vec{A})$

2. $\qquad \vec{A} \times (\vec{B} \times \vec{C}) = -(\vec{A} \cdot \vec{B})\vec{C} + (\vec{A} \cdot \vec{C})\vec{B}$

3. $\quad (\vec{A} \times \vec{B}) \cdot (\vec{C} \times \vec{D}) = (\vec{A} \cdot \vec{C})(\vec{B} \cdot \vec{D}) - (\vec{A} \cdot \vec{D})(\vec{B} \cdot \vec{C})$

4. $\qquad\qquad \nabla(fg) = f\nabla g + g\nabla f$

5. $\qquad\quad \nabla \cdot (f\vec{A}) = (\nabla f) \cdot \vec{A} + f(\nabla \cdot \vec{A})$

6. $\qquad\quad \nabla \times (f\vec{A}) = (\nabla f) \times \vec{A} + f(\nabla \times \vec{A})$

7. $\qquad \nabla \times (\nabla \times \vec{A}) = \nabla(\nabla \cdot \vec{A}) - \nabla^2 \vec{A}$

8. $\qquad \nabla \times (\vec{A} \times \vec{B}) = (\vec{B} \cdot \nabla)\vec{A} - (\vec{A} \cdot \nabla)\vec{B} + (\nabla \cdot \vec{B})\vec{A} - (\nabla \cdot \vec{A})\vec{B}$

9. $\qquad \nabla \cdot (\vec{A} \times \vec{B}) = \vec{B} \cdot (\nabla \times \vec{A}) - \vec{A} \cdot (\nabla \times \vec{B})$

10. $\qquad \nabla(\vec{A} \cdot \vec{B}) = (\vec{B} \cdot \nabla)\vec{A} + (\vec{A} \cdot \nabla)\vec{B} + \vec{B} \times (\nabla \times \vec{A}) + \vec{A} \times (\nabla \times \vec{B})$

11. $\qquad \vec{A} \times (\nabla \times \vec{B}) = -(\vec{A} \cdot \nabla)\vec{B} + \sum_{i=1}^{3} A_i(\nabla B_i)$

12. $\qquad \vec{v} \times (\nabla \times \vec{B}) = \nabla(\vec{v} \cdot \vec{B}) - (\vec{v} \cdot \nabla)\vec{B}, \quad \text{if} \quad \vec{v} = \dfrac{d\vec{x}}{dt}$

Various integral theorems:

1. $$\oint_{\partial S} \vec{A} \cdot d\vec{l} = \int_S (\nabla \times \vec{A}) \cdot d\vec{S}$$

2. $$\oint_{\partial V} \vec{A} \cdot d\vec{S} = \int_V (\nabla \cdot \vec{A})\, dV$$

3. $$\int_V (\nabla \times \vec{A})\, dV = -\oint_{\partial V} \vec{A} \times d\vec{S}$$

4. $$\int_V \nabla f\, dV = \oint_{\partial V} f\, d\vec{S}$$

5. $$\int_S (\nabla f) \times d\vec{S} = -\oint_{\partial S} f\, d\vec{l}$$

6. $$\int_V (\phi \nabla^2 \psi - \psi \nabla^2 \phi)\, dV = \oint_{\partial V} (\phi \nabla \psi - \psi \nabla \phi) \cdot d\vec{S}$$

7. $$\int_V (\phi \nabla^2 \psi + \nabla \phi \cdot \nabla \psi)\, dV = \oint_{\partial V} (\phi \nabla \psi - \psi \nabla \phi) \cdot d\vec{S}$$

Some Bessel function relations for $Z \in \{J_\nu(x), Y_\nu(x), H_\nu^{(1)}(x), H_\nu^{(2)}\}$:

1. $$x^2 \frac{d^2 Z}{dx^2} + x \frac{dZ}{dx} + (x^2 - \nu^2)Z = 0$$

2. $$\int Z_1(x)\, dx = -Z_0(x)$$

3. $$\int x\, Z_0(x)\, dx = x\, Z_1(x)$$

4. $$\int x^{p+1} Z_p(x)\, dx = x^{p+1} Z_{p+1}(x)$$

5. $$\int x^{1-p} Z_p(x)\, dx = x^{1-p} Z_{p-1}(x)$$

6. $$\int_0^1 t\, J_\nu(X_{\nu p} t)\, J_\nu(X_{\nu q} t)\, dt = \frac{1}{2} \left[J_\nu'(X_{\nu p}) \right]^2 \delta_{pq}, \quad \nu > -1$$

For power loss and related calculations:

1.
$$\bar{P}_{\text{loss}} = \frac{1}{2} R_s \int_S |H_\parallel|^2 \, dS$$

2.
$$R_s = \sqrt{\frac{\mu\omega}{2\sigma}}$$

3.
$$\delta \simeq \sqrt{\frac{2}{\mu\omega\sigma}}$$

4.
$$Q = \frac{U\omega_0}{\bar{P}_{\text{loss}}}$$

Index